GALLIUM ARSENIDE DIGITAL
INTEGRATED CIRCUITS

GALLIUM ARSENIDE DIGITAL INTEGRATED CIRCUITS:
A Systems Perspective

NICK KANOPOULOS

*Center for Digital Systems Research,
Research Triangle Institute, North Carolina*

PRENTICE HALL
Englewood Cliffs, New Jersey, 07632

Library of Congress Cataloging-in-Publication Data

Kanopoulos, Nick
    Gallium arsenide digital integrated circuits : a systems
perspective / Nick Kanopoulos
        p.   cm.
    Includes bibliographies and index.
    ISBN 0-13-346214-5
    1. Digital integrated circuits—Design and construction.
2. Gallium arsenide semiconductors—Design and construction.
I. Title
TK7874.K345   1989   89-3558
621.3815'2—dc19   CIP

Editorial/production supervision: Hartley Ferguson
Cover design: Wanda Lubelska Design
Manufacturing buyer: Bob Anderson

© 1989 by Prentice-Hall, Inc.
A Division of Simon & Schuster
Englewood Cliffs, New Jersey 07632

The publisher offers discounts on this book when ordered
in bulk quantities. For more information, write:

    Special Sales/College Marketing
    Prentice Hall
    College Technical and Reference Division
    Englewood Cliffs, NJ 07632

All rights reserved. No part of this book may be
reproduced, in any form or by any means,
without permission in writing from the publisher.

Printed in the United States of America

10 9 8 7 6 5 4 3 2 1

ISBN   0-13-346214-5

Prentice-Hall International (UK) Limited, *London*
Prentice-Hall of Australia Pty. Limited, *Sydney*
Prentice-Hall Canada, Inc., *Toronto*
Prentice-Hall Hispanoamericana, S.A., *Mexico*
Prentice-Hall of India Private Limited, *New Delhi*
Prentice-Hall of Japan, Inc., *Tokyo*
Simon & Schuster Asia Pte. Ltd., *Singapore*
Editoria Prentice-Hall do Brasil Ltda., *Rio de Janeiro*

*To my teachers:
Pete Marinos,
Vasili Makios,
and Kreon Keramitsopoulos,
who led me to new paths in the quest for learning.*

CONTENTS

PREFACE xi

ACKNOWLEDGMENTS xiii

1 GaAs COMPONENTS 1

    1.1   Description of Fabrication Steps 1

    1.2   JFET and MESFET Principles of Operation 9

        1.2.1 The Junction FET, 9
        1.2.2 The Metal Semiconductor FET, 16
        1.2.3 The Normally-Off FET, 19

    1.3   High Electron Mobility Transistor (HEMT) 22

    1.4   Heterojunction Bipolar Transistor (HBT) 23

    1.5   Modeling of GaAs Devices 24

        1.5.1 The Diode Model, 24
        1.5.2 The JFET Model, 27
        1.5.3 Noise Models, 30
        1.5.4 Device Models for Computer Simulation, 32

    1.6   Device Scaling 35

    Projects 37

    References 37

2 GaAs LOGIC GATE DESIGN   40

- 2.1 D-MESFET Logic Approaches   42
- 2.2 E-MESFET Logic Approaches   45
- 2.3 Propagation Delays and Power Dissipations   49
    - 2.3.1 Depletion-Mode Logic Gates, 50
    - 2.3.2 Enhancement-Mode Logic Gates, 56
- 2.4 Noise Margins   59
- 2.5 Interconnection Lines   63
    - 2.5.1 Electrical Properties of On-Chip Interconnection Lines, 64
    - 2.5.2 Proposed Solutions to Interconnection Line Problems, 66
- 2.6 Logic Gate Configurations   68
- 2.7 Logic Family Selection Trade-Offs for Gate Implementation   76

    Projects   77

    References   78

3 GaAs LOGIC CIRCUITS   80

- 3.1 Flip-Flops   80
- 3.2 Shift Registers   87
- 3.3 Counters   98
    - 3.3.1 Ripple Counters, 102
    - 3.3.2 Synchronous Counters, 105
    - 3.3.3 Nonbinary Counters, 111
    - 3.3.4 Combinations of Counter Modules, 114
    - 3.3.5 Variable Modulus Counters, 115
    - 3.3.6 Lockout, 118
- 3.4 Sequence Generators   118
- 3.5 Arithmetic Circuits   123
    - 3.5.1 Adders, 123
    - 3.5.2 Multipliers, 131
- 3.6 Multiplexers/Demultiplexers   141
- 3.7 Memory Circuits   144
    - 3.7.1 Random-Access Memory, 144
    - 3.7.2 First-In, First-Out Memory, 153
- 3.8 Programmable Logic Arrays   156
- 3.9 Data Conversion Circuits   159
    - 3.9.1 D-A Converters, 161
    - 3.9.2 A-D Converters, 163

Contents ix

- 3.10 I/O Drivers 171
- 3.11 Miscellaneous Circuits 174
  Projects 178
  References 178

4 GaAs DIGITAL INTEGRATED CIRCUIT DESIGN PRINCIPLES 182

- 4.1 Chip Design Methodologies 183
  - *4.1.1 Custom Design, 184*
  - *4.1.2 Standard Cells, 184*
  - *4.1.3 Gate Arrays, 187*
- 4.2 Chip Design Sequence and Computer-Aided Design Tools 189
  - *4.2.1 Schematic Capture, 189*
  - *4.2.2 Logic Simulation, 190*
  - *4.2.3 Circuit Simulation, 192*
  - *4.2.4 Layout, 193*
  - *4.2.5 Design Rule Checking, 201*
  - *4.2.6 Net List Extraction, 202*
  - *4.2.7 Switch Level Simulation, 202*
  - *4.2.8 Circuit Extraction, 203*
- 4.3 Selection of Circuits to Meet Design Requirements 203
- 4.4 Areas for Special Consideration 211
- 4.5 Interface with Si Technologies 214
- 4.6 A Chip Design Example 218
- 4.7 Architectures Attractive for GaAs Implementation 227
  Projects 229
  References 232

5 PACKAGING 234

- 5.1 Package Design Considerations 235
  - *5.1.1 Package Size, 235*
  - *5.1.2 Package Loading, 235*
  - *5.1.3 Crosstalk, 236*
  - *5.1.4 Power Supply Decoupling, 237*
  - *5.1.5 Temperature Effects, 237*
- 5.2 Chip-Package Designs 238
  - *5.2.1 Multilayer Ceramic Packages, 238*
  - *5.2.2 Silicon-Based Package, 242*
  - *5.2.3 Hybrid Packages, 243*

Projects 244
References 244

6 HIGH-SPEED TESTING AND DESIGN FOR TESTABILITY   245

6.1 Testing Issues 246
- 6.1.1 Test Generation, 247
- 6.1.2 Test Application, 250
- 6.1.3 Test Evaluation, 252

6.2 Testing Approaches 253
- 6.2.1 An Example of Testing a GaAs Chip, 255

6.3 Design for Testability 257
- 6.3.1 Practical Guidelines for Enhancing Circuit Testability, 258
- 6.3.2 Structured Design for Testability Techniques, 262

6.4 Built-In Self-Test (BIST) 272
- 6.4.1 Signature Analysis, 274
- 6.4.2 Built-In Logic Block Observer, 278
- 6.4.3 Built-In Self-Test at the Board Level, 280

Projects 292
References 292

7 GaAs INSERTION INTO SYSTEM DESIGN   294

7.1 System Applications to Be Strongly Impacted by GaAs Insertion 295
- 7.1.1 High-Throughput Signal/Image Processing, 295
- 7.1.2 Biomedical Research, 297
- 7.1.3 Communications Applications, 298
- 7.1.4 Scientific Computers, 299

7.2 Insertion Issues and Trade-Offs 300
- 7.2.1 Design Partitioning, 300
- 7.2.2 Intercomponent Connection Problems, 302
- 7.2.3 System Timing, 310
- 7.2.4 System Testing, 313
- 7.2.5 A Final Note on Insertion Issues, 320

7.3 Architectural Approaches at the System Level 321
- 7.3.1 Pipeline Techniques, 322
- 7.3.2 Systolic Arrays, 327
- 7.3.3 Special Purpose Architectures, 333

Projects 349
References 353

INDEX   355

PREFACE

Gallium Arsenide (GaAs) has been hailed the semiconductor material of the future by those people who have been developing GaAs technologies since the early 1960s. The early efforts, focused mainly on developing bipolar transistors, were frustrated by processing difficulties in introducing dopants through diffusion and the brief lifetime of minority carriers. Since then, progress in fabrication technologies and the development of metal-semiconductor field-effect transistors and junction field-effect transistors has made possible the design and implementation of digital integrated circuits that can be mass produced. The integration level for the first circuits was limited to a few gates, but further development of fabrication processes and material quality made circuits of several thousand gates commercially available by 1988. It is believed that this trend toward higher integration levels will continue in the future. The available GaAs components offer very high throughput while requiring moderate to low power. This feature opens a new era for applications where system implementation previously was considered impractical.

The work on the development of GaAs technologies and circuits has been, for the most part, confined to industry and government laboratories. Research also has been performed on materials and devices at a small number of leading universities. Unfortunately, few formal programs teach new engineers design aspects of GaAs circuits and the impact of these circuits on system design. However, in the late eighties there is a trend in university programs to include courses dealing with GaAs technologies and circuit design in graduate programs. This book is written to facilitate the teaching of these concepts to graduate students as well as to educate Si circuit designers and system designers about different aspects of GaAs circuit technology, and its impact on system design.

The book is concerned almost exclusively with the design of digital circuits. Only in chapter 3 is there a brief description of data acquisition circuits (i.e., analog-to-digital and digital-to-analog converters). The material selection for the book is based on my belief that a new circuit technology is best utilized for implementing integrated circuits—and, eventually, systems—when there is an understanding of all pertinent design-related aspects, spanning from the operating principles of the circuit components to the impact of circuit insertion into system design. Based on this objective, the book gives a broad view of the technology and design issues, the design methodologies, and the trade-offs that must be made to incorporate GaAs digital circuits into systems. Some of the circuit design and design for testability methodologies naturally apply both to Si and GaAs technologies, but the trade-offs the designer must make are valid only for GaAs circuit design. To my knowledge, this is the first attempt to treat the design of digital GaAs circits in this manner. Three earlier books on GaAs technology present an excellent overview and focus mainly on the material and device aspects of the technology and its application to the design of microwave circuits.

The book contains seven chapters that address different technological aspects and problems related to the design of digital GaAs circuits. Chapter 1 presents a description of fabrication and operating principles of GaAs components. Chapter 2 discusses the basic principles for logic gate design in different GaAs logic families. The design of GaAs logic circuits using basic logic gates is presented in chapter 3. These circuits are most commonly used as parts of complex integrated circuits or as standalone, off-the-shelf components. Chapter 4 presents integrated circuit design principles (including different design methodologies) and traces the sequence that the designer typically follows, starting from a functional circuit description to producing a pattern generation tape that is used to fabricate a chip. Chapter 5 addresses packaging issues and considerations that can limit circuit performance. Chapter 6 presents high-speed testing issues and techniques and emphasizes the need to consider testing as part of the design activity. Finally, chapter 7 deals with the impact of GaAs technology insertion into system design. After all, integrated circuits are used eventually in designing systems, and maximum advantage of a circuit technology can be achieved only when the circuit designer understands the impact on system design and the system designer understands the capabilities, limitations, and inherent characteristics of the circuits used.

Material in these chapters is condensed to provide a basic understanding of both principles and techniques. In fact, some of the chapters could form subjects for whole books. The reader more interested in the material of specific chapters should consult the cited references, where the subject is treated in much greater detail. Also, keep in mind that GaAs technology is a fast-evolving one; debate continues among researchers on many aspects of the technology. The principles and techniques presented here should be considered always under the assumptions made and the design and application requirements presented. This text is meant to provide a broad picture of a wide range of technological aspects related to the design and utilization of GaAs digital integrated circuits. It is my hope that you will readily profit from reading it.

Nick Kanopoulos
Research Triangle Park, North Carolina

ACKNOWLEDGMENTS

This book was made possible with the direct and indirect contributions of many people to whom I wish to express my gratitude. The Research Triangle Institute (RTI), through its professional development award program, supported the beginning of this effort, and my manager James Clary encouraged me to continue and supported this work until the end. Professor Pete Marinos from Duke University encouraged the idea of writing a book on the subject and reviewed the outline and part of the work. Professors Craig Casey and Apostolos Dollas, also from Duke, reviewed some chapters and gave me helpful feedback. My old-time friend Panos Argyroudis, who was designing GaAs chips for GigaBit Logic at the time I was writing, reviewed parts of the book from a Si-designer-converted-to-GaAs-designer point of view and gave me very helpful comments. Dr. Agit Rode from TriQuint was of great help in obtaining permissions for material related to TriQuint's technology and in discussing different issues of the fabrication and testing of GaAs circuits. I also had long discussions with Arthur Frasier, also from TriQuint, on reliability issues and design approaches that may help increase the circuit reliability.

The secretarial and technical assistant staff in the Center for Digital Systems Research at RTI did a superb job in typing and editing the manuscript and drafting the figures, all with computer-aided document production tools. Sally Pilson, Jan Muller, Bernie Taylor, and Betsy Shay worked on the manuscript at some point while Kristi Simmons and Susan Rogers carried most of the weight until the end.

My editor at Prentice Hall, Elizabeth Kaster, was instrumental in getting four technical reviews that had significant, highly appreciated contributions to the final shape of the manuscript. Ms. Kaster's knowledge of the publishing business and her timely responses to all my questions made this part of the project an easy one.

Last, but not least, my wife and best friend Kimberley read and edited the first manuscript from the graduate student point of view, which was very valuable in shaping the discussion on the various topics in the book. Her continuous emotional support and understanding, undivided even after our little Nichole came to life, kept this effort alive and on schedule.

All these people I thank for their involvement and contribution to the development of this book.

GALLIUM ARSENIDE DIGITAL INTEGRATED CIRCUITS

1

GaAs COMPONENTS

The GaAs components used for implementation of digital integrated circuits (ICs) are mainly Field-Effect Transistors (FETs), Heterojunction Bipolar Transistors (HBTs), and Schottky Diodes (SDs), while implementation of Monolithic Integrated Microwave Circuits (MIMIC) includes resistors, capacitors, and inductors. This chapter presents the fabrication steps involved in manufacturing these GaAs components, their electrical characteristics, their modeling for circuit simulation, and finally, the effect of scaling process parameters on the electrical behavior of these components.

1.1 DESCRIPTION OF FABRICATION STEPS

Several GaAs process techniques have been developed; their differentiation depends on the technology different manufacturers utilize and on the trade-offs they make in terms of circuit performance, level of integration, and process yield. All GaAs process techniques used, however, have two basic requirements:

1. that bulk resistivity be sufficiently high, even after the ion implantation anneal, to assure acceptable low leakage current between circuit elements, and
2. that residual impurities be low enough to ensure uniform and repeatable implanted-layer sheet resistance and acceptably low backgating.[1]

The process steps to be discussed here correspond to a GaAs Depletion Mode Process.[2] This process is used by a particular GaAs foundry (i.e., TriQuint

Semiconductor) and illustrates the typical steps used in the fabrication of GaAs devices. It should be noted here that the process steps may vary, depending on the manufacturer and the type of device the process is meant to fabricate. Published information on different processes addresses only the process steps and the resulting device characteristics. Process control parameters and details of the techniques used to achieve the device's characteristics are proprietary to the manufacturer. This process is based on 1 µm gate length, depletion mode, metal semiconductor FET (MESFET) devices. In addition to MESFETs, the process can fabricate diodes, implanted resistors, thin film resistors, capacitors, and inductors. Two levels of metalization are provided for interconnection between different devices on a circuit. Gate metal is used for the first level of interconnect; the second level is airbridge metal. The airbridge metal provides low parasitic capacitance per unit length of the metal run, since the air is the surrounding dielectric layer. This is an important characteristic since low resistance and low capacitance interconnections are critical for achieving high-speed IC operation.[2]

The first step in this GaAs IC process is a series of wafer cleaning and etching steps which are performed in order to provide an appropriate surface for device fabrication. Following this step, an N^- ion implantation is used to form the MESFET channel and Schottky diodes and may also be used to form ion-implanted resistors. N^+ ion implantation is used for forming the source and drain contact regions. Similar to the N^- implant, N^+ can also be used to form Schottky diodes and implanted resistors. The ion implantation steps are performed using Si^+ selectively implanted into the semi-insulating GaAs wafer. This is accomplished by depositing a thin layer of SiO_2 using a Chemical Vapor Deposition (CVD) step. Photoresist is used as the implant mask. A high-temperature anneal is performed to repair the implant damage and to activate the implanted ions. During the implant anneal, the SiO_2 layer provides the necessary cap to prevent dissociation of the GaAs at the surface. After annealing, the SiO_2 is removed and a layer of silicon nitride (SiN) is deposited. The two ion implantation steps are illustrated in Figure 1.1. Thin film resistors can be made at this point by using Ni-Cr resistor metal, which is deposited on the SiN. Contact metal (Figure 1.2a) is deposited at the ends of the resistors to ensure high-quality, stable electrical contacts. Alloyed Au/Ge/Ni (Figure 1.2b) is used to form ohmic contacts with low contact resistance. The gate metal (Figure 1.2c) makes a Schottky barrier contact with GaAs and is defined using a lift-off technique. The gate metal (Figure 1.2d) is Ti/Pd/Au (i.e., Ti = Schottky metal, Pd = antidiffusion barrier, Au = low-resistance component) and the smallest feature on this mask is 1 µm. This is the most critical mask for alignment. Due to its low sheet resistance (0.090 $\Omega/\Box$), gate metal is also used for interconnections in the first-level metalization. This metal layer, which is placed on top of SiN, connects to ohmic metal, gate metal, and contact metal by physically overlapping it. Therefore, in these cases, there is no need for a via to define the contact areas. On the other hand, contact to second-level metal is achieved through a dielectric via and an airbridge via. This metal layer (i.e., first metal) is also used to form one electrode of capacitors

Sec. 1.1 Description of Fabrication Steps

fabricated on the GaAs substrate. Although the same material is used for gate metal and first metal, it takes two separate processing steps to form these structures. Figure 1.2 illustrates the metalization steps, through the first metal layer, following the implantation steps.

Following the gate metalization, as Figure 1.3a illustrates, a layer of silicon nitride is deposited as a dielectric for capacitors. Dielectric vias are defined and etched in the SiN when there is a requirement to form a contact between the first and second layers of metal interconnect. Airbridge vias (Figure 1.3b) are used to make contact between the first and second levels of metal and also to form capacitors. The via defines the capacitor area where the airbridge metal (second metal layer) "lands" and becomes the top electrode of the capacitor (the bottom electrode being the first metal). Airbridge Ti/Au metal (Figure 1.3c) is used to interconnect the logic gates of an integrated circuit. It is also used to form the top electrode of a capacitor. The first metal layer is the only metal that can be connected to the airbridge metal through the two sets of vias.

Figure 1.3 illustrates the vias and second-level metalization steps. Follow-

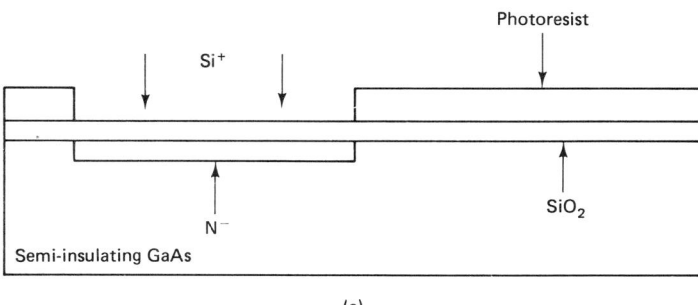

(a)

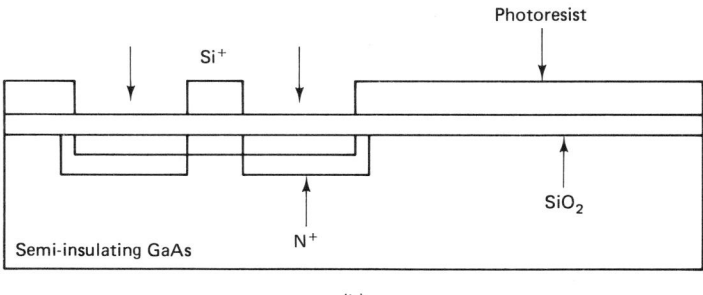

(b)

Figure 1.1 Ion implementation steps in the fabrication flow of GaAs MESFETS: (a) N^{-A} ion implantation and (b) N^+ ion implantation (from Triquint's GaAs Short Course Notes; by permission).

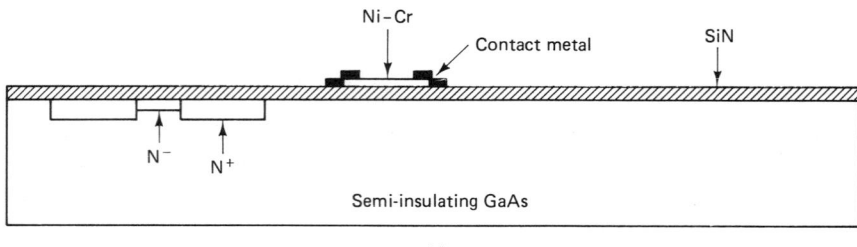

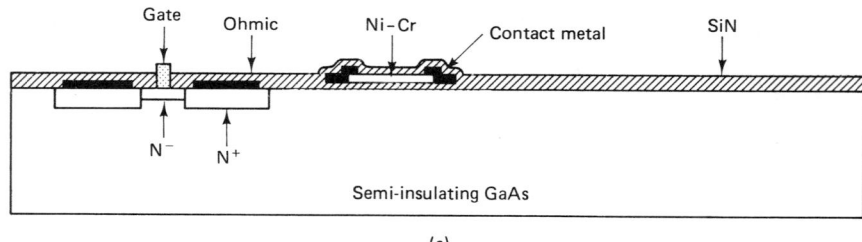

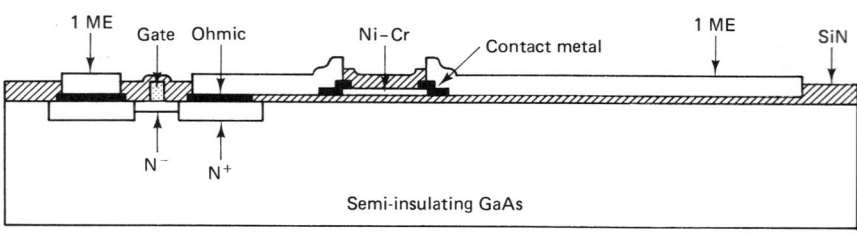

Figure 1.2 Metalization steps in the fabrication flow of GaAs MESFETS: (a) Ni_1-Cr resistor metal; (b) ohmic metalization; (c) gate metalization; and (d) first-level metal.

Sec. 1.1   Description of Fabrication Steps   5

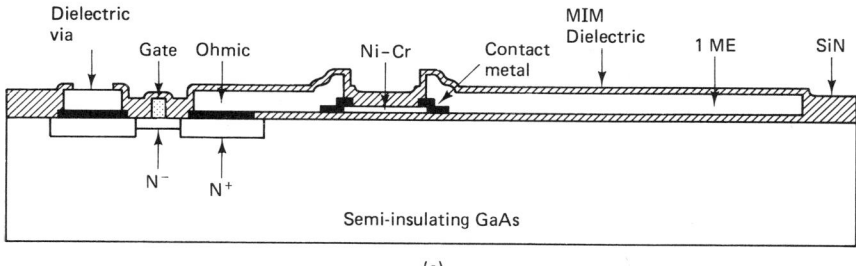

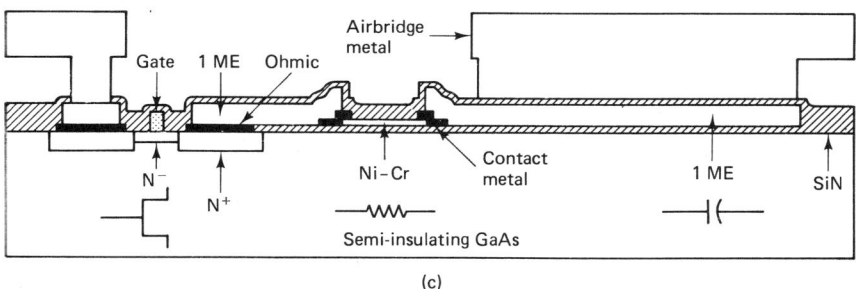

Figure 1.3 Vias and second-level metalization steps in the fabrication flow of GaAs MESFETS: (a) dielectric via; (b) airbridge via; and (c) airbridge metalization.

ing the airbridge metalization, a layer of silicon nitride is deposited to passivate all metalization. The airbridge is coated with a thin layer of silicon nitride, thick enough to passivate the metal without significantly changing the air-gap.

Contact photolithography (i.e., the mask being put in contact with the GaAs substrate) is used for the entire process. Typical component parameters are listed in Table 1.1. The significance and relationships of these parameters are discussed

TABLE 1.1 Typical Component Parameters

Component	Parameter	Value
FET	I_{DSS} (V_{DS}=2.5V, V_{GS}=0V) V_p(V_{DS}=2.5V, I_{DS} = 1%I_{DSS}) g_m(V_{DS}=2.5V, V_{GS}=0V) f_T(V_{DS}=2.5V, V_{GS}=0V)	120 mA/mm -1.5V 130 mS/mm 12.4 GHz
Schottky Diode	$C_{junction}$ J_{sat}	2fF/μm^2 1.6X10^{-14}A/μm^2
Resistor	R_{sheet}	50 Ω/▫
Thin-Film Capacitor	C $V_{breakdown}$	0.25fF/μm^2 40V
Spiral inductors	L Q (nH at 1 GHz)	1nH–20nH 1

later in this chapter. The parameters for the FET transistor are obtained from DC measurements on a 1 × 300 μm MESFET.[2]

The depletion-mode process steps were described in detail because this process is considered the most mature today. However, the enhancement-mode MESFET offers many advantages (i.e., simpler logic gates, single power supply) over depletion-mode devices and a manufacturable process for these devices is also available today.

Figure 1.4 illustrates a cross section of a self-aligned-gate E-MESFET. The "normally off" condition of E-MESFETs requires a very thin, lightly doped channel region. Such thin, active layers are very surface-sensitive, highly resistive, and difficult to control during fabrication. In order to solve the problem of high resistance between the gate and the source or drain, recessed gates were proposed or regions of enhanced doping next to the edges of the gates. Recessed gates are metal gates placed physically below the GaAs surface using a shallow channel etch. Early E-MESFET process techniques that used etching to recess the gates did not permit fine control of the threshold voltages, so they were not attractive for successful fabrication of ICs with a significant scale of integration.

More recently, semiconductor manufacturers constructed E-MESFETs using techniques borrowed from the self-aligned-gate (SAG) fabrication used for silicon MNOS.[3] The SAG technique uses a Schottky gate as a mask for implanting the source and drain regions of a device. The problem in this case is finding a gate metal system that will not be affected by the 800°C temperatures during annealing after ion implantation. One successful approach used by Fujitsu[4] employs tungsten silicide gate metalization.

The SAG process produces high carrier concentration regions immediately adjacent to the gate depletion region that affects the device threshold voltage. The

Sec. 1.1   Description of Fabrication Steps

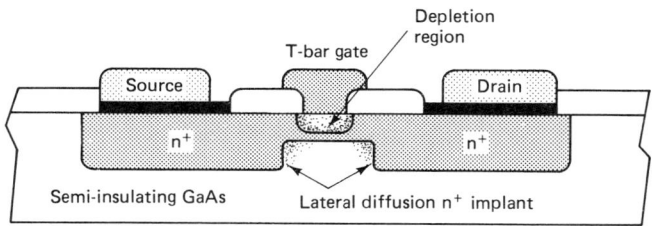

Figure 1.4   Cross section of E-MESFET T-bar gate structure.

threshold voltage is further affected by the high stress prevalent in refractory metals (such as tungsten silicide) and by the inherent lateral diffusion of n-type GaAs dopants. In order to overcome these problems, a T-bar gate structure (Figure 1.4) was developed by several laboratories (i.e., Nippon Telephone and Telegraph, Hughes Research Labs, and Cornell University). The T-bar gate structure places the heavily doped region 0.1−0.2 μm away from the critical gate channel depletion region. This minimizes the effects on threshold voltages while maintaining lower series source-to-drain resistance.

High electron mobility transistors (HEMTs) offer significant performance advantages over enhancement or depletion MESFETs. These heterojunction GaAs devices are usually fabricated using molecular beam epitaxy (MBE), although metalorganic chemical vapor deposition (MOCVD) is an alternative method which was under development in the middle 1980s. Figure 1.5 illustrates a cross section of an HEMT device. Processing steps for HEMTs are very similar to those for D-MESFETs, with the exception of the MBE layers, and that device isolation is provided by proton implantation rather than undoped semi-insulating GaAs.[5] The thin layer (approximately 700 Å) of aluminum gallium arsenide ($Al_{0.3}Ga_{0.7}As$) deposited over the undoped GaAs channel provides a two-dimensional gas of electrons that flows through the undoped GaAs where it can attain very high velocities.

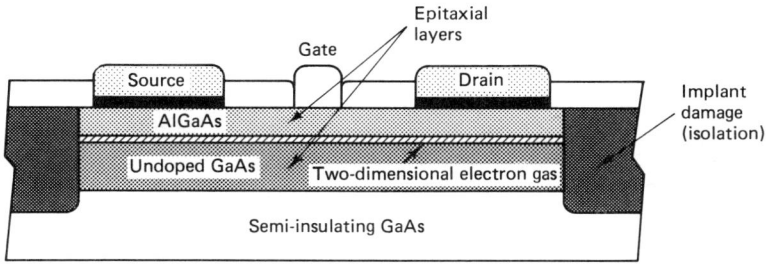

Figure 1.5   HEMT cross section.

The main difficulty in producing circuits using HEMT technology is the lack of production MBE equipment in the early 1980s and the low throughput of the existing MBE equipment. Additional problems exist with the high-channel resistance, voltage uniformity and control, and the thinness of the active layer, which is sensitive to process steps associated with heat and plasma kinetics. Variations in the thickness of the active layer due to these processing steps can result in threshold voltage shifts which are highly undesirable in IC fabrication. The superior control over layer thickness offered by MBE may overcome these limitations.

The most sophisticated (process-wise) GaAs device fabricated so far is the heterojunction bipolar transistor (HBT). HBTs are fabricated using MBE or metallorganic chemical vapor deposition (MOCVD) techniques to form the layer structure shown in Figure 1.6. The HBT structure is based on very thin layers of GaAs and AlGaAs formed on a semi-insulating undoped GaAs substrate. The proportions of aluminum and gallium are strictly controlled in these layers, leading to a desired wide-energy bandgap for the AlGaAs emitter. Since it is a vertical device, the HBT's speed characteristics are mainly dependent on the thin-base region and the high-electron mobility of GaAs and AlGaAs, rather than the gate length of a MESFET. Another important characteristic of the HBT device is that its threshold voltage is determined in part by the bandgap of the AlGaAs and GaAs layers, which is relatively constant, providing a built-in threshold voltage control for the device in contrast to MESFET devices, where pinch-off voltage is affected by channel doping and layer thickness.

Presently, there are several fabrication difficulties in commercially producing HBT circuits. Their fabrication requires sophisticated MBE techniques to grow the heterojunction layers—the main fabrication difficulty. In addition, ion implantation is used for base region contacts and for providing electrical isolation between devices by damaging the crystal lattice.

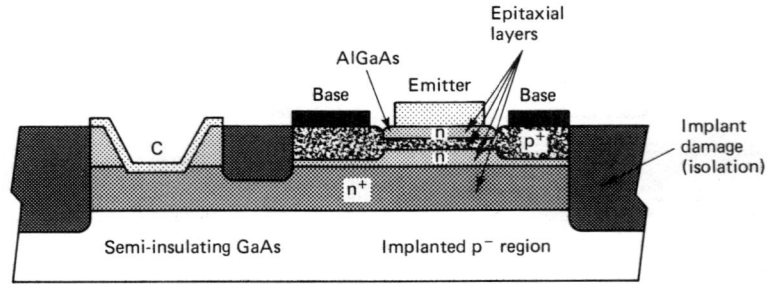

Figure 1.6 HBT cross section.

1.2 JFET AND MESFET PRINCIPLES OF OPERATION

The junction field-effect transistor (JFET) is basically a voltage-controlled resistor. Its conduction process is dominated by one kind of carrier that designates the JFET as a "unipolar" device to distinguish it from bipolar devices where both types of carriers are involved in the conduction process. The first working JFET was reported by Dacey and Ross in 1955.[6]

The first metal-semiconductor field-effect transistor (MESFET) using an epitaxial layer of GaAs on semi-insulated GaAs substrate was fabricated by Hooper and Lehrer in 1967.[7] The MESFET operates similarly to the JFET; however, the gate electrode in the MESFET is a metal semiconductor rectifying contact instead of a p-n junction.

Field-effect transistors are square-law or linear devices and therefore have much smaller intermodulation and cross-modulation products than those of bipolar transistors, which have generally exponential characteristics. Since FETs are unipolar devices, they do not experience problems with minority-carrier storage effects; thus, they have high switching speeds and high cut-off frequencies. At high current levels, FETs have a negative temperature coefficient; this implies that current decreases as temperature increases. This characteristic leads to a more uniform heat distribution over the device area and prevents the FET from experiencing thermal runaway or second breakdown associated with bipolar transistors.

1.2.1 The Junction FET

A JFET schematic diagram is illustrated in Figure 1.7. This device consists of a conductive n-type channel (a p-type channel may also be used) with two ohmic contacts that act as a source and a drain. Under positive bias between drain and source, electrons flow from source to drain—thus the direction of conventional current is from drain to source. The third electrode of the device is the p-type gate, which forms a rectifying junction with the channel. The junction voltage-variable depletion region width is used to control the effective cross-sectional area of the conductive channel. If a reverse bias is applied between the p^+ gate and the channel, the depletion region will extend deeper into the n-material and, therefore, the effective thickness of the channel will decrease. Since the resistivity of the channel region is fixed by its doping, the channel resistance varies with changes in its cross-sectional area. Therefore, varying the thickness of the depletion region changes the effective cross-section of the channel and, hence, the resistance of the device. This operating characteristic leads to the claim that the JFET operates as a voltage-controlled resistor in the nonsaturated current region. Since the p^+ regions are heavily doped, their conductivity is high and one can assume that the potential throughout each gate is uniform. The channel material, however, is less heavily doped and the potential varies with position. If we consider the channel as a distributed resistor carrying a current I_D, then the

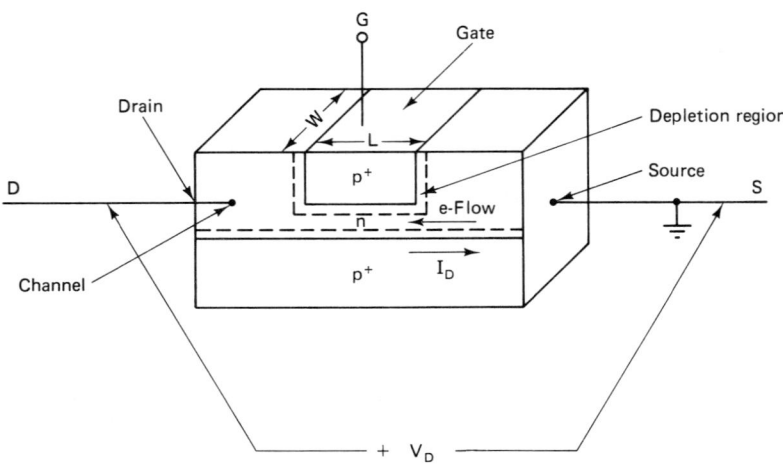

Figure 1.7 Simplified view of a JFET.

voltage in the channel will vary between a maximum value V_D at the drain end and 0 at the source end. For low current values, we can assume that this voltage varies linearly with the distance from the drain end. Figure 1.8 illustrates this voltage variation.

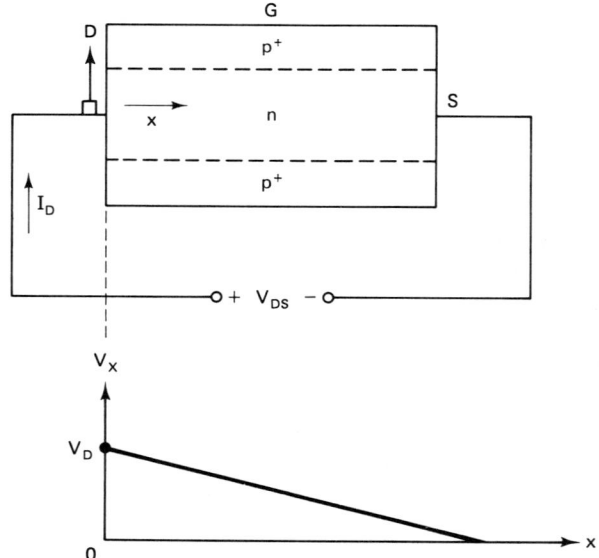

Figure 1.8 Voltage variation along the channel with gate voltage $V_G = 0$ and small I_D.

Sec. 1.2 JFET and MESFET Principles of Operation

If the channel were p-type, then the gate regions would be n$^+$ type and holes would flow from the source to the drain in the same direction as the current flow.

The current-voltage characteristics of a typical JFET are shown in Figure 1.9. The drain current I_D is plotted against the drain voltage V_D for various gate voltages V_G. There are three regions of operation that can be defined for these characteristics: the linear region where V_D is small and I_D is proportional to V_D; the saturation region where I_D is essentially constant and independent of V_D; and the breakdown region where I_D increases rapidly beyond a maximum voltage BV_D. The variation of the JFET depletion regions during operation at different current levels is illustrated in Figure 1.10. For very small values of I_D, the widths of the depletion regions are close to the equilibrium values (Figure 1.10a). However, as I_D increases, V_x becomes larger near the drain end than near the source end of the channel. For $V_G = 0$, the reverse bias across each point in the gate-to-channel junction is V_x; this can be used to estimate the shape of the depletion regions as shown in Figure 1.10b. The reverse bias is $V_{GD} = -V_D$ near the drain end and decreases to zero near the source. At this point, the I-V plot for the channel departs from the linear region, which was valid at low current levels because the resistance of the constricted channel is higher than the channel resistance near equilibrium. As V_D and I_D are further increased, the depletion regions and the channel resistance continue to increase, and the channel becomes further restricted. Further increase of V_D reaches a bias point at which the depletion regions meet near the drain end of the device and essentially pinch off the channel (Figure 1.10c). At this point, further increase of V_D does not result in significant increase of I_D. Beyond pinch-off, I_D is said to be saturated and it maintains

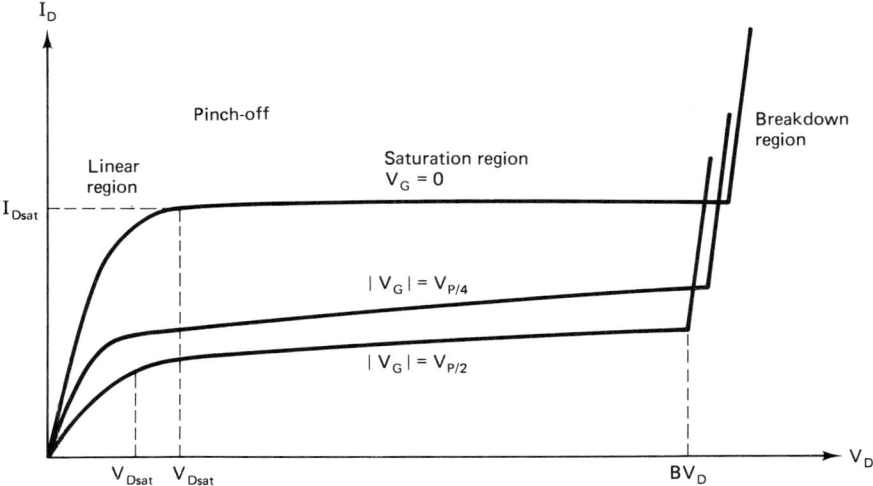

Figure 1.9 I-V characteristics of a JFET. For a given gate voltage V_G, the V_D and I_D at the pinch-off point are designated as V_{Dsat} and I_{Dsat}, respectively.

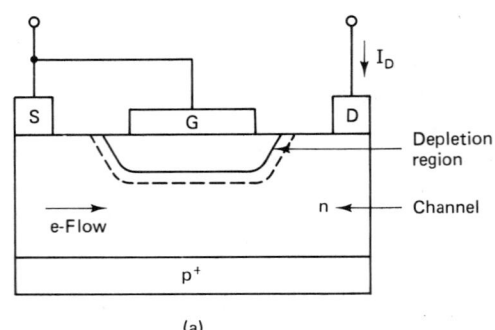
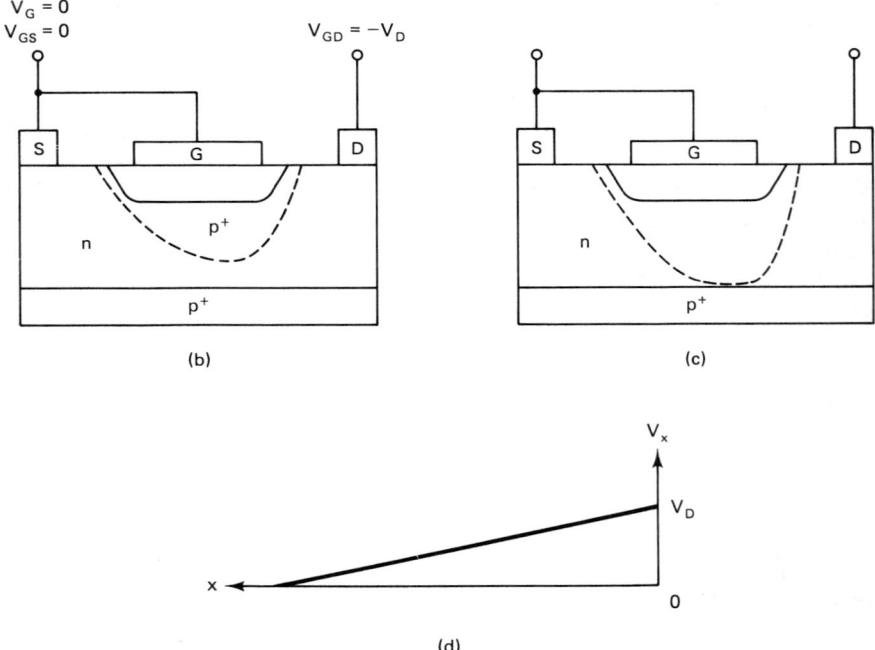

Figure 1.10 Depletion regions in the channel of a JFET with $V_G = 0$ and several values of V_D: (a) linear region; (b) near pinch-off, and (c) beyond pinch-off.

approximately its value at pinch-off. The nature of this region is complicated in that the limiting factor for the saturation current may be mainly electron velocity saturation rather than the effect of electron injection into a depletion region. As the reverse gate bias increases (Figure 1.9), both V_{Dsat} and the corresponding I_{Dsat} decrease. This is because larger negative V_G results in larger depletion regions which render the effective channel thickness smaller. Since a smaller channel

Sec. 1.2 JFET and MESFET Principles of Operation

thickness yields a higher channel resistance, pinch-off is induced at lower I_D values. This is illustrated in Figure 1.9 where the curve with the larger V_G value has the smallest slope (or the least current).

A simplified calculation of the pinch-off voltage and the saturation current can be derived by considering the approximate form of the channel illustrated in Figure 1.11. Calculations are performed based on the assumption that the region is uniformly doped and the gradual channel approximation is valid. This approximation assumes that the source and the drain are far enough apart so that the depletion region is not affected by the electric field at the source and the drain. The pinch-off voltage can be found by calculating the reverse bias between the p^+ gate and the n-channel at $x = 0$, which is the drain end of the device. The width of the depletion region at $x = 0$ is given by the abrupt junction expression:

$$W_{x=0} = \left[2\varepsilon_S \frac{(-V_{GD} + V_{bi})}{qN_d} \right]^{1/2} \tag{1.1}$$

where ε_S is the material permitivity and V_{bi} is the built-in potential between the $p^+ n$ junction and is given by $\left[\dfrac{kT}{q}\right] \ln \left[\dfrac{N_a N_d}{n_i^2}\right]$.

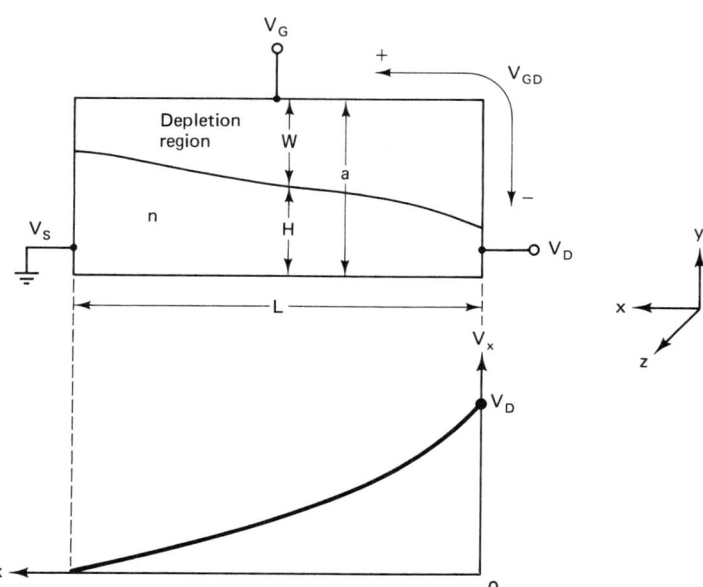

Figure 1.11 Approximation of the channel, with definition of parameters and dimensions used for pinch-off voltage and saturation current calculations.

According to the device operation discussed above, pinch-off occurs at the drain end of the channel when:

$$h_{x=0} = a - W_{x=0} = 0 \tag{1.2}$$

or when $W_{x=0} = a$. The value of $-V_{GD}$ at pinch-off is the pinch-off voltage defined as V_p. Using Equation (1.2),

$$a = \left[\frac{2\varepsilon_s (V_P + V_{bi})}{qN_d}\right]^{1/2}$$

or

$$V_P = \frac{qN_d}{2\varepsilon_s}(V_{bi} - a^2) \tag{1.3}$$

The relation of V_P to V_G and V_D is

$$V_P = V_D - V_G = -V_{GD} \tag{1.4}$$

where V_{GD} is zero or negative for proper device operation for an n-channel JFET. Equation (1.4) indicates that pinch-off results from a combination of gate-to-source and drain-to-source voltage. The same equation also indicates that pinch-off is reached at lower values of V_D, and hence I_D, when a negative gate bias is applied—as Figure 1.9 illustrates.

Calculation of the channel current is based on the assumption that I_D saturates at pinch-off and the saturation current, I_{Dsat}, beyond pinch-off remains fairly constant at its pinch-off value. We should note, however, that in small devices used in logic circuits, it is common to find the output current saturated by carrier velocity saturation rather than by pinch-off of the channel.[8]

The current density along the channel (x direction) is given by the ohmic law equation:

$$J_x = \sigma E_x \tag{1.5}$$

or

$$J_x = qN_d\mu E_x \tag{1.6}$$

where J_x is the current density, σ the conductivity of the n-channel material, E_x the electric field along the x direction, and μ the carrier mobility. It is assumed here that the mobility is field independent. It is also noted that $E_x = \frac{-dV_x}{dx}$, the minus sign indicating that V_x decreases as x increases along the channel (Figure 1.11). Assuming that the channel is symmetrical with respect to the x axis (Figure 1.11) and the effects of the gates in each half of the channel are the same, then the channel (drain) current for the upper half-channel region is given by

Sec. 1.2 JFET and MESFET Principles of Operation

$$I_D = qN_d\mu \left[\frac{-dV_x}{dx}\right](a - W_x)Z \tag{1.7}$$

where Z is the channel depth in the z direction.

The channel half-width at any point x depends on the reverse bias between the gate and channel at this point, denoted $-V_{GX}$. This can be expressed as

$$h(x) = a - W(x) = a - \left[\frac{2\varepsilon_S(V_{bi} - V_{GX})}{qN_d}\right]^{1/2} \tag{1.8}$$

Using $V_{Gx} = V_G - V_x$ and the expression for V_p given in Equation (1.3), we can rewrite Equation (1.8) in terms of V_P:

$$h(x) = a\left[1 - \left(\frac{V_{bi} + V_x - V_G}{V_P}\right)^{1/2}\right] \tag{1.9}$$

Substitution of Equation (1.9) into Equation (1.7) yields

$$-I_D\, dx = qN_d\,\mu a\left[1 - \left(\frac{V_{bi} + V_x - V_G}{V_P}\right)^{1/2}\right]dV_x \tag{1.10}$$

By integrating along with channel, one can obtain

$$I_D = g_D V_P\left[\frac{V_D}{V_P} + \frac{2}{3}\left(\frac{V_{bi} - V_G}{V_P}\right)^{3/2} - \frac{2}{3}\left(\frac{V_D - V_G + V_{bi}}{V_P}\right)^{3/2}\right] \tag{1.11}$$

where $g_D = 2Z\mu qN_d\frac{a}{L}$ is the channel conductance (also called drain conductance), assuming negligible $W(x)$ and a symmetrical channel, and V_G is negative. This expression is valid up to pinch-off where $V_D - V_G = V_P$. Based on the assumption that the saturation current remains essentially constant at its pinch-off value, an expression for I_{Dsat} can be written as follows:

$$\begin{aligned}I_{Dsat} &= g_D V_P\left[\frac{V_D}{V_P} - \frac{2}{3}\left(\frac{V_G + V_{bi}}{V_P}\right)^{3/2} - \frac{2}{3}\right] \\ &= g_D V_P\left[\frac{V_G}{V_P} - \frac{2}{3}\left(\frac{V_G + V_{bi}}{V_P}\right)^{3/2} + \frac{1}{3}\right]\end{aligned} \tag{1.12}$$

where $\frac{V_D}{V_P} = 1 + \frac{V_G}{V_P}$

As Equation (1.12) suggests, the saturation current is greatest when V_G is zero and reduces as V_G becomes negative. For drain voltages beyond V_{Dsat}, the drain current is assumed to sustain its saturation value. As the drain voltage increases significantly beyond V_{Dsat}, avalanche breakdown of the gate-to-channel junction will occur and the I_{Dsat} will increase abruptly. This is illustrated in

Figure 1.9. The breakdown occurs at the drain end of the channel where the reverse voltage is the highest and is normally destructive if I_d is not limited.

A very important device parameter that can be derived from Equation (1.12) is the device transconductance that relates changes of I_D to variations of V_G. The device transconductance is expressed as

$$g_{msat} = \frac{\partial I_{Dsat}}{\partial V_G} = g_D \left[1 - \left(\frac{V_G}{V_P}\right)^{1/2}\right] \qquad (1.13)$$

The assumption of a constant electron mobility in the channel leads to constant channel conductivity in the g_D term. As mentioned, however, electron velocity saturation at high fields may make this assumption invalid. This is particularly likely for devices with short channels.

The assumptions made concerning the JFET operation are valid in an ideal situation but rarely in cases of real devices. These assumptions often were made to produce closed-form mathematical models that approximate the operation of real devices and to give the reader an idea of the basic principles upon which these devices operate.

1.2.2 The Metal Semiconductor FET

The operation of a MESFET is very similar to the operation of the JFET, discussed above. The difference is that the depletion of the channel in a MESFET is accomplished by the use of a reverse-biased metal-semiconductor Schottky barrier instead of a p-n junction as in a JFET. MESFETs are presently the predominantly used device for the design of GaAs integrated circuits. This is mainly due to the simplicity of the Schottky barrier gate, which allows device fabrication to close geometrical tolerances.

Although the electrical behavior of MESFETs can be described in general with the formulas discussed for JFETs, significant discrepancies are encountered between theory and experiment when short channel devices are considered. A major reason for the discrepancies is field-dependent mobility which leads to high field velocity saturation. Figure 1.12 illustrates the dependence of the drift velocity on the electric field for GaAs at room temperature.[9] At low fields, the velocity increases linearly with the field and the slope corresponds to a constant mobility $\mu = \dfrac{d\upsilon}{d\varepsilon}$. The drift velocity first reaches a peak value and then decreases toward a constant value of about $6 \sim 8 \times 10^6 \ cm/sec$, at high fields. The field value at which the drift velocity reaches its peak is designated as the critical field E_c. For short MESFET gates, a saturated velocity model has been proposed[10] and is expected to be valid for very short gates ($L < 2$ μm) in which current saturation is assumed under the gate. This saturated current, I_{Dsat}, is directly modulated by the difference between the depletion width, W, and the channel depth, a. Assuming uniformly doped channels, this model yields:

Sec. 1.2   JFET and MESFET Principles of Operation   17

$$I_{Dsat} = q\upsilon_{sat} ZN_d (a - W) \tag{1.14}$$

As the MESFET gate length becomes shorter and the drain voltage becomes larger, two-dimensional effects will dominate the operation of the device. Two-dimensional effects are not included in the approximations made for deriving closed form expressions for the I-V characteristics of JFET and MESFET devices. The key features of a MESFET operated in the saturation region are shown in Figure 1.13.[11] The narrowest opening of the channel is at the drain end. At low field, the drift velocity rises to a peak value and falls to the low saturated-velocity value under the drain edge of the gate. Current continuity is preserved by heavy electron accumulation in this area because of channel narrowing and progressive slowing of the electrons as x increases. As the channel widens and electrons move faster, a strong depletion layer results. This is exactly the opposite of what happens between points x_1 and x_2 in Figure 1.13. The charges in the accumulation and depletion layers are approximately equal, and this stationary dipole layer is responsible for most of the drain voltage drop.

For very short gates, the electrons may not reach equilibrium transport conditions in the high field region of the channel.[12] The nonequilibrium situation is illustrated in Figure 1.14. When the field has a value below the critical value E_c at which the drift velocity reaches its peak, the electrons are in equilibrium. As soon as the electrons enter the high-field region where $E > E_c$, they are accelerated to a higher velocity before reaching the equilibrium velocity. Figure 1.14 shows the expected velocity overshoot to be more than twice the peak velo-

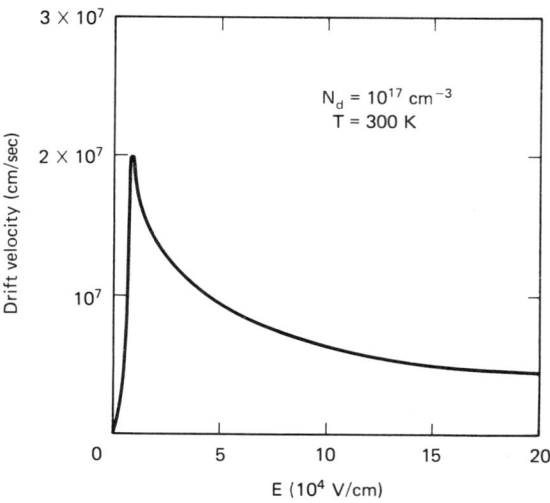

Figure 1.12 Drift velocity versus electric field (after P. Smith, Imoue, Frey, Reference 9).

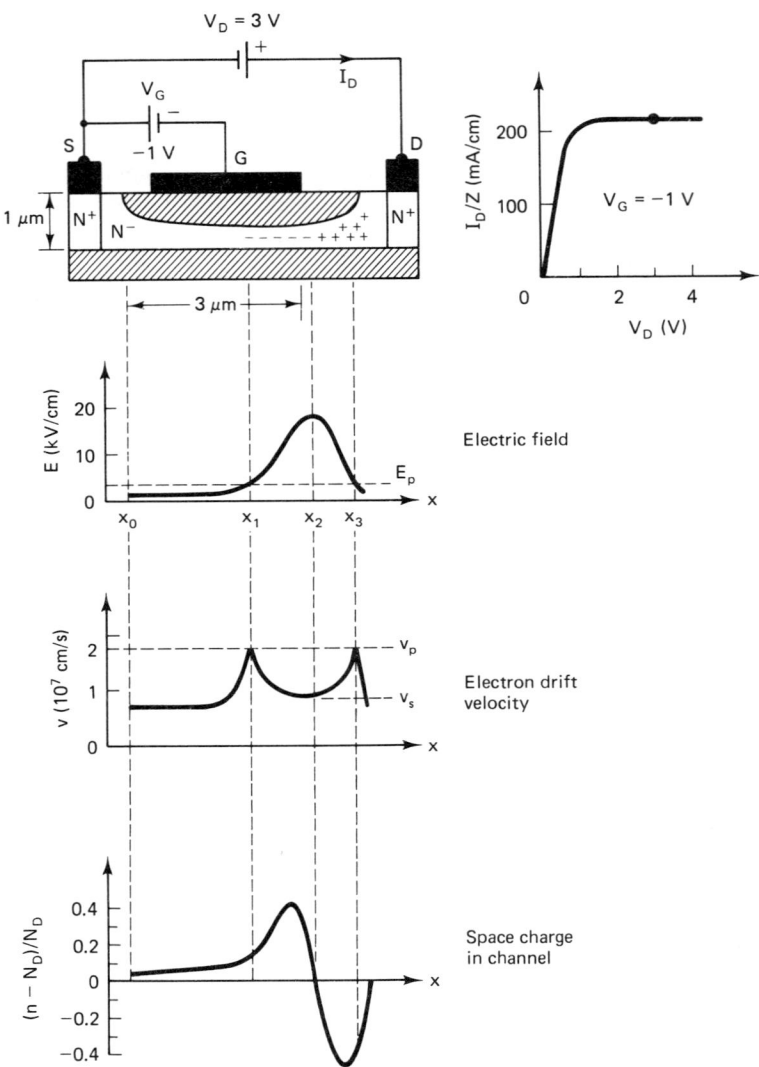

Figure 1.13 Channel cross section, electric field, drift velocity, and space-charge distribution in the channel of a GaAs MESFET operated in current-saturation region (after Leichti, Reference 11).

city v_c and the relaxation to the equilibrium after traveling a distance of 1 μm. The overshoot will result in shorter electron transit times through the high-field region and is expected to improve the high-frequency response of MESFET.

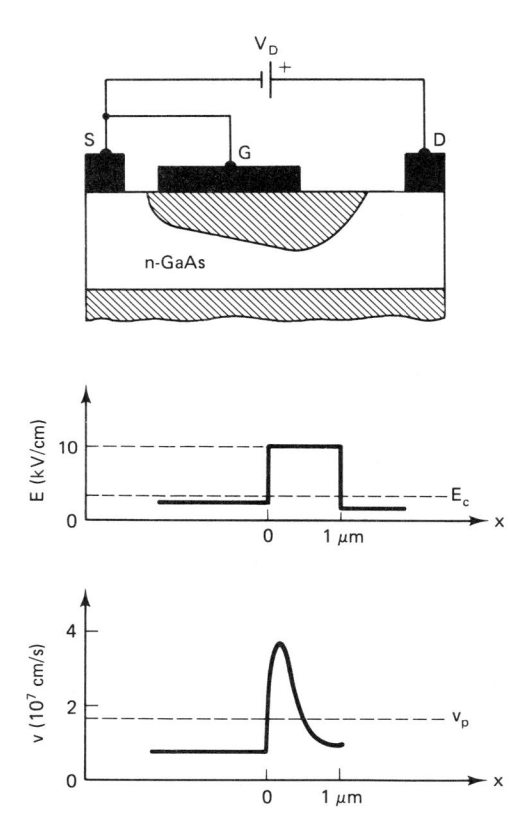

Sec. 1.2 JFET and MESFET Principles of Operation   19

Figure 1.14 Velocity overshoot of electrons as they enter the high field region ($E > E_c$) under the gate (after Liechti, Reference 11).

1.2.3 The Normally-Off FET

The operation of JFETs and MESFETs discussed so far concerns only the n-type normally-on (depletion-mode) device. This device, as described, has an undepleted n-type conductive channel at zero gate bias. An n-type normally-off (enhancement-mode) device is one that does not have a conductive channel at $V_G = 0$, because the built-in potential, V_{bi}, of the gate junction is sufficiently high to deplete the channel region totally. This device is attractive for high-speed, low-power circuits. The device turns on and current flows through the channel when a positive bias greater than the device threshold is applied to the gate. The threshold voltage V_{th} is approximately given by

$$V_{th} = V_{bi} - V_P \tag{1.15}$$

or

$$V_P = V_{bi} - V_{th} \qquad (1.16)$$

where V_P is the internal pinch-off voltage given by $V_P = \dfrac{qN_d}{2\varepsilon_s}(V_{bi} - a^2)$. The drain saturation current near the threshold can be obtained by substituting V_P from Equation (1.16) to Equation (1.13) and using Taylor series expansion around the point $V_G = V_{th}$. This procedure yields:[13]

$$I_{Dsat} = \dfrac{Z\mu\varepsilon_s}{2aL}(V_G - V_{th})^2 \qquad (1.17)$$

The basic current-voltage characteristics of normally-on and normally-off devices are similar, the main difference being the shift of threshold voltage on the V_G axis. The normally-off device has $I_D = 0$ for $V_G = 0$. Current conduction begins at $V_G = V_{th}$ and the current varies according to Equation (1.17) for $V_G > V_{th}$. The results obtained for normally-on devices are applicable to normally-off devices with appropriate change of gate voltages. Figure 1.15 illustrates the two modes of device operation.

The operation of the JFET and MESFET and the discussion of their characteristics were based on the following assumptions:

1. The electron mobility is constant.
2. There is a sharp transition between the charge neutral and fully depleted regions (i.e., depletion layer approximation).
3. Channel approximation is gradual.

As already mentioned, for short channel FETs or FETs with large bias between drain and source, assumptions (1) and (3) become invalid. Assumption (2) is always an approximation at best. Removing these assumptions makes the derivation of quantitative results, in a format easy to interpret, a very difficult process. A qualitative description of some general consequences of velocity saturation (which defies the above assumptions) is given below:

1. The transition from the nonsaturated region of operation to saturation is much more abrupt than in the classical case described for the JFET and MESFET.
2. The drain current at the onset of saturation is smaller due to velocity saturation. This also applies to the drain saturation voltage.
3. The drain voltage at the onset of saturation is a weaker function of gate-to-source voltage than is implied by the classical FET treatment.
4. When velocity saturation dominates the operation of the FET, then $g_m \sim V_{max} W$, where V_{max} is the electron saturated velocity and W the channel width.
5. When velocity saturation dominates, g_m no longer increases with decreasing

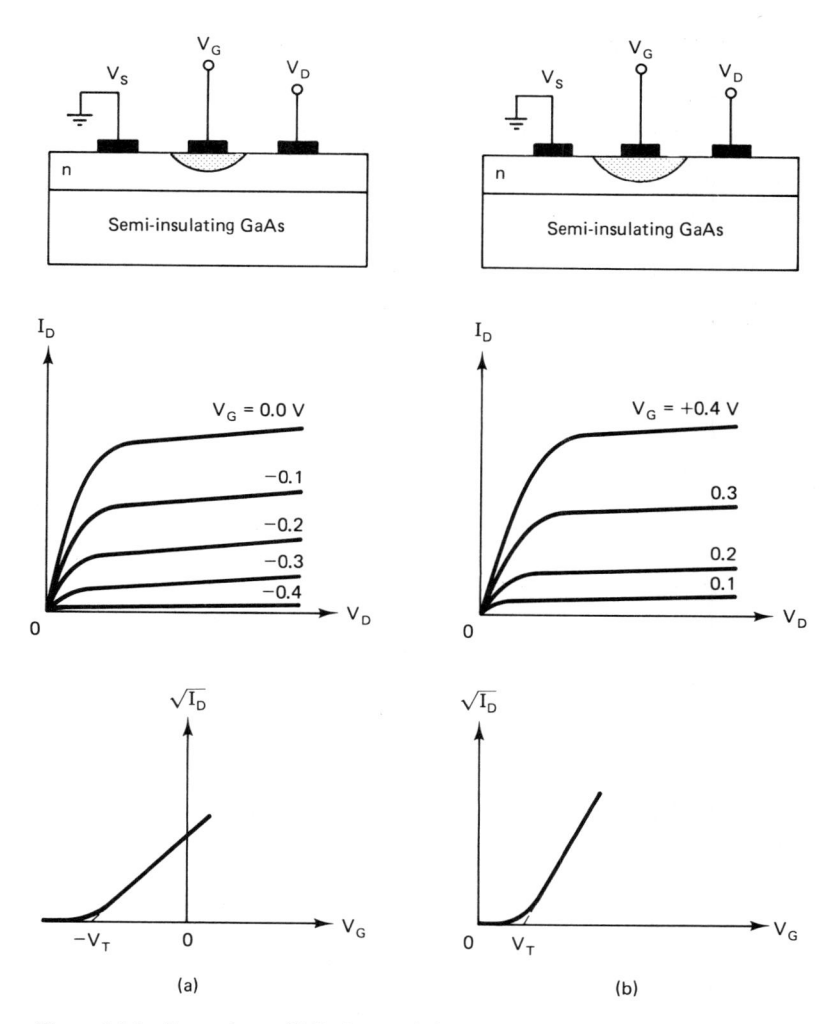

Sec. 1.2   JFET and MESFET Principles of Operation   21

Figure 1.15 Comparison of I-V characteristics: (a) normally-on (depletion) MESFET, and (b) normally-off (enhancement) MESFET.

L (as Equation [1.14] implies); it is independent of L for very short channels.

In general, these features due to velocity saturation become more prominent as the gate length becomes shorter. For circuit design with small geometry devices, we should note the importance of items (4) and (5) above. The reader interested in a two-dimensional analysis of the FET behavior, with and without velocity saturation, should consult references.[14,15]

1.3 HIGH ELECTRON MOBILITY TRANSISTOR (HEMT)

The HEMT device is a variation of an E-MESFET but offers improved performance over a standard E-MESFET, particularly at low temperature. This device is also known as two-dimensional electron gas FET (TEGFET), modulation doped FET (MODFET), and selectively doped heterojunction transistor (SDHT). The concept behind these modulation doped structures is to place donor atoms in an AlGaAs (a wide bandgap material) layer adjacent to an undoped GaAs (lower bandgap material) channel layer which receives the free electrons from the ionized donors due to the electron's inherent affinity to move to the lower bandgap region.[16]

The most widely used form of this device is composed of a thick undoped layer of GaAs covered with a thin (approximately 700 Å) layer of AlGaAs. A number of properties are unique to HEMT devices. The gate metal forms a Schottky barrier with the AlGaAs donor layer. The higher bandgap of that donor layer makes the HEMT similar to a Si-MOSFET, and this physical correspondence allows the HEMT to be analyzed using standard MOS theory.[17] However, there are also very important differences between the MOSFET and the HEMT because the high bandgap material is not only a dielectric but also a supplier for channel electrons. These differences become apparent when the gate is forward biased. After the channel is fully turned on, parallel conduction is induced in the AlGaAs donor layer. Since charges in the highly doped donor layer have a low mobility, this action causes a reduction of the transconductance. The device is turned off for negative gate voltages. As the gate voltage increases, channel conduction rises rapidly due to the high mobility of the electrons in the channel and the maximum density. The application of additional positive gate voltage induces charges into the AlGaAs layer and results in a reduction of the transconductance.

The principal performance advantage of the HEMT over a MESFET is due to the fact that the electron mobility in the channel of a HEMT is higher than in a MESFET, because there are no doping ions in the channel to scatter carriers. Electron mobilities of 8500 $cm^2v^{-1}s^{-1}$ at 300K and 50000 $cm^2v^{-1}s^{-1}$ at 77K can be achieved in HEMT devices compared to channel mobilities of 4500 $cm^2v^{-1}s^{-1}$ at 300 K achievable in typical GaAs MESFETs. This allows a HEMT to develop nearly its full transconductance with a small gate logic voltage swing (~ 200 mV) above threshold. Furthermore, the thin gate insulator region gives these devices comparatively high transconductance. The reported switching speeds for minimally loaded HEMT inverters ($L = 1$ µm) range from 12 to 17 psec.[18,19,20]

Since HEMT devices belong to the E-MESFET family of technologies, they enjoy the inherent advantages and disadvantages of such technologies. For example, using similar ground rules for threshold uniformity control, a variance of threshold voltage of about 10 mV must be obtained for HEMT VLSI circuits. The superior control offered by MBE is inherently capable of attaining this degree of uniformity; however, making such control repeatable is necessary for producing HEMT circuits. Other problems to be overcome for the HEMT technology to realize its potential in the commercial IC market are light sensitivity, threshold voltage shift with temperature, and instability in the I-V characteristic which manifests in a hysterisis loop of I_{DS} curves.[21] An important advantage of HEMT devices is that they achieve their high speed at lower logic voltage swings than conventional E-MESFETs, which may result in dynamic power dissipation one to two orders of magnitude lower. Although the small logic swing may compromise the circuit noise margin, low power is of great importance for HEMT VLSI applications—and this is the key motivating factor for continuing research in this technology in the late 80s.

1.4 HETEROJUNCTION BIPOLAR TRANSISTOR (HBT)

The main feature of the HBT relies upon a wide bandgap emitter, wherein part of the energy bandgap difference between the emitter and base, in addition to electric fields, is used to control the flow and distribution of electrons and holes. The wide bandgap emitter allows the base to be more heavily doped than the emitter, leading to a low base resistance and emitter-base capacitance (both factors are essential for high-frequency operation) while maintaining a high emitter injection efficiency.[22]

Most efforts on MBE-grown HBTs have been focused on the AlGaAs/GaAs system. By using a wide bandgap n-AlGaAs emitter the doping level of the p^+ GaAs base region can be made very high without degrading the current gain, since hole injection from the base into the wider bandgap emitter is essentially impossible. Another advantage of HBT technology is determination of the threshold voltage strictly by the inherent bandgap of the AlGaAs and GaAs, which varies very little compared to controlling the channel doping and layer thickness of MESFET structures.

The majority of the research and development work in HBT technology, in the late 80s, has been concentrated on the material and fabrication requirements. Molecular beam epitaxy (MBE) has been used almost exclusively (MOCVD is considered an alternative) for this work. A wide range of improvements in MBE material, device structures, and fabrication technologies must be achieved to make HBT technology a practical reality. The HBT is very difficult to fabricate due to its use of complicated MBE heterojunction layer structures in conjunction with ion implantation steps for contacting critical p^+ base regions as well as for device isolation created by implemented damage. As can be observed from this brief discussion, GaAs HBTs have all the lithographic difficulties of the most advanced GaAs and/or Si device technologies. They have more difficult MBE heterojunction layer requirements than HEMT, and all the demanding requirements of small, low-resistance p- and n-type ohmic contacts.

Performance of HBT devices ultimately is projected to be in the 100–200 GHz range with gate delays in the range of 10 psec.[23] Furthermore, the high current drive capability of these devices coupled with their threshold voltage insensitivities makes HBTs the prime candidate device for ultra-high-performance integrated circuits. Commercial application of this technology, however, will have to follow a wide range of improvements needed in materials and fabrication technologies.

1.5 MODELING OF GaAs DEVICES

The discussion of the operation of GaAs devices provides the means for understanding the device's behavior and how it produces its I-V characteristics. However, the formulae given neglect resistance and capacitance effects inherent to the devices. For a realistic description of GaAs devices, these effects must be incorporated into the device models. In addition, noise models are needed for estimating the noise characteristics of a device.

This section presents GaAs diode and FET models suitable for computer simulation of GaAs circuits. Parameters of the model equations can be used with the SPICE[24] circuit simulator to determine the dc and ac behavior of GaAs circuits.

1.5.1 The Diode Model

The diode model to be described is generally applicable to either junction diodes or Schottky barrier diodes. The equivalent circuit for the diode model is illustrated in Figure 1.16. The ohmic resistance of the diode is modeled by the linear resistor R_S.

The dc characteristics of the diode are modeled by the nonlinear current source I_D whose value is given by:

$$I_D = I_S \left[\exp\left[\frac{V'}{\eta V_t}\right] - 1 \right] \tag{1.18}$$

where I_S is the saturation current, η the ideality factor, and V_t the thermal voltage. V_t is given by:

$$V_t = \frac{kT}{q} \tag{1.19}$$

where k is Boltzmann's constant, T is the absolute temperature in degrees Kelvin, and q is the electronic charge. V_t is about 25.85 mV at room temperature. A graph of log (I_D) versus V' is shown in Figure 1.17. This graph corresponds to the forward bias region. In the ideal region of operation ($V' \leq 600$ mV), the diode characteristics can be determined by the equation:

$$\log_{10}(I_D) = \log_{10}(I_S) + 2.3 \frac{V'}{NV_t} \tag{1.20}$$

The saturation current I_S can be experimentally estimated by considering several

Sec. 1.5 Modeling of GaAs Devices

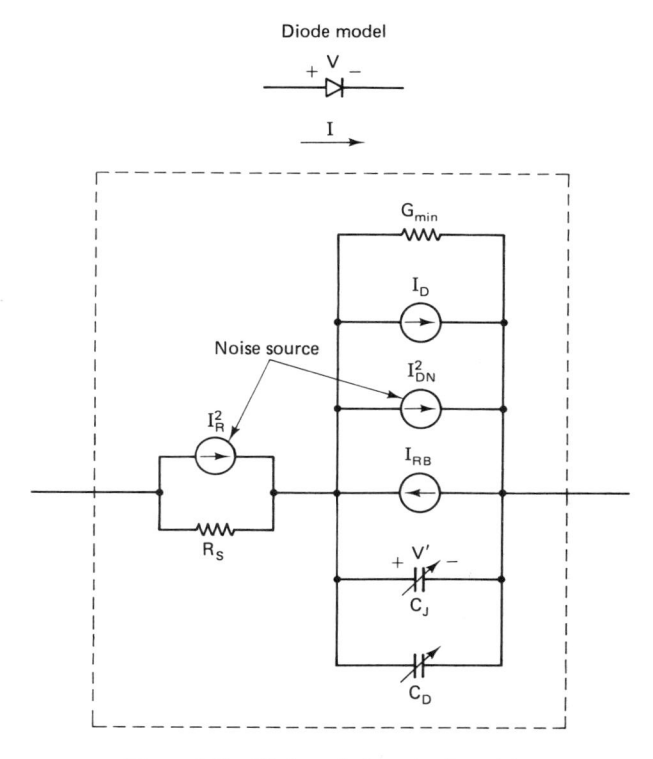

Figure 1.16 Diode equivalent-circuit model.

(V', I_D) points in this region of operation. The ideality factor η is determined from the slope of the diode characteristics in this region.

In practice, at higher levels of bias the diode current deviates from the ideal exponential characteristic given by Equation (1.18). This deviation is due to the presence of ohmic resistance in the diode. A small conductance $G_{\min}$ has been added to the small signal equivalent circuit to help the convergence of the I_D calculation in a computer environment. Therefore, at higher levels of bias, I_D is given by:

$$I_D = I_S \left[\exp\left(\frac{V'}{\eta V_t} \right) - 1 \right] + G_{\min} V' \qquad (1.21)$$

Diode breakdown at reverse bias is modeled as an exponential increase in the reverse diode current I_{RB} and is determined by the breakdown voltage V_B and the value of current at V_B, I_{BV}. I_{RB} is given by:

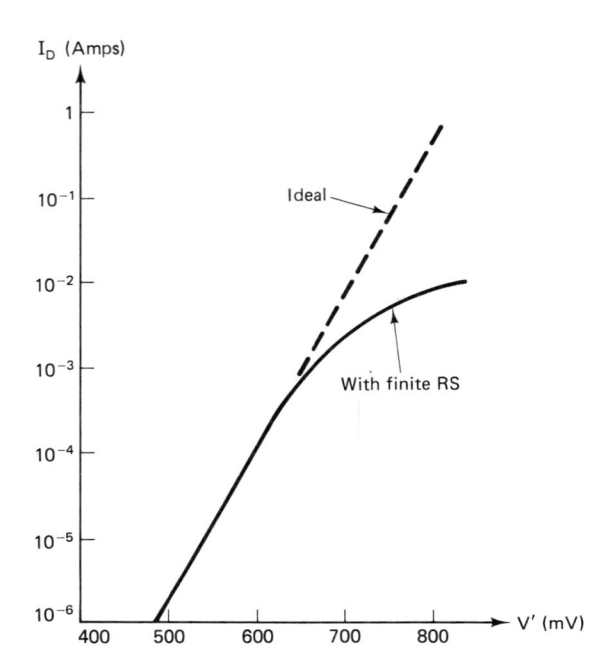

Figure 1.17 Graph of log (I_D) versus V'_D in the forward-bias region of a diode.

$$I_{RB} = I_{BV} \exp\left[-\frac{q(V' + V_B)}{KT}\right] \quad (1.22)$$

In general, there are two distinct charge storage mechanisms in a diode: charge storage in the junction depletion region, which is modeled by the C_J capacitance, and charge storage due to injected minority carriers, modeled by the C_D capacitance. C_D is zero for GaAs Schottky diodes because there is no minority carrier injection. The C_J capacitance depends on the zero bias junction capacitance C_{jo}, the built-in potential of the diode V_b, and a grading coefficient M. These parameters are obtained from capacitance bridge measurements at several values of reverse bias. Typically, V_b is 0.7–0.8 volts for a GaAs Schottky diode and M is 0.3–0.5 for both junction diodes and Schottky Barrier diodes. C_J is given by:

$$C_J = C_{jo}\left[1 - \frac{V'}{V_b}\right]^M \quad (1.23)$$

The C_D capacitance of a p-n junction diode is basically determined by the transit time t_t and is given by:

Sec. 1.5 Modeling of GaAs Devices

$$C_D = t_t \left[I_S \frac{1}{\eta V_t} \exp\left(\frac{V'}{\eta V_t}\right) + G_{min} \right] \quad (1.24)$$

The temperature dependence of the saturation current is defined by the energy gap E_g and the saturation current temperature coefficient P_{ti}. If I_S is the saturation current at T_1, then the saturation current I'_S at T_2 is given by

$$I'_S = I_S \left(\frac{T_2}{T_1}\right)^{\frac{P_{t2}}{N}} \exp\left[\frac{q E_g}{\eta k}\left(\frac{T_2 - T_1}{T_1 T_2}\right)\right] \quad (1.25)$$

1.5.2 The JFET Model

The equivalent circuit for an n-channel JFET (and similarly for a MESFET) is illustrated in Figure 1.18. (For a p-channel device, both the polarities of the voltages V'_{GD}, V'_{GS}, and V'_{DS} and the direction of the I_{GD}, I_{GS}, and I_D current sources must be reversed.) The ohmic resistance of the drain and source regions of the FET are modeled by the two linear resistors R_D and R_S, respectively. The dc characteristics of the device are represented by the nonlinear current source I_D. The value of I_D can be computed using the following equations:

$$I_D = \begin{cases} 0 \\ \quad \text{for } V'_{GS} < V_T \\ \beta(V'_{GS} - V_T)^2 (1+\lambda V'_{DS}) \tanh(a\, V'_{DS}) \\ \quad \text{for } V_T < V'_{GS} < V'_{DS} + V_T \\ \beta V'_{DS}\,[2\,(V'_{GS} - V_T) - V'_{DS}](1 + \lambda V'_{DS}) \tanh(a\, V_{DS}) \\ \quad \text{for } V'_{GS} > V'_{DS} + V_T \end{cases} \quad (1.26)$$

The drain current is modeled by a "square-law" characteristic defined by the zero-bias threshold voltage V_T and the transconductance parameter β. V_T and β determine the variation of drain current with gate voltage; they can be defined from a graph of $\sqrt{I_D}$ versus V'_{GS}. An example of such a graph for an n-channel device is illustrated in Figure 1.19. V_T is the x-axis intercept of this graph, while β is defined as the slope of the $\sqrt{I_D}$ versus V'_{GS} characteristic.

The parameter, λ, is the channel length modulation parameter and determines the output conductance which, in the forward saturation region, is given by

$$g_{Dsat} = \beta\lambda(V'_{GS} - V_T)^2 \approx \lambda I_D. \quad (1.27)$$

The parameter, a, determines the voltage at which the drain current characteristics saturate. The I_{GD} and I_{GS} current sources that model the two gate junctions and their values are given by the diode equations

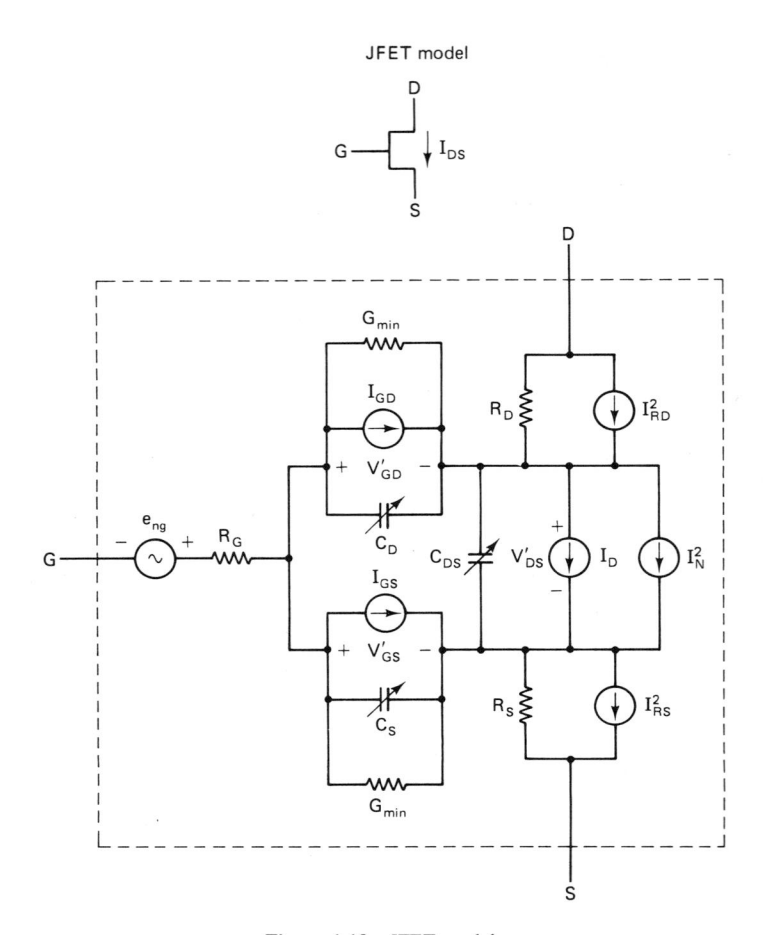

Figure 1.18 JFET model.

$$I_{GD} = I_S \left[\exp\left(\frac{V'_{GD}}{V_t}\right) - 1 \right] + G_{min} V'_{GD}$$

(1.28)

$$I_{GS} = I_S \left[\exp\left(\frac{V'_{GD}}{V_t}\right) - 1 \right] + G_{min} V'_{GS}$$

where I_S is the saturation current, and G_{min} is a small conductance added in parallel to the junctions to help computation convergence when the model is used in a computer environment. Since the device is normally operated with the gate junctions reverse-biased, I_S should be chosen to correspond to measured values of gate-junction leakage current.

Charge storage in a JFET (MESFET) occurs mainly in the gate junctions.

Sec. 1.5 Modeling of GaAs Devices

The charge storage of the gate junctions is modeled by two voltage dependent capacitors whose values are determined by

$$CS = C_{GS}\left(1 - \frac{V'_{GS}}{V_{bi}}\right)^{-1/2} \quad (1.29)$$

and

$$CD = C_{GD}\left(1 - \frac{V'_{GD}}{V_{bi}}\right)^{-1/2} \quad (1.30)$$

(C_{GS} and C_{GD} are the zero-bias gate-to-source and gate-to-drain capacitances, respectively.) The device transconductance g_m and output conductance g_D can be defined by the equations

$$g_m = \left.\frac{\partial I_D}{\partial V_{GS}}\right|_{\text{operating point}} \quad (1.31)$$

$$g_D = \left.\frac{\partial I_D}{\partial V_{DS}}\right|_{\text{operating point}} \quad (1.32)$$

The temperature effect on the saturation current of the device is similar to that of a diode. It is described by the same expression as in Equation (1.25).

It should be noted at this point that the equations modeling the current source I_D and the capacitances C_s and C_D are approximations, which may result in significant differences between the behavior of a device predicted by the model and the behavior of an actual device. For example, Equation (1.26) was found to be a good representation of the device behavior for a given V_{GS} for FETs with large pinch-off voltage.[25] Also, Expressions (1.29) and (1.30) are based on the

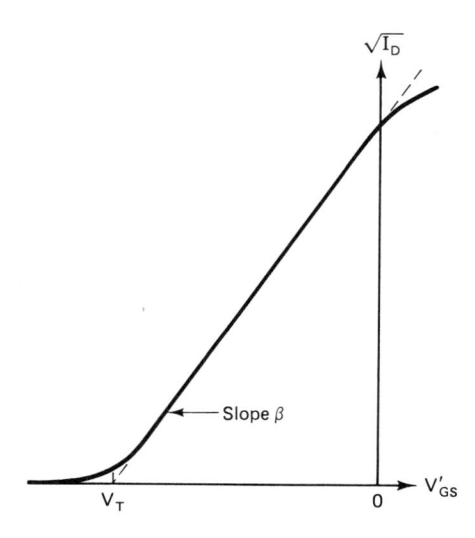

Figure 1.19 Typical JFET drain current characteristics.

assumption that the channel thickness is infinite in extent—this is obviously not valid for a real device.

The accurate modeling of the behavior of a MESFET (JFET) is a complicated problem and the subject of ongoing research. Although the equations given in this section can be used as a basis for modeling the behavior of a device, they have to be augmented to produce highly accurate results for devices fabricated with a particular process and exhibiting certain measurable characteristics. In fact, device modeling used for circuit design is driven by laboratory measurements. These, in turn, are used to modify the basic expressions discussed here so as to represent better the behavior of the devices the designer will be using to implement circuits. Another important issue when developing device models to be used in a computer-aided design environment is the convergence performance of the model in the numerical analysis. This often prompts the inclusion of nonphysical parameters in the model solely for the purpose of helping conversion. The reader who seeks more accurate descriptions of the behavior of FETs under different conditions should consult References 25, 26, 27, 28, 29, and 30.

1.5.3 Noise Models

During the AC analysis of a circuit, it is important to determine its noise characteristics. The noise characteristics of a device have little effect on the design of digital circuits, but they are a major performance factor for microwave circuits. Although the subject of this book is the design of digital GaAs circuits, a brief discussion of noise models is presented because these models characterize certain aspects of device behavior and are employed by most circuit simulators. The reader interested in detailed treatment of noise models should consult Reference 31 and Chapter 5, Reference 8.

Every device in a circuit generates both shot and flicker noise in addition to the thermal noise generated by the ohmic resistance of the device. The generated noise can be modeled as stochastic electrical excitation to the device's equivalent small signal circuit. In developing a noise model, it is assumed that each noise source in the circuit is statistically uncorrelated to the other noise sources of the circuit; the contribution of each noise source to the total noise at a specified output is determined separately. The total circuit noise is defined as the RMS sum of the individual noise contributions, assuming the noise is white.

Noise is specified with respect to a specific noise bandwidth and is expressed in $\frac{V}{\sqrt{Hz}}$ or $\frac{A}{\sqrt{Hz}}$ units. The output noise of a circuit is the RMS level of output noise divided by the square root of bandwidth. The equivalent input noise is the output noise divided by the transfer function of the output with respect to the specified input.

Thermal noise generation in the ohmic region of a diode is modeled by the current source I_R^2 in Figure 1.16. The value of I_R^2 is determined by

Sec. 1.5 Modeling of GaAs Devices

$$I_R^2 = \frac{4kT}{R_S} \tag{1.33}$$

where k is Boltzmann's constant, and T is the absolute temperature in degrees Kelvin.

The shot and flicker noise of the diode are modeled by the current source I_{DN}^2 in Figure 1.16. The value of I_{DN}^2 is determined by

$$I_{DN}^2 = \underbrace{2qI_D}_{\text{(shot noise)}} + \underbrace{\frac{K_F I_D^{AF}}{f}}_{\text{(flicker noise)}} \tag{1.34}$$

where K_F and AF are the flicker noise coefficient and exponent respectively, and f is frequency. These model parameters are estimated from measurements of the diode noise at low frequency.

The thermal noise generation in the drain and source regions of a MESFET is modeled by the two sources I_{RD}^2 and I_{RS}^2 in the equivalent circuit of the device shown in Figure 1.18. The values of these sources are determined by

$$\begin{aligned} I_{RS}^2 &= \frac{4kT}{R_S} \\ I_{RD}^2 &= \frac{4kT}{R_D} \end{aligned} \tag{1.35}$$

Shot and flicker noise is modeled by the current source I_N^2 in Figure 1.18. The value of I_N^2 is determined by

$$I_N^2 = \underbrace{\frac{8}{3}kTg_m}_{\text{(shot noise)}} + \underbrace{\frac{K_F I_D^{AF}}{f}}_{\text{(flicker noise)}} \tag{1.36}$$

where g_m is the small signal transconductance of the MESFET.

These device noise models are employed by the SPICE circuit simulator[24] to perform noise analysis of a circuit. Another measure of internally generated noise in a MESFET is the noise figure. The noise figure represents a ratio of signal-to-noise level at the device input compared to that at the output. At high frequencies (i.e., microwave), the primary noise contribution is thermal. Thermal noise generated in the channel is capacitively coupled through the gate and is resistively coupled to the input via the source-to-gate series resistance. The noise is amplified along with the input signal and appears at the output. It is desirable to reduce the magnitude of noise that is amplified by the device. This can be achieved by appropriate choices of design, process, and material, so that the gate capacitance, series resistance, and other parasitic elements of the device are minimized. At the same time, the device transconductance must be maximized. A trade-off is, of course, implied here. A process used for microwave circuits may offer a great noise figure, but a digital GaAs process will more likely emphasize optimization of other device parameters such as transconductance.

An empirical model for the device noise figure was developed by Fukui[31] and relates the noise figure to equivalent circuit elements:

$$F_m = 1 + KfC_s \sqrt{\frac{R_S + R_G}{g_m}} \qquad (1.37)$$

where F_m is the optimum noise figure, f is the frequency of operation, and K is Fukui's material quality parameter. K is essentially a fitting factor; so, Expression (1.37) is in close agreement with results obtained through wafer measurements. C_S, R_S, and R_G are elements in the equivalent circuit of Figure 1.18; namely, the gate capacitance, source series resistance, and gate metalization resistance. Typical values for F_m obtained through measurements of MESFETs with gate length of 1.85 µm and gate width of 500 µm were found to be between 1.5 dB and 1.8 dB.[31] The measurements were conducted at 1.8 GHz. The measured devices had different channel thicknesses and different free-carrier concentrations in the active channel.

Equation (1.37) is commonly used to assist understanding the relationships between noise figure and device parameters. Although the equation is generally used to analyze a single device, successful use of it to compare devices from the same wafer has been reported (cf. Reference 8, Chapter 5).

Note that the equivalent circuit elements used in Equation (1.37) are obtained with zero gate bias, although the noise figure is measured using a typical gate bias. In spite of such a difference in the gate-bias conditions, the above relationship was found to be valid.[31] If the equivalent circuit elements were measured at different than zero gate bias, then the material quality factor K must be modified.

1.5.4 Device Models for Computer Simulation

Circuit design performance is determined by circuit simulation at the transistor level. This type of simulation also provides information about power dissipation of the circuit, temperature impact on performance, and noise analysis. The most widely used circuit simulator is the SPICE simulator, which was originally developed at the University of California at Berkeley. The accuracy of the simulation results, of course, greatly depends on the device models used and the values given to the model parameters. These values depend on the fabrication process and are obtained through either process simulation or actual measurements of device characteristics fabricated with this process. The parameters that are sensitive to process variation are specified on a SPICE simulation, while others are used in their default value.

As an example, the device models for a 1 µm GaAs D-Mode are given. The models are for diodes and MESFETs. Table 1.2 lists the parameters for the diode model utilized by SPICE, while Table 1.3 lists the parameters for the MESFET. An example of the model parameters specified in the SPICE (JFET) model

Sec. 1.5 Modeling of GaAs Devices

TABLE 1.2 SPICE Diode Model Parameters

No.	Keyword	Meaning	Units	Default
1	AREA	area factor	—	1.0
2	STATE	ON/OFF flag	—	ON
3	VDO	initial voltage	V	0.0
4	IS	saturation current	A	10^{-14}
5	RS	ohmic resistance	Ω	0.0
6	N	emission coefficient	—	1.0
7	TT	transit time	S	0.0
8	CJO	zero-bias junction capacitance	F	0.0
9	PB	junction built-in potential	V	1.0
10	M	grading coefficient	—	0.5
11	EG	energy gap	eV	1.11
12	XTB	saturation current temperature exponent	—	3.0
13	KF	flicker noise coefficient	—	0.0
14	AF	flicker noise exponent	—	1.0
15	FC	forward-bias nonideal junction-capacitance coefficient	—	0.5
16	BV	reverse breakdown voltage	V	∞
17	IBV	current at BV	A	10^{-3}

TABLE 1.3 SPICE MESFET Model Parameters

No.	Keyword	Meaning	Units	Default
1	AREA	area factor	—	1.0
2	STATE	ON/OFF flag	—	ON
3	VGSO	initial VGS	V	0.0
4	VGDO	initial VGD	V	0.0
5	POL	NJF/PJF polarity flag	—	NJF
6	VTO	zero-bias threshold voltage	V	−2.0
7	BETA	transconductance parameter	A/V	10^{-4}
8	LAMBDA	channel-length modulation parameter	A/V	0.0
9	RD	drain ohmic resistance	Ω	0.0
10	RS	source ohmic resistance	Ω	0.0
11	CGS	zero-bias gate-source junction capacitance	F	0.0
12	CGD	zero-bias gate-drain junction capacitance	F	0.0
13	PB	gate junction potential	V	1.0
14	IS	gate junction saturation current	A	10^{-14}
15	KF	flicker noise coefficient	—	0.0
16	AF	flicker noise exponent	—	1.0
17	FC	forward-bias nonideal junction-capacitance coefficient	—	0.5

card, which simulate the behavior of devices fabricated with a 1 μm GaAs process, are as follows:

- MODEL NP2 D (IS = 1.6E–13 N = 1.27 RS = 950 BV = 2 IBV = 0.1UA CJO = 2.8FF PB = 1.0 M = 0.5 FC = 0.0)
- MODEL J1Q NJF (LAMBDA = 0.01 VTO = –0.90 BETA = 1.95E–4 IS = 1E–30)

The measured I-V characteristics of a MESFET with gate width of 300 μm are shown in Figure 1.20. This MESFET was fabricated with the process described in Section 1.1. In Figure 1.21, the measured I-V characteristics are compared with I-V characteristics obtained through simulation using the model presented. The close agreement of simulated and measured results indicates that this model can be used to simulate fairly accurately the behavior of circuits to be fabricated with this specific process. The difference in the I-V characteristics in Figures 1.20 and 1.21 is due to the frequency dependency of the FET output impedance.[32] The output impedance decreases as the frequency increases, up to a point where it becomes constant, independent of any further increase in the frequency.

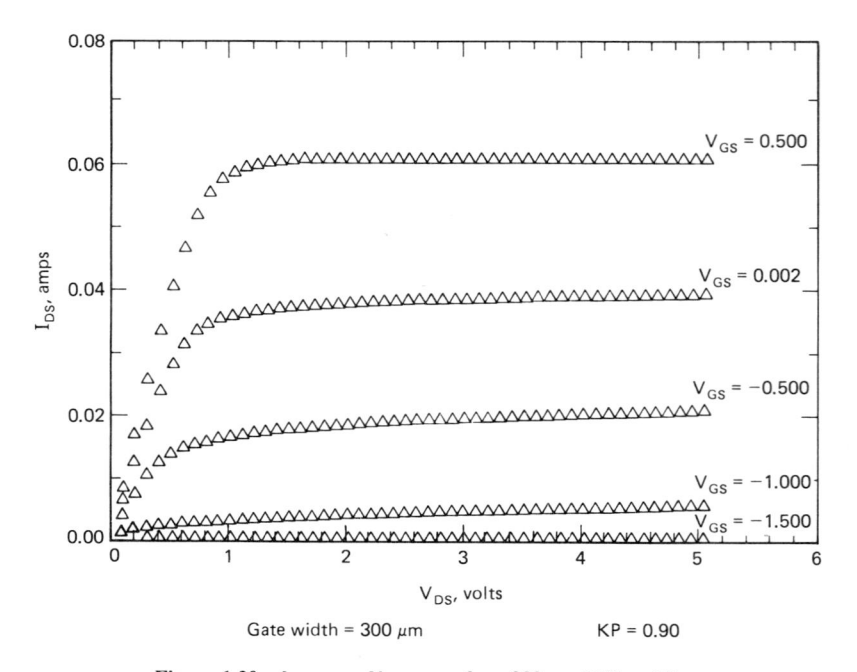

Figure 1.20 I_{ds} versus V_{ds} curves for a 300 μm FET at dC.

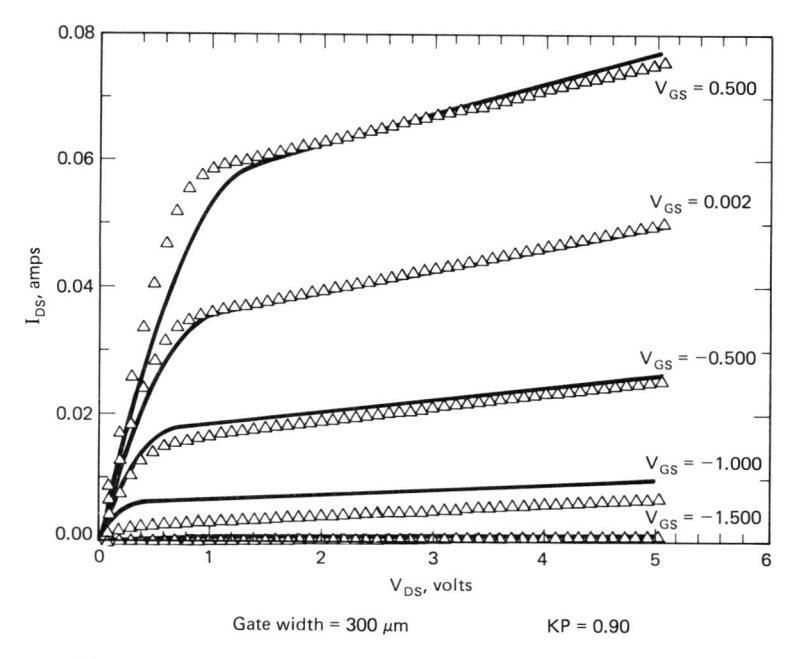

Figure 1.21 I-V curves for a 300 μm FET at high frequency. The triangles signify measured data; the solid lines signify simulated data.

1.6 DEVICE SCALING

This section discusses the manner in which individual devices are scaled down to small geometrical sizes. Scaling results in increased packing density and therefore increased functionality on a monolithic IC.

Scaling theory has been extensively used as a guide to increasing MOSFET integration density. According to this theory, if the device's physical dimensions and applied potentials are scaled by a common factor $\frac{1}{K_{sf}}$ (where $K_{sf} > 1$) and the impurity concentration is increased by K_{sf}, the shape of the electric-field pattern within the scaled device remains constant. As a result, two-dimensional effects such as drain-source punch-through, threshold sensitivity to channel length, and drain voltage can be controlled. However, for submicron devices, straightforward application of this theory is limited because of the temperature variation of the threshold voltage and the nonscalability of the junction built-in potential. Both these properties require the threshold and supply voltages to be reduced less than the conventional scaling theory would indicate.

Because of these considerations and since scaling of currently available MESFETs will result in submicron devices, it is desirable to generalize any scaling theory. We must identify the design criteria that are crucial in maintaining both the shape of the electric field and the potential distributions constant, while still allowing the local fields to increase, if desired. In order to achieve this, the physical dimensions and the potentials have to be scaled by different factors. This yields a considerably increased flexibility, while still maintaining some control of two-dimensional effects.

For any particular device geometry and set of boundary conditions, the electric field configuration in the channel results from a solution of Poisson's equation:

$$\nabla^2 V = \frac{-q(p - n + N_D - N_A)}{\varepsilon_S} \tag{1.38}$$

and the current continuity equation

$$\nabla \vec{J} = -\frac{\partial \rho}{\partial t} \tag{1.39}$$

Our prime concern is with Equation (1.38), since this equation determines the scaling conditions for the potential and the shape of the electric field.

Let us consider the variable transformation:[33]

$$V' = \frac{V}{K_{sf}} \tag{1.40a}$$

$$(x', y', z') = \frac{(x, y, z)}{\lambda_s} \tag{1.40b}$$

$$(n', p', N_D', N_A') = (n, p, N_D, N_A) \frac{\lambda_s^2}{K_{sf}} \tag{1.40c}$$

by which Expression (1.39) can be rewritten as

$$\nabla^2 V' = -\frac{q}{\varepsilon_S}(p' - n' + N_D' - N_A') \tag{1.41}$$

This is identical to Equation (1.38) and indicates that Poisson's equation remains unchanged by the scaling, which implies that the shape of the potential solutions to the equation will retain their prescaled form if the boundary conditions are also scaled. In Equation (1.40a), K_{sf} is the potentials scaling factor, while λ_s is the linear dimensions scaling factor in Equation (1.40b).

The earlier forms of constant-field scaling are based on the assumption that $\lambda_s = K_{sf}$, although several other forms of constant-voltage scaling and quasi-constant-voltage scaling have been proposed.

Table 1.4 shows the scaling factors associated with the most important physical quantities in the general case of $K_{sf} \neq \lambda_s$. The most important circuit limitations resulting from the choice $\lambda_s > K_{sf}$ are represented by the increase in power density by $\frac{\lambda_s^3}{K_{sf}^3}$ and in current density within the interconnection lines by $\frac{\lambda_s^3}{K_{sf}^2}$. For a given die size, the former effect causes increased power dissipation

TABLE 1.4 Scaling Factors and Their Association to Physical Quantities

Physical Parameter	Expression	Scaling Factor
Linear dimensions		
Gate length	L	$1/\lambda_s$
Gate width	Z	$1/\lambda_s$
Depletion region width	W	$1/\lambda_s$
Potentials	V_G, V_D, V_S	$1/K_{sf}$
Impurity concentrations	N_A, N_D	λ_s^2/K_{sf}
Electric field	E	λ_s/K_{sf}
Capacitances	Device and interconnection	$1/\lambda_s$
Current (saturated velocity)	I_D	$1/K_{sf}$
Power	$I_D V_{DD}$	λ_s/K_{sf}^3
Power density	$I_D \dfrac{V_{DD}}{Z \cdot L}$	λ_s^3/K_{sf}^3
Gate delay	$C_{GS} \dfrac{V_{DD}}{I_D}$	K_{sf}/λ_s^2
Power delay product	$I_D V_{DD} \cdot t_d$	$1/\lambda_s K_{sf}^2$
Line resistance	R_l	λ_s
Current density	$\dfrac{I_D}{\text{line area}}$	λ_s^3/K_{sf}^2
Time constant	$R \cdot C$	1

and, thus, heat removal problems. The latter is of special concern due to electromigration, which affects long-term reliability.

PROJECTS

1.1. Perform a SPICE simulation on a D-MESFET with $L = 1\mu m$ and W varying between 10 μm and 300 μm. Use the device model J1Q discussed in Section 1.5.4. Set $V_{DD} = 2$ V, $V_{SS} = -2$ V, and V_G a 1 GHz pulse between −1 and 1 V. Simulate for ten different values of W and provide a discussion for the obtained results.

1.2. Repeat Exercise 1.1 by using a constant value for W and by increasing L at 0.5 μm increments.

1.3. Perform I-V measurements on a process-control MESFET (i.e., transistors used on wafers to monitor the process characteristics) and use the measured values to obtain a device model similar to J1Q, so that measured and simulated curves are almost identical.

REFERENCES

1. R. L. Van Tuyl et al., "A Manufacturing Process for Analog and Digital Gallium Arsenide Integrated Circuits," *IEEE Trans. on Circuits and Systems*, p. 1031, July 1982.

2. A. Rode et al., "A High-Yield GaAs MSI Digital IC Process," *Proceedings IEEE Intl. Electron Device Modeling Conf. (IEDM)*, p. 162, 1982.
3. R. C. Eden, A. R. Livingston, B. M. Welch, "Integrated Circuits: The Case for Gallium Arsenide," *IEEE Spectrum*, p. 30, December 1983.
4. Y. Nakayama et al., "An LSI GaAs DCFL Using Self-Aligned MESFET Technology," *Proceedings IEEE GaAs IC Symposium*, p. 6, 1982.
5. C. P. Lee et al., "Ultra High Speed Digital Integrated Circuits Using GaAs/GaAlAs HEMTs," *Proceedings IEEE GaAs IC Symposium*, p. 162, 1983.
6. G. C. Dacey, I. M. Ross, "The Field-Effect Transistor," *Bell System Tech. J.* 34, p. 1149, 1955.
7. W. W. Hooper, W. I. Lehrer, "An Epitaxial GaAs Field-Effect Transistor," *Proceedings IEEE* 55, p. 1237, 1967.
8. D. K. Ferry, *Gallium Arsenide Technology* (ch. 9). Indianapolis: Howard W. Sams & Co., 1985.
9. P. Smith, M. Imoue, J. Frey, "Electron Velocity in Si and GaAs at Very High Electric Fields," *Appl. Phys. Lett.* 37, p. 797, 1980.
10. R. E. Williams, D. W. Shaw, "Graded Channel FET's Improved Linearity and Noise Figure," *IEEE Trans. Electron Devices*, ED-25, p. 600, 1978.
11. C. A. Liechti, "Microwave Field-Effect Transistors," *IEEE Trans. Microwave Theory and Tech.* MTT-24, p. 279, 1976.
12. J. Ruch, "Electron Dynamics in Short Channel Field Effect Transistors," *IEEE Trans. Electron Devices*, ED-19, p. 652, 1979.
13. R. Zuleeg, J. K. Notthoff, K. Lehovec, "Femtojoule High-Speed Planar GaAs E-JFET Logic," *IEEE Trans. Electron Devices* ED-25, p. 628, 1978.
14. D. P. Kennedy, R. R. O'Brien, "Computer Aided Two-Dimensional Analysis of the Junction Field-Effect Transistor," *IBM J. of R/D*, 14, p. 95, March 1970.
15. K. Yamaguchi, S. Asai, H. Kodera, "Two-Dimensional Numerical Analysis of Stability Criteria of GaAs FETs," *IEEE Trans. Electron Devices*, vol. ED-25, no. 6, p. 612, June 1978.
16. T. Mimura et al., "A New Field-Effect Transistor with Selectively Doped GaAs/m-$Al_x Ga_{1-x}$as Heterojunctions," *Japan J. Appl. Phys.*, vol. 19, no. 5, p. 225, May 1980.
17. P. F. Pierret, M. S. Lundstrom, "Correspondence Between MOS and Modulation-Doped Structures," *IEEE Trans. Electron Devices*, vol. ED-31, p. 383, February 1984.
18. R. C. Eden, "Comparison of GaAs Device Approaches for Ultrahigh Speed VLSI," *Proceedings IEEE*, vol. 70, no. 1, p. 5, January 1982.
19. C. P. Lee et al., "Ultra High Speed Digital Integrated Circuits Using GaAs/GaAlAs High Electron Mobility Transistors," *Proceedings GaAs IC Symposium*, p. 162, 1982.
20. T. Mimura et al., "High Electron Mobility Transistor Logic," *Japan J. Appl. Phys.*, vol. 20, no. 8, p. 598, August 1981.
21. K. H. Duh et al., "Instabilities in MODFETs and MODFET Circuits," *IEEE Trans. Electron Devices*, vol. ED-31, p. 1345, October 1984.
22. H. Kroemer, "Heterostructure Bipolar Transistors and Integrated Circuits," *Proceedings IEEE*, vol. 70, no. 1, p. 13, January 1982.

Chap. 1 References

23. P. M. Asbeck et al., "4.5 GHz Frequency Dividers Using GaAs/GaAlAs Heterojunction Bipolar Transistors," *Tech. Digest*, ISSCC, p. 50, 1984.
24. L. W. Nagel, "SPICE 2: A Computer Program to Simulate Semiconductor Circuits," Electronics Research Laboratory, Univ. of California at Berkeley, *Memorandum ERL-MS20*, May 1975.
25. H. Statz et al., "GaAs FET Device and Circuit Simulation in SPICE," *IEEE Trans. Electron Devices*, ED-34, p. 160, February 1987.
26. H. C. Ki, et al., "A Three-Section Model for Computing I-V Characteristics of GaAs MESFETs," *IEEE Trans. Electron Devices*, ED-34, p. 1929, September 1987.
27. P. A. Sandborn, J. R. East, G. I. Haddad, "Quasi Two-Dimensional Modeling of GaAs MESFETs," *IEEE Trans. Electron Devices*, ED-34, p. 985, May 1987.
28. T. Chen and M. A. Shur, "Analytical Models of Ion-Implanted GaAs FETs," *IEEE Trans. Electron Devices*, ED-32, p. 18, January 1985.
29. M. S. Shur, "Analytical Models of GaAs FETs," *IEEE Trans. Electron Devices*, ED-32, p. 70, January 1985.
30. T. Hariu, K. Takahashi, Y. Shibata, "New Modeling of GaAs MESFETs," *IEEE Trans. Electron Devices*, ED-30, p. 1743, December 1983.
31. H. Fukui, "Design of Microwave GaAs MESFETs for Broad-Band Low Noise Amplifiers," *IEEE Trans. Microwave Theory and Tech.* MTT-27, p. 643, July 1979.
32. N. Scheinberg, R. Bayruns, R. Goyal, "A Low-Frequency GaAs MESFET Circuit Model," *IEEE J. Solid State Circuits*, SC-23, no. 2, p. 605, April 1988.
33. G. Baccarani, M. R. Wordeman, R. H. Dennard, "Generalized Scaling Theory and Its Application to a 1/4 Micrometer MOSFET Design," *IEEE Trans. Electron Devices*, ED-31, p. 452, 1984.

2
GaAs LOGIC GATE DESIGN

The choice of a particular FET device for implementing integrated circuits is dependent on the circuit performance requirements of the integrated circuit and the fabrication process of the device. Table 2.1 relates the circuit and consequent device requirements for high-speed, low-power ICs, while Table 2.2 relates the device characteristics and consequent device physical parameters for high-speed, low-power ICs.[1]

The depletion-mode MESFET (D-MESFET) was the most widely used device in the early 1980s for implementing GaAs ICs. Circuits employing D-MESFETs pose the least fabrication problems, since Schottky barriers on GaAs ICs are easier to fabricate than p-n junctions. Furthermore, the large voltage swings associated with D-MESFET circuits relax the requirements for FET

TABLE 2.1 Circuit and Consequent Device Requirements for Very High-speed, Low-power ICs

Circuit Requirements	Consequent Device Requirements
1. Small logic voltage swings (low-power ~ $\frac{1}{2}C\Delta V^2$)	Very uniform threshold voltages for active devices (particularly for VLSI)
2. Low device and parasitic capacitances	Low input capacitance devices and semi-insulating substrate for low parasitics
3. High switching speeds with reasonable fanout loadings at low switching voltages	Very high current gain bandwidth, very high power gain bandwidth, and fast increase in transconductance above threshold (strong nonlinearity)

Chap. 2   GaAs LOGIC GATE DESIGN

TABLE 2.2 Device Characteristics Desired for High-speed, Low-power Switching with the Consequent Device Physical Parameters and Structural Characteristics

Desired Device Electric Characteristics	Consequent Physical Parameters
1. High transconductance at control voltages low above threshold	High carrier mobilities. Very short channel (e.g., gate length)
2. Very uniform threshold voltages	Very low threshold voltage sensitivity to horizontal geometry variations. Low threshold voltage sensitivity to vertical geometry variations and doping variations
3. Very low input capacitances	Small geometries and low carrier storage effects
4. High current and power gain bandwidths	High carrier mobilities and saturation velocities, small geometries, and good thermal design

threshhold voltage uniformity. The logic voltage swing in a D-MESFET can extend below $-V_{th}$ to the onset of gate conduction in spite of the forward-biased gate-source junction. In D-MESFETs any regions of the source-drain channel not under the gate are conductive. This eases the requirements for precise gate alignments and special gate-recess and etch processes that are necessary for avoiding parasitic source and drain resistances. The relative simplicity of fabricating D-MESFET circuits results in acceptable yields for commercial production of D-MESFET GaAs ICs. However, for proper logic switching of circuits designed with depletion-mode active devices, a voltage level shifting between FET drains and gates is required. This level shift is necessary to meet device turn-off requirements, and it is achieved by employing two power supplies. The two power supplies impose some penalty in terms of circuit area overhead and chip interface overhead, since most circuit logic families require only one power supply.

Enhancement-mode MESFET circuits (E-MESFETs) avoid the requirement for a dual power supply and level-shifting circuitry because the E-MESFETs have positive threshold voltages. However, E-MESFETs are restricted to small logic voltage swings (typically 0.5 V) because their gates cannot be forward-biased above 0.6 to 0.8 V without drawing excessive current. Since the difference between voltages representing logical 0 and 1 states must be approximately 20 times the standard deviation of threshold voltages to allow for adequate noise margins for implementing integrated circuits, E-MESFET threshold voltage must be uniform to within 25 millivolts. This degree of control is difficult to achieve consistently, and it was the main reason for the difficulty in mass production of commercial E-MESFET ICs in the early 80s. However, refinements in fabrication technology made possible commercial production of E-MESFET ICs by the late 1980s.

2.1 D-MESFET LOGIC APPROACHES

Logic circuits implemented with D-MESFETs have three basic characteristics:

1. they employ two power supplies
2. they employ one inverting stage and one level-shift stage
3. logic is implemented with diodes or transistors

Buffered FET logic (BFL)[2] and Schottky Diode FET Logic (SDFL)[3] circuit design approaches have been extensively employed for the design of depletion-mode GaAs ICs. Design variations of BFL and SDFL result in other, less frequently used, logic families such as unbuffered FET logic (UFL), Capacitor-Diode FET Logic (CDFL), and Buffered Diode FET Logic (BDFL).

The circuit configuration of a basic BFL inverter is shown in Figure 2.1. The logic is implemented with transistors in the inverting stage, while the output is driven by a source follower with level shifting diodes to restore the required logic levels to +0.7 V (high) to $-V_{th}$ (low) or below, voltages required by the input FETs. The source follower output driver has relatively low sensitivity to fanout loading and loading capacitance. Also, no dc current is required to drive subsequent BFL gate inputs. However, because the circuit operation relies on the use of forward-biased, level shifting diodes, the BFL approach to logic circuit design results in relatively high power consumption. BFL power consumption

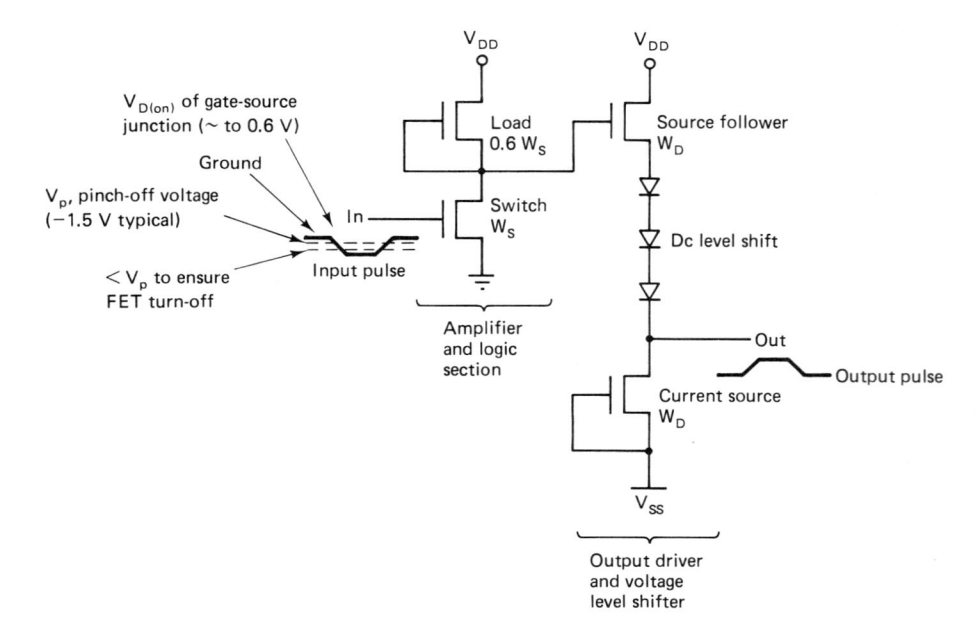

Figure 2.1 BFL basic inverter circuit.

Sec. 2.1   D-MESFET Logic Approaches   43

can be reduced by using a somewhat different circuit structure, shown in Figure 2.2. This circuit configuration, known as UFL, consumes less power by omitting the load driver source follower. In this case, however, the circuit is sensitive to high fanout because there is no buffer between the switching transistor and the output node. The output node is loaded by the input impedance at connected stages and its capacitance increases (for multiple fanout) linearly with the fanout number as do the rise, fall, and delay times.

A different approach to reducing BFL power consumption is taken by the CDFL logic family, which was introduced as Feed-Forward Static Logic (FFSL).[4] Figure 2.3 illustrates the basic inverter configuration in CDFL. The Schottky diode added in the voltage-shift section of the circuit is always reverse-biased and acts as a capacitor providing capacitive coupling between stages through which the high-frequency signal is transmitted. The transmission of the high-frequency signal through the capacitor diode (CD) allows for smaller device width of the current source in the voltage shift section and wider tolerance to pinch-off variation than BFL without performance degradation. While BFL circuits have to retain internal logic levels to achieve certain performance, in CDFL the voltage shift section need not be biased for optimal speed but only for correct dc levels. Two factors must be considered in designing CDFL circuits. First, the CD must not be punched through under operating conditions, which implies that charge flow to the depletion edge of the capacitor must not be excessively hindered. Second, the capacitance of the CD must be significantly greater (at least 4 to 5 times) than the maximum capacitive load of the output node. Since the two capacitors act as a divider network, this provision will ensure that the high-frequency signal is not excessively attenuated.

Another approach to minimizing circuit area and power requirements is to

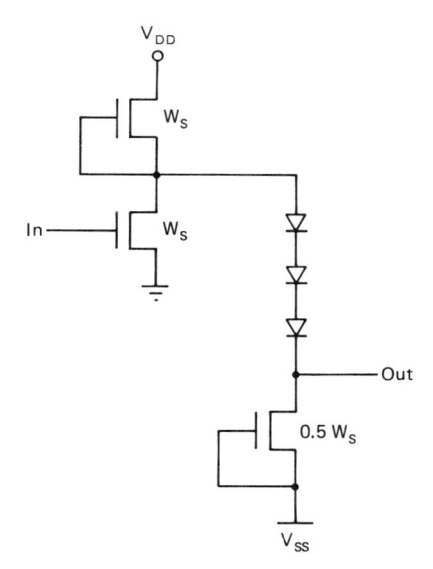

Figure 2.2   UFL basic inverter circuit.

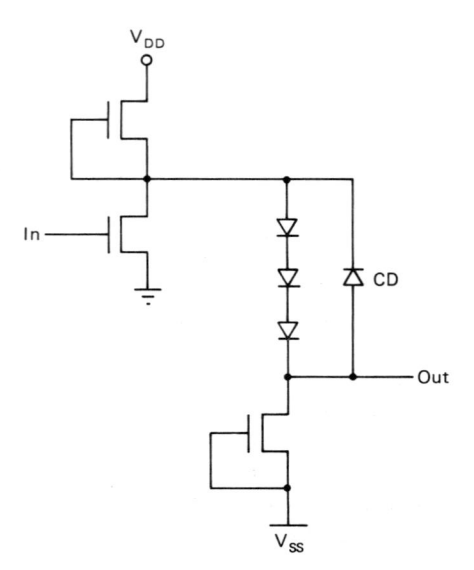

Figure 2.3 CDFL basic inverter circuit.

use diodes to perform the logic function of a gate and a single driven D-MESFET for output inversion and buffering. This circuit approach is called SDFL, and the design of a basic SDFL inverter is illustrated in Figure 2.4. The SDFL approach offers savings in power (the input diodes are not always forward-biased) and in circuit area, since the logic is implemented using diodes that occupy a smaller area than FETs. The fact that diodes are 2-terminal devices also significantly reduces the number of vias and cross overs required in most circuits as compared to the vias and crossings needed when the 3-terminal FETs are used to perform

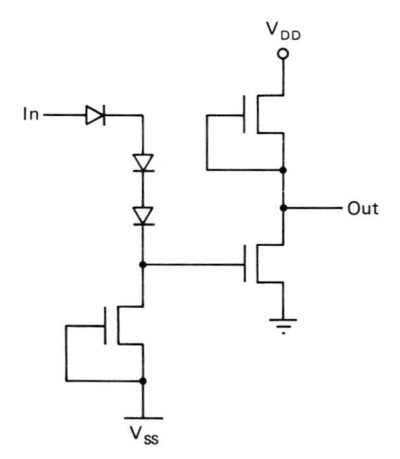

Figure 2.4 SDFL basic inverter circuit.

Sec. 2.1　D-MESFET Logic Approaches

the logic function of the circuit. The use of small, low-capacitance diodes for switching and input level shifting allows for high fan-in logic gates to be constructed in SDFL. There are, however, limitations to SDFL when used for large fanouts (greater than 3). The gate delay is dependent on fanout and capacitance loading, as is the noise margin. The noise margin may also be limited by high fan-in due to division of pull-down current among many driver pull-ups. To overcome fanout problems, a push-pull output driver for SDFL has been suggested.[5] The SDFL inverter with the push-pull output driver is illustrated in Figure 2.5. The driver includes a source follower plus a switched, pull-down FET. The source follower has very good current sourcing capabilities and is relatively insensitive to loading, as has been indicated in the discussion for the BFL approach. The pull-down current available for a given gate depends on the input states of the gates loading it, but the parasitic capacitance does not. The switched pull-down is added to the driver to address this problem. When the output is high, this device is off. Therefore, it does not add more loading or draw power. When turned on, it helps to pull down the output node and interconnect lines independent of the state of the inputs of the loading gates. The push-pull driver results in SDFL circuits that are relatively insensitive to fanout at the expense of extra area and power requirements.

Buffering the inputs of the SDFL basic inverter results in the BDFL basic inverter circuit illustrated in Figure 2.6. This circuit configuration features buffered inputs, low input capacitance, and no dc current loading on the output. However, as in SDFL, the output (if not buffered) is sensitive to fanout.

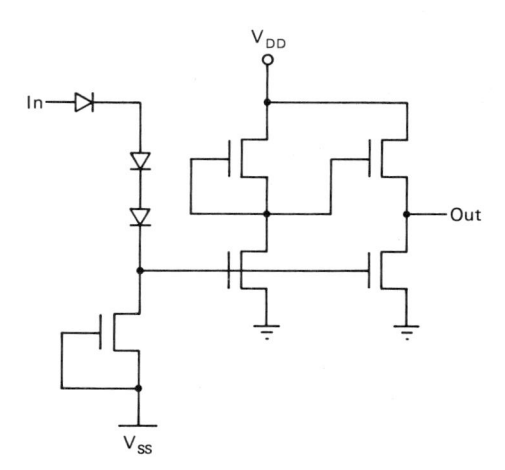

Figure 2.5 SDFL basic inverter with push-pull driver.

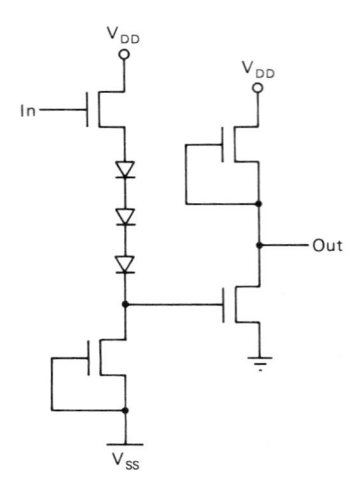

Figure 2.6 BDFL basic inverter circuit.

2.2 E-MESFET LOGIC APPROACHES

The most common circuit design approach based on E-MESFETs (normally-off) is called Direct-Coupled FET Logic (DCFL). DCFL gate structures utilize E-MESFETs in conjunction with resistors or depletion loads. Figure 2.7(a) illustrates a basic DCFL inverter circuit. In this approach, a logic "0" corresponds to a voltage near zero. A logic "1" corresponds to a positive voltage capable of driving to saturation the E-MESFET. The value of this voltage is usually limited by the onset of gate conduction in the E-MESFET, which is typically on the order of 0.6 V. Since the switching occurs between zero and a positive voltage, only one power supply is required for DCFL circuit configurations. On the other hand, the small logic swing attributed to the E-MESFET requires strict uniformity control of the device threshold in a logic structure. An improvement of the DCFL inverter of Figure 2.7(a) is the substitution of the load resistor R_L with a D-MESFET shown in Figure 2.7(b). The D-MESFET with its gate tied to its source acts as an active nonlinear current source. Such a nonlinear load improves the transfer characteristic, the speed, and speed-power products of logic circuits. A small width D-MESFET suffices to provide the same load current as in the resistor load configuration, thus reducing the area required per gate. Speed is also enhanced because of decreased load capacitance. The saturation current supplied by the D-MESFET load of small negative threshold voltage is very sensitive to threshold variations. This sensitivity can be reduced by making the threshold more negative. The fabrication of the D-MESFET active load requires a default concentration profile different from that required by the E-MESFET. This requirement increases the fabrication complexity of E-MESFET circuits when an active load is used.

From a static point of view, DCFL has very good fanout capability determined by the very low leakage currents. From a dynamic point of view, however,

Sec. 2.2 E-MESFET Logic Approaches   47

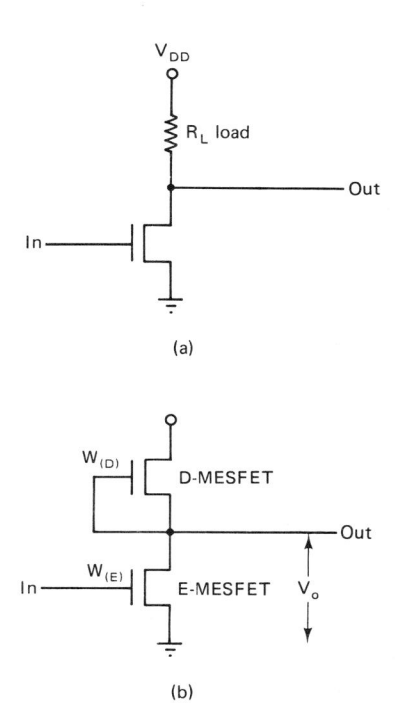

Figure 2.7 DCFL inverter circuit: (a) basic and (b) with active load.

the switching speed of DCFL gates is reduced by the gate capacitance loading of the output node. The factor of switching speed reduction is approximately $\frac{1}{N}$, where N is the number of loading gates. The current through the load resistor R_L or the active D-MESFET load is kept fairly low in order to reduce static power dissipation and improve noise margins by reducing the output logic "low" of the E-MESFET. As a consequence, the output rise time of the circuit with high fanout is slow. This can be greatly improved with the pseudo complementary output buffer configuration shown in Figure 2.8.[6] The added buffer to the inverter of Figure 2.7(a) is called pseudo complementary because the load transistor Q_2 turns off when its switching transistor Q_3 is on and vice versa (Figure 2.8). (The gate voltage of Q_2 is the inverted gate voltage of Q_1 and Q_3.) Based on the operation of the buffer circuit, there is negligible dc power dissipation in the second inverter provided that no dc current is drawn from its output node. The trade-off for good fanout performance in this case is the area overhead due to the two additional E-MESFETs of the buffer.

The concern over the limitations of FET threshold control in DCFL has provided impetus for the development of a logic family with a wide allowable threshold voltage range. This logic family is called Source-Coupled FET Logic (SCFL) and it has been used to design circuits which have demonstrated a wide range of tolerance to threshold voltage and a partial immunity to temperature variation.[7] The basic SCFL inverter consists of a differential amplifier and two source-

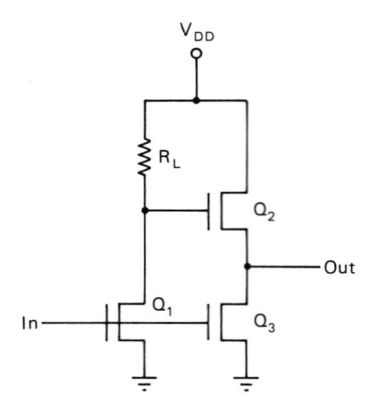

Figure 2.8 DCFL inverter with pseudo complementary output buffer.

follower buffers with diode level shifters, as shown in Figure 2.9. When the input voltage V_{IN} is applied to Q_1 in the differential amplifier, the voltage V_{IN} is compared to the fixed reference voltage V_{REF} applied to Q_2 so that either Q_1 or Q_2 can turn "on" depending on whether V_{IN} is higher or lower than the reference voltage V_{REF}. Switching of the differential occurs nearly independently of the switching transistors since the critical level in the transfer characteristic is equal to the externally applied reference voltage V_{REF}. This wide allowable threshold range is from −0.6 V to 0.4 V, with logic swings as large as 1.2 V. Typical values for the R_1 and R_s resistors are 12K ohms and 4K ohms repectively, with V_{SS} at −5 V. As Figure 2.9 shows, both the true and the complementary outputs are obtainable

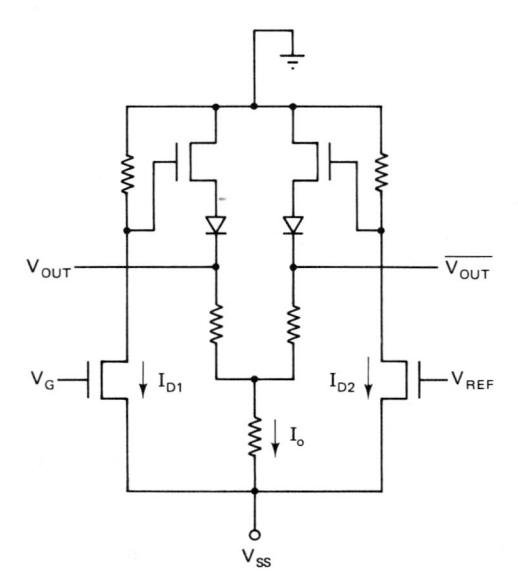

Figure 2.9 SCFL basic inverter circuit.

from the differential amplifier, and this may be advantageous for certain circuit applications.

There are several features of the SCFL family that make it suitable for implementing high-speed, low-power circuits. First, the gate-drain capacitance is small because the drain voltage at the ON state may be higher than any other logic family by design. The discharge time of the differential amplifier outputs is short because the discharging current is dominated by the saturation current of the switching transistors. The SCFL has also good fan-out capability because the source follower buffers enable fast charging of the load capacitance.

Although SCFL overcomes the tight threshold control requirement for DCFL, it requires more power consumption than DCFL due to the configuration of the SCFL basic inverter which is composed of two transistors for the differential amplifier and two output buffers.

2.3 PROPAGATION DELAYS AND POWER DISSIPATIONS

The charge required to move the output voltage of a logic gate by half of the logic swing, $\frac{V_L}{2}$, is $\frac{C_N V_L}{2}$ in a linear circuit with a constant output node capacitance C_N. Experimental measurements on fabricated MESFETs have shown a weak and similar dependence of the device transconductance g_m and gate-to-source capacitance C_{GS} on operating current I_{DS}.[8] This dependence is attributed to nonlinear charge storage and low and high drain currents in the gate-to-source capacitance.[9] It was found that this excess charge is only 27 percent of the nominal charge and therefore its influence, although significant, is not dominant. An additional mechanism of nonlinear charge is present in the drain-to-gate capacitance C_{GD}, but its contribution to excess charge is small since it is only 5 percent of the nominal charge.[9] The contribution of nonlinear charges to propagation delay can be taken into account as a correction after the propagation delay of a logic gate is computed with constant capacitances and transconductances. The propagation delay (t_{pd}) of a gate is proportional to $\frac{C_N \Delta V_L}{\Delta I_{DS}}$.

Two important characteristics of MESFET logic gates have been obtained through ring oscillator measurements:[2]

1. Propagation delay of a logic gate depends strongly on the pinch-off voltage of the switching MESFETs, because ΔV_L is a function of V_P. Pinch-off voltage is a function of the conductive layer thickness, as discussed in Chapter 1.
2. For constant gate length, the propagation delay increases when the width of the switching MESFETs is reduced, because of the consequent reduction of ΔI_{DS} and the ability to charge C_{load} ($\frac{dv}{dt} = \frac{I}{C}$). Circuits with very narrow gate width (and correspondingly low power dissipation) approach a state

where propagation delay is inversely proportional to power, leading to a constant minimum power-delay product. Circuits with wide MESFETs approach a constant minimum propagation delay at higher power dissipation levels.

2.3.1 Depletion-Mode Logic Gates

The propagation delay and power dissipation for D-mode logic gates is computed for the BFL structure of Figure 2.1 and the UBFL structure of Figure 2.2. Certain assumptions are made for computing these propagation delays;[9] some of these assumptions are based on an actual gate layout.[2] The following assumptions are used:

1. The effects of nonlinear storage charge are approximated by increasing by 25 percent the propagation delay computed by considering only linear charge storage.
2. The inverter switching transistor has a current gain–bandwidth product $f_T = \dfrac{g_m}{2\pi C_{GS}} = 16$ GHz which implies that the characteristic time constant τ for this device is $\tau = \dfrac{1}{2\pi f_T} = \dfrac{C_{GS}}{g_m} \sim 10$ psec.
3. Capacitances C_{GS} and C_{GD} are proportional to gate width W.
4. Capacitance C_{DS} is proportional to $W + W_e$, where W_e is an experimentally determined constant.[2]
5. The voltage-independent values of capacitances for a device with gate width $W = 500$ μm are: $C_{GS} = 0.7$pF, $C_{DS} = 0.1$pF, and $C_{DG} = 0.03$pF.
6. Every level-shifting Schottky diode has a stray capacitance to ground approximately equal to C_{DS} of a MESFET with an I_{DSS} equal to the current flowing in the Schottky diode.
7. Capacitances of interconnections running in parallel with C_{GS}, C_{DS}, and C_{DG} are approximately equal to 0.15 C_{GS}, 0.4 C_{DS}, and C_{DG}, respectively.
8. The ratio of power supply voltages is $RV = \dfrac{|V_{SS}|}{V_{DD}}$. Typical values for V_{SS} and V_{DD} are -3 and 4 volts, respectively.
9. When the output is high, the source follower source current is 1.5 times its I_{DSS}.

A linear circuit equivalent to the basic inverter of Figure 2.1 is shown in Figure 2.10.[9] The difference between the constant current of the load and the current of the switch in the inverter is modeled by the controlled current source I_i. The controlled current source, I_D, represents the transconductance of the source

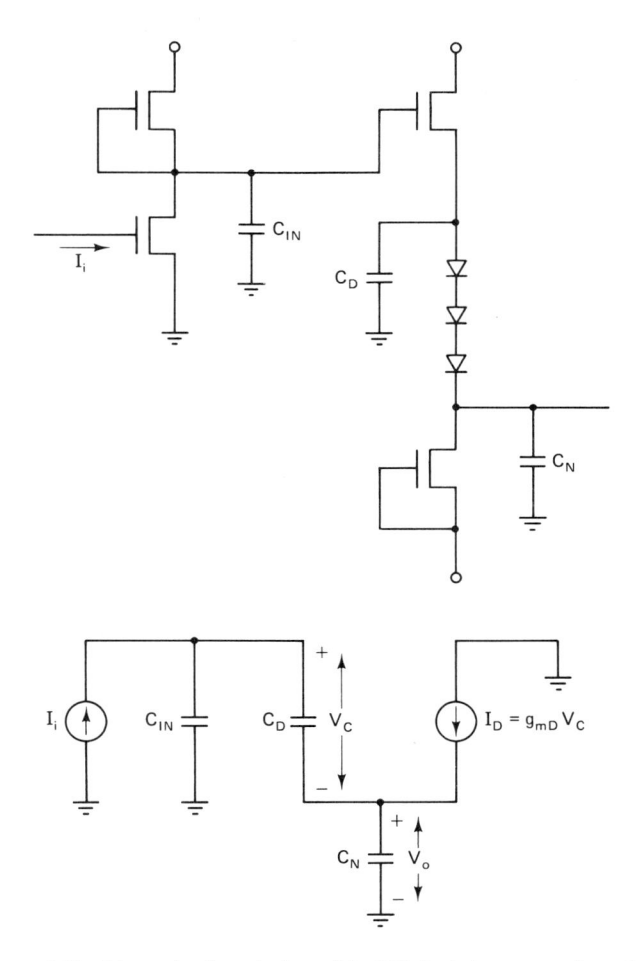

Sec. 2.3   Propagation Delays and Power Dissipations   51

Figure 2.10 Linear circuit equivalent of the BFL basic inverter configuration.

follower MESFET. The C_{DS} and C_{DG} capacitances of the active load and the switch MESFET along with the C_{DG} capacitance of the source follower MESFET and stray capacitances have been lumped into the input capacitance, C_{IN}. The output node capacitance C_N includes the C_{DS} capacitance of the source follower MESFET, the C_{DG} and C_{DS} capacitances of the current source MESFET, the stray capacitances to ground of the level-shifting Schottky diodes, and the input capacitances of FO subsequent logic gates for a fan-out of FO.

Assuming a step current input, $I_i = I_o u(t)$, it can be shown that the output voltage, V_o, is given by:[9]

$$V_o = \begin{cases} 0 & \text{for } t < 0 \\ \dfrac{I_o T_o}{C_{IN}} \left[\dfrac{t}{T_o} - \left(1 - \dfrac{C_t}{C_N}\right)\left(1 - e^{-t/T_o}\right) \right] & \text{for } t \geq 0 \end{cases} \quad (2.1)$$

where

$$C_t = \dfrac{1}{\dfrac{1}{C_{IN}} + \dfrac{1}{C_N} + \dfrac{1}{C_D}} \quad (2.2)$$

and

$$T_o = \dfrac{C_N\, C_D}{g_{mD}\, C_t} \quad (2.3)$$

In Expression (2.3), g_{mD} is the transconductance of the source follower MESFET.

Our primary interest is the computation of the time at which the output voltage V_o equals half the logic swing V_L. However, Equation (2.1) cannot yield, in closed form, this time and therefore it must be determined graphically or numerically. A graphical solution obtained in reference 9 is shown in Figure 2.11. The abscissa $\dfrac{t}{T_o}$ is determined by the intersection of the appropriate $\dfrac{C_t}{C_N}$ curve for

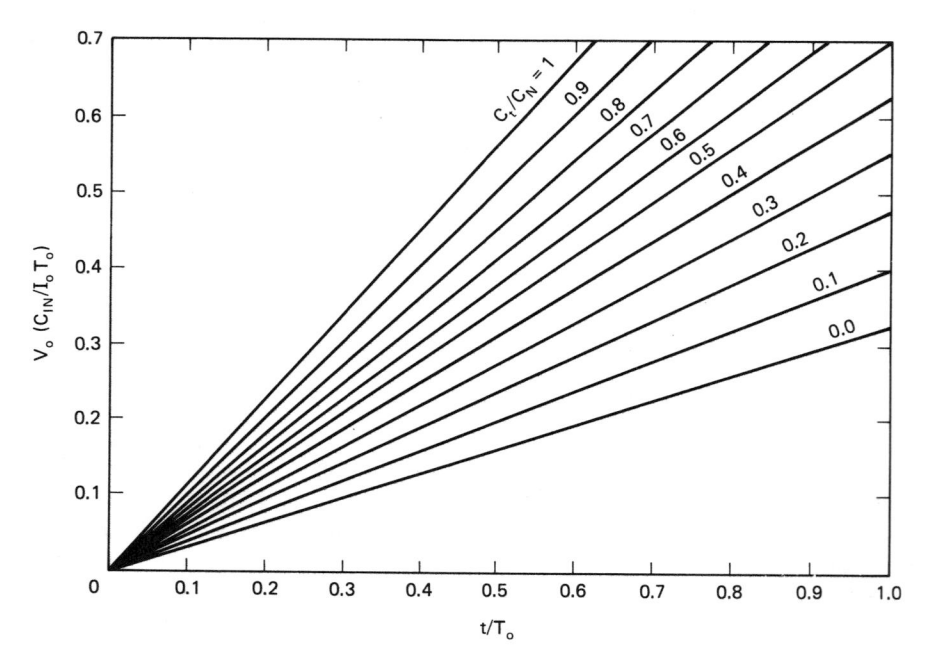

Figure 2.11 Transient response of the circuit of Figure 2.4 for various values of C_t/C_n (After Barna, Reference 9).

Sec. 2.3   Propagation Delays and Power Dissipations   53

the circuit under consideration, with a horizontal line of $\dfrac{V_L}{2} \cdot \dfrac{C_{IN}}{I_o T_o}$. The evaluation of the circuit capacitances is based on the assumptions made above. Therefore, for a MESFET with gate width W, we have $C_{DG} = \left[\dfrac{0.03}{0.7}\right] C_{GS}$ and $C_{DS} \sim \left[\dfrac{0.1}{0.7}\right]\left[1 + \dfrac{W_e}{W}\right] C_{GS}$. By considering all capacitances contributing to C_{IN} we have:

$$C_{IN} = C_{GS(S)}\left[\frac{15}{70} + \frac{2}{7}FI + (0.2FI + 0.12)\frac{W_e}{W_S} + \frac{3}{35}\frac{W_D}{W_S}\right] \quad (2.4)$$

$$C_D = \frac{W_D}{W_S} 1.15 C_{GS(S)} \quad (2.5)$$

and

$$C_N = C_{GS(S)}\left[\frac{92.5}{70} FO + \frac{32}{35}\frac{W_D}{W_S} + \frac{5.8}{7}\frac{W_e}{W_S}\right] \quad (2.6)$$

where the subscripts refer to the logic switch MESFET, FI is the fan-in ratio (for the general case of a logic gate with fan-in ratio FI), FO is the fan-out ratio, W_S is the gate width of the MESFETs in the inverter, and W_D is the gate width of the MESFETs in the source follower. The implicit assumption in the capacitance expressions is that the FETs have equal gate lengths.

The following two variables are introduced in order to evaluate the remaining parameters in Equation (2.1):

$$K = \frac{I_o}{I_{DSS(S)}} \quad (2.7)$$

and

$$\Lambda = \frac{g_{m(S)} V_L}{I_{DSS(S)}} \quad (2.8)$$

where $I_{DSS(S)}$ is the drain current of the switch FET at $V_{GS} = 0$ V.

Since the current available from a MESFET is proportional to its gate width for the circuit of Figure 2.1, we can assume that the current through the active load MESFET is equal to $0.6 I_{DSS(S)}$. In addition, the logic switch MESFET is driven to $V_{GS} > 0$ with a resulting $I_D \geq 1.2 I_{DSS(S)}$. Therefore, during the transient, we have $I_o \geq 0.6 I_{DSS(S)}$ and if we use $I_o = 0.6 I_{DSS(S)}$ then K has a constant value of 0.6.

Equation (2.8) implies that Λ has a constant value independent of gate width if V_L is constant. This assertion is valid because both variables $g_{m(S)}$ and $I_{DSS(S)}$ in Equation (2.8) are proportional to gate width. Measurements (with ±15% error margin) have shown that I_{DSS} is proportional to the device pinch-off voltage $|V_P|$ when $1 \text{ V} \le |V_P| \le 3 \text{ V}$.[9] Also V_L is approximately proportional to V_P, and $g_{m(S)}$ is largely independent of V_P and therefore, from Equation (2.8) we can assume that Λ remains constant as V_P varies.

Assuming $K = 0.6$ and $\Lambda = 2$, we have

$$\frac{C_{IN} V_o}{I_o T_o} = \frac{2}{0.69} \frac{C_{IN} C_t}{C_N C_{GS(S)}} \frac{V_o}{V_L} \tag{2.9}$$

and for $V_o = \dfrac{V_L}{2}$

$$\frac{C_{IN} V_o}{I_o T_o} = \frac{2}{1.38} \frac{C_{IN} C_t}{C_N C_{GS(S)}} \tag{2.10}$$

From Equation (2.10) and Figure 2.11 a numerical solution can be obtained for $\dfrac{t_o}{T_o}$ where t_o is the linear delay. Using some algebraic operations and including the correction factor for nonlinear charges, we find the equation for the propagation delay is:[9]

$$t_{pd} \sim 0.5 \text{ (psec)} + 2 \frac{C_N}{C_t} \frac{t_o}{T_o} \tag{2.11}$$

The positive-going and the negative-going propagation delays are nominally identical in this circuit.

The average power dissipation for the circuit of Figure 2.1 can be calculated as:[9]

$$P = V_{DD} I_{DSS(S)} \left[0.3 + 1.25 \frac{W_D}{W_S} + RV \frac{W_D}{W_S} \right] \tag{2.12}$$

Equation (2.12) can be written in terms of the pinch-off voltage when V_{DD} is chosen to be proportional to V_P, and $I_{DSS(S)}$, which is proportional to V_P and W_S, is approximated as a function of these two parameters. Therefore, if we use the approximations

$$V_{DD} \sim 1.8 |V_P| \tag{2.13}$$

and

$$I_{DSS(S)} \sim 0.08 (\text{mmho/}\mu\text{m}) \, V_P \, W_s \tag{2.14}$$

then Equation (2.12) yields

$$P = 1.8 \times 0.08 \cdot V_P^2 \, W_S \left[0.3 + (1.25 + RV) \frac{W_D}{W_S} \right] \tag{2.15}$$

The coefficients in Equations (2.13) and (2.14) are obtained from Reference 2.

Sec. 2.3   Propagation Delays and Power Dissipations

A similar analysis for the UFL basic inverter of Figure 2.2 leads to a propagation delay given by:[9]

$$t_{pd} = 0.5 \text{ (psec)} + N\left[0.86 + 0.36FI + 1.65FO + (0.25FI + 0.64)\frac{W_e}{W_S}\right] \quad (2.16)$$

and average power dissipation given by[9]

$$P = 1.8 \cdot 0.08 \cdot V_P^2 \, W_S \, (1 + 0.5RV) \quad (2.17)$$

The expressions for propagation delay and power dissipation of the BFL and UFL basic inverters are very important in determining device sizes for achieving required circuit performance in terms of speed and power. Although these expressions were developed based on certain layout rules and layout topology, they describe generally valid dependencies of propagation delay and power dissipation on device geometry and pinch-off voltage. Furthermore, by combining Equations (2.11) and (2.12) we can determine gate propagation delays, for a given power allocation, by manipulating device geometries of BFL gates. The same procedure can be applied to UFL gates. Also, for given propagation delay requirements and different fan-in and fan-out values, the choice for using a BFL or a UFL gate can be determined as the one resulting in the lowest power dissipation. The design of a logic gate must address two basic issues: (1) which circuit configuration should be used for the gate, and (2) what is the "optimal" ratio of the load-driver gate width W_D and the logic-switch gate width W_S leading to minimal propagation delay for a given power allocation. Expressions 2.11 through 2.17 indicate the interdependencies of all relevant parameters (assuming gate length of 1 µm for all FETs) for resolving these issues. It should be noted that, for a given process, a circuit designer can affect only the device geometry, the circuit bias, and the fan-in and fan-out ratios of a logic gate.

Although Expressions 2.11 through 2.17 illustrate a quantitative procedure for designing a logic gate (given a specific process) according to a set of propagation delay, power dissipation, and fan-in/fan-out requirements, these equations have to be modified (in terms of their proportionality coefficients) to be used for accurate design of a gate using a different process. In general, an integrated circuit designer uses a circuit simulation program, such as SPICE, to determine the circuit characteristics that will meet design requirements. The circuit simulation program employs a device model with parameters adjusted for the process intended for the circuit fabrication. The major benefit of studying 2.11 through 2.17, however, is the understanding of the way circuit parameters have to be determined by the designer in order to meet the circuit design requirements.

In general, for a given gate length, power dissipation and propagation delay are inversely proportional for constant fan-in and fan-out values. The increase of power dissipation is limited by the maximum permitted dissipation for the chip. One way to reduce power dissipation is to reduce the pinch-off voltage, since the power dissipation is proportional to V_P^2. However, reduction of V_P leads to increasing requirements of V_P uniformity across both a chip and a wafer—this implies strict control on the fabrication process that is difficult to achieve.

In UFL, propagation delay can be reduced by reducing W_e. This reduction, however, requires changes in the device layout and fabrication process.

The effect of reducing the gate length to 0.5 μm from the 1 μm that was assumed for the discussion in this chapter was studied in Reference 10. The changes considered by this reduction of the gate length were increases of g_m and I_{DSS} by 7.7 percent and decrease of C_{GS} by 37.5 percent. For a value of $FI = FO = 2$, it was found that the propagation delay of a BFL gate is improved by 5 percent, while the propagation delay of a UFL gate is improved by 25 percent.[10] These results offer an indication of the relative performance improvement in different logic families when the gate length of the FETs is reduced below 1 μm.

The sensitivity of logic gate propagation delay to temperature variations was measured in Reference 2 by monitoring the frequency of a ring oscillator. The temperature for coefficients was found to be $-1.84\frac{\text{MHz}}{C^o}$ for measured oscillation frequencies corresponding to $\frac{0.22\text{psec}}{C^o}$ of propagation delay. These measurements yield propagation delay variations of $\frac{0.2\%}{C^o}$. A similar percentage change also occurs in power supply currents. From measurements on discrete MESFETs, the saturation current was found to be inversely proportional to temperature while pinch-off voltage remains constant with temperature variations. The resulting decrease in transconductance must be mainly responsible for increasing logic gate propagation delays as temperature increases. It should be noted that these measurements offer a relative indication of the dependency of propagation delay on temperature. This dependency may not accurately apply to all logic families and/or circuit topologies.

2.3.2 Enhancement-Mode Logic Gates

The propagation delay and power dissipation for the E-Mode basic inverter with a depletion load is computed in this section. This inverter structure, shown in Figure 2.8, is used as the basic building element for DCFL logic gates. Its operation is identical to an NMOS inverter and similar approximations are used to derive formulas for both NMOS and GaAs circuits. In this section we find the propagation delay $t_{pd(p)}$ of a positive going output transition.

The propagation delay $t_{pd(p)}$ for a positive going output transition is defined as the time it takes the voltage to reach the gate threshold voltage, V_T, when the input is switched from the high logic level V_{HIGH} to the low logic level V_{LOW}. The computation of $t_{pd(p)}$ assumes the output node is loaded with a constant capacitive load C_N.

During this transient, the E-Mode device is cut off while the D-Mode device is in its saturated region. This is because, in general, V_T is chosen so that $V_T \leq V_{DD} - |V_{P(D)}|$ where $V_{P(D)}$ is the pinch-off voltage of the D-Mode device.

Sec. 2.3　　Propagation Delays and Power Dissipations　　57

Considering that $V_{GS} = 0$ for the depletion device, the saturation current in the depletion device can be approximated as

$$I_{DS(D)} = K'_{(D)} \frac{W_{(D)}}{L_{(D)}} |V_{P(D)}|^2 (1 + \lambda V_{DS(D)}) \qquad (2.18)$$

where the subscript (D) signifies the D-Mode device, K' is the process transconductance parameter, and λ is the channel-length modulation parameter that represents the small influence of $V_{DS(S)}$ on $I_{D(D)}$ in saturation.[11] Assuming, as in the D-Mode case, that we are working with 1 µm gate technology, then Equation (2.18) can be written as

$$I_{DS(D)} = K'_{(D)} W_{(D)} |V_{P(D)}|^2 (1 + \lambda V_{DS(D)}) \qquad (2.19)$$

With $V_{DS(D)} = V_{DD} - V_o$, the above expression becomes:

$$I_{DS(D)} = K'_{(D)} W_{(D)} |V_{P(D)}|^2 (1 + \lambda V_{DD} - \lambda V_{OUT}) \qquad (2.20)$$

or

$$I_{DS(D)} = \frac{V_G - V_O}{R_G} \qquad (2.21)$$

where

$$V_G \equiv \frac{\lambda \cdot V_{DD} + 1}{\lambda} \qquad (2.22)$$

and

$$R_G \equiv \frac{1}{|V_{P(D)}|^2 K_{(D)} W_{(D)} \lambda} \qquad (2.23)$$

Equations (2.21) through (2.23) imply that the depletion device can be represented by a voltage source V_G in series with a source resistance R_G.

Assuming the initial value of the output voltage is V_{LOW}, then the output voltage as a function of time can be written as

$$V_o = V_{LOW} + (V_G - V_{LOW})[1 - e^{-t/(R_G C_N)}] \qquad (2.24)$$

and therefore $V_o = V_T$ at time $t_{pd(p)}$ given by

$$t_{pd(p)} = R_G C_N \ln \frac{V_G - V_{LOW}}{V_G - V_T} \qquad (2.25)$$

For most cases it can be assumed that $V_{LOW} = 0$, and therefore

$$t_{pd(p)} = R_G C_N \ln \frac{V_G}{V_G - V_T} \qquad (2.26)$$

The total effective load capacitance, C_N, of the output node is a result of the inherent device capacitances in a logic gate, the number of inputs (fan-in ratio) *FI* to the logic gate, and the number of identical logic gates (fan-out ratio) FO connected to the output node. Therefore, for a logic gate of fan-in FI and fan-out FO and by assuming one input is switched at a time, C_N can be written as

$$C_N = (FI + 1) C_{GD(E)} + FIC_{DS(E)} + \\ C_{DS(D)} + C_{GD(D)} + FO(C_{GS(E)} + 2C_{GD(E)}) \quad (2.27)$$

where subscripts (E) and (D) refer to enhancement and depletion devices, respectively, and a factor of 2 has been included for the Miller effect. The above equation does not include any stray capacitance which, in general, will be loading the circuit. The stray capacitance depends on the circuit geometry and process. It should be calculated and added to C_N given by Equation (2.27).

The propagation delay $t_{pd(n)}$ for a negative going output transition is defined as the time required for the output voltage to switch from V_{HIGH} to the logic gate threshold, when the input is switched from V_{LOW} to V_{HIGH}. In this case also, a constant capacitive load C_N is assumed for the output node.

This transient is not obtainable analytically—unless $\lambda \to 0$, which would be an unrealistic assumption for the short channel devices considered here. This transient is computed numerically through computer simulations.

An indirect way to determine average propagation delay of a gate (i.e., $\left[\frac{t_{pd(p)} + t_{pd(n)}}{2} \right]$) is by measuring the oscillation frequency, f_o, of an odd number N_R of inverters connected to form a ring oscillator. The average delay time, t_D, is then calculated as:

$$t_D = \frac{t_{pd(n)} + t_{pd(p)}}{2} = (2N_R f_o)^{-1} \quad (2.28)$$

This expression, however, assumes that all inverters have the same geometry and the interconnection between stages contributes very little to the propagation delay. This assumption results in more optimistic results than the expected performance of an inverter in a logic circuit. Also, in a ring oscillator, all inverters have $FO = FI = 1$, which is rarely the case in a logic circuit. Therefore, results obtained from Expression (2.28) should be treated as relative indications of an upper bound that should be expected from a logic gate fabricated with a given process.

The power dissipation of a DCFL inverter switching at frequency f, has a static component P_{st} which is independent of f and a dynamic component P_d which is proportional to f. The static component

$$P_{st} = \frac{I_{DS(D)} (V_{DD} - V_{LOW}) V_{DD}}{2} \quad (2.29)$$

results from resistive heating of the switching device and the load during the periods when the switching device is on, which is assumed to be half of the time. Additional static power is dissipated by current flow through the gates of circuits

Sec. 2.4   Noise Margins   59

connected to the output node in case $V_{DD} > V_{GM}$, where V_{GM} is the forward gate bias at which the device gate junction becomes highly conductive. This power dissipation can be expressed as

$$P'_{st} = \frac{I_{DS(D)}(V_{DD} - V_{GM})V_{DD}}{2} \qquad (2.30)$$

The dynamic power dissipation results from the periodic charge and discharge of the output node capacitance C_N, and is given by

$$P_d = C_N V_o^2 f \qquad (2.31)$$

where V_o is the output voltage.

An important parameter often used as a figure of merit for various circuit implementations is the product of delay time and dynamic power dissipation:

$$t_d P_d \sim C_N V_o^2 \qquad (2.32)$$

The delay time-dynamic power product, often called speed-power product, or dynamic switching energy, is the minimum energy a gate requires to change states during a logic transition, assuming two transitions per cycle.

2.4 NOISE MARGINS

Noise margins can be defined as the voltage difference, measured on the input voltage axis of a logic gate transfer characteristic, between the operating point (i.e., high or low) and the nearest unity gain point. The unity gain point is defined as the point on the transfer characteristic where the slope $\dfrac{\partial V_o}{\partial V_{IN}} = 1$.
Almost identical values for noise margins can be obtained by using, instead of unity gain points, those points where the output voltages reach 10% and 90% of the logic swing, or the intersection points of the tangent in the inflection point with the "low" and "high" levels as illustrated in Figure 2.12 for a DCFL inverter.[6] For this circuit, the noise margin for logic "0" input is 0.06 V and the noise margin for logic "1" input is 0.07 V.

A quantity often used when comparing different circuits is noise immunity, which is defined as the ratio of noise margin to logic swing. Comparisons using this figure are quite relevant since internally generated noise increases with the logic swing.

For reliable circuit operation, the design of logic gates should guarantee positive noise margins. This implies that for any gate combination it is required that $V_{oh} > V_{INh}$ and $V_{ol} < V_{INl}$, where V_o is the output voltage, V_{IN} is the input voltage, and the subscripts h and l refer to logic "high" and logic "low," respectively. In order to fulfill these requirements the designer should investigate the conditions that give $\min V_{oh}/\max V_{ol}$, and $\max V_{INh}/\min V_{INl}$ for worst case of power supplies, process variation, gate combinations, and temperature. Consider,

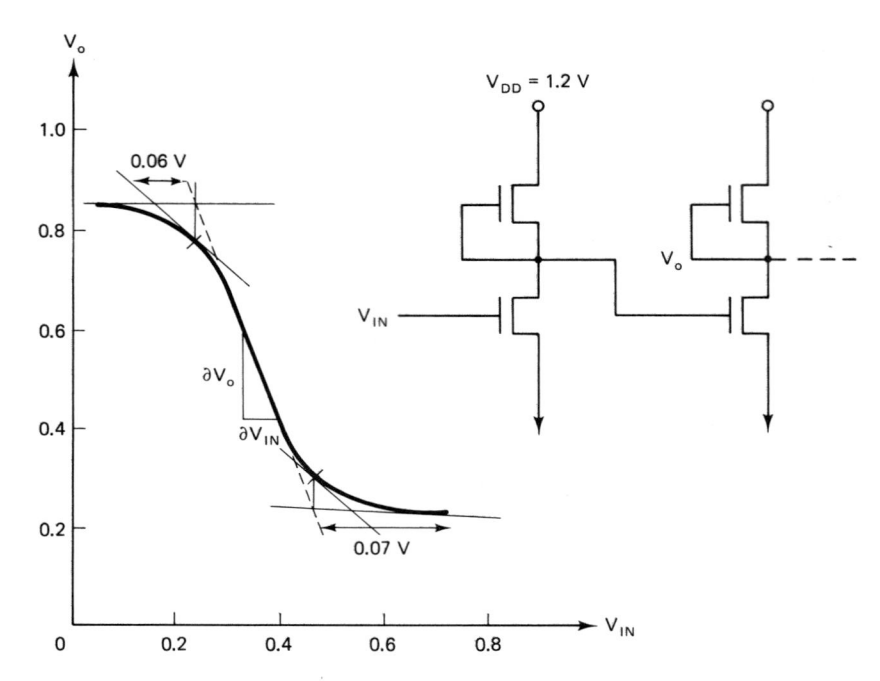

Figure 2.12 Transfer function of CDFL ring oscillator section, with noise margins indicated.

for example, the BFL inverter of Figure 2.1. When the output is "high" then the input to the source follower is near V_{DD}. Therefore, the minV_{oh} level occurs for minimum V_{DD}. The output "low" level increases as the current supplied by the source follower current source decreases. The value of source follower current source current is inversely proportional to V_{SS}. Therefore, the maxV_{ol} occurs for maximum V_{SS}.

The conditions for worst case process are more negative pinch-off voltage V_P and low diode voltage drops. More negative V_P results in minimum V_{INl}. Also, as V_P goes more negative the input device is not fully turned off with V_{INl}; this results in decrease of V_{oh} since the pull-up device has finite V_{DS}. Low diode drops, on the other hand, result in maximum V_{ol}. As V_{ol} increases, the input device driven by it turns on, thus decreasing V_{oh} of a subsequent gate.

The highest temperature specified in the design requirements (typically, 75°C for commercial specifications and 125°C for military specifications) is the worst case of temperature, since it results in low diode drops and more negative pinch-off voltage resulting in the effects discussed for worst case process.

Gates with high fan-in ratio produce high V_{INl} currents resulting in maximum values for V_{INl}. On the other hand, high V_{INl} currents produce lower V_{oh}, thus resulting in minimum values for V_{oh}. However, a high fan-in gate is

Sec. 2.4   Noise Margins   61

equivalent to a single fan-in gate if one input is "high" and the other inputs are completely turned off, thus resulting in maximum values for V_{ol}. Figure 2.13 illustrates these conditions. Special design considerations should also be given to logic gates where the input devices are connected in series. Figure 2.14 shows such an example. In this case V_{ol1} is greater than V_{ol2} unless the width of device Q_1 is twice the width of device Q_2—that is, $W_{Q1} = 2W_{Q2}$. The reason for this requirement is that when high level inputs are applied simultaneously to the gates of the Q_2 devices, the same resistance ratio must exist between the load and input devices as in the case of the single gate inverter in order to have the same V_{ol}. Since the Q_1 devices are in series between the output terminal and ground, the effective resistance of each device must be half of the resistance of the single input device, or the device gate width must be twice as wide. It is assumed here that both devices Q_1 and Q_2 have the same gate length and also the load devices Q_L are identical. Another effect that must also be considered is that, while the circuit is in the conducting state, the source of the upper input device Q_1 is at a potential higher than ground, thus placing a somewhat higher level requirement on

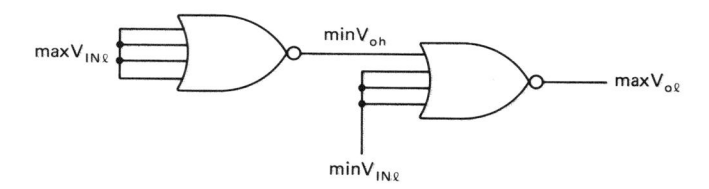

Figure 2.13   Worst-case gate configuration.

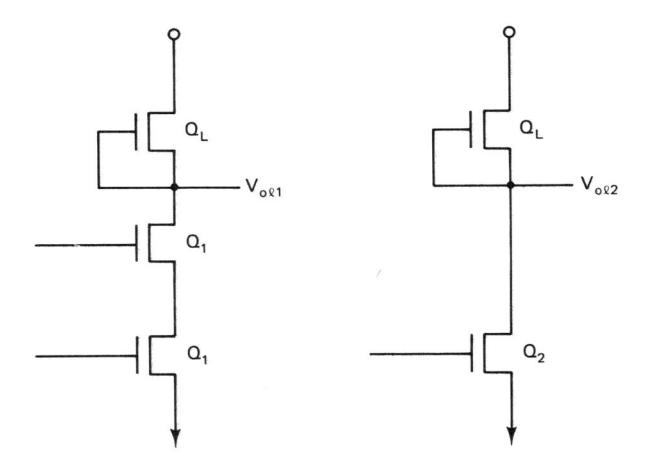

Figure 2.14   Logic gate with input devices connected in series versus a gate with a single input device.

the input to maintain a specific on-resistance for the device. These considerations place a practical limit on the number of input devices that may be connected in series between the output node and ground and still maintain reliable function of logic gates.

A circuit used to determine worst case noise margins through computer simulation is shown in Figure 2.15. The noise margin, based on this circuit configuration, is defined as the maximum value of V_{n1} or V_{n2}, at which the latch is at the ultimate edge of switching to the wrong state.[12] To determine the noise margin via computer simulation we ramp the voltage of the noise sources V_{n1} and V_{n2} and monitor the state of the latch. The value of the noise sources that causes the latch to toggle is the noise margin. An example of such a simulation is illustrated in Figure 2.16.

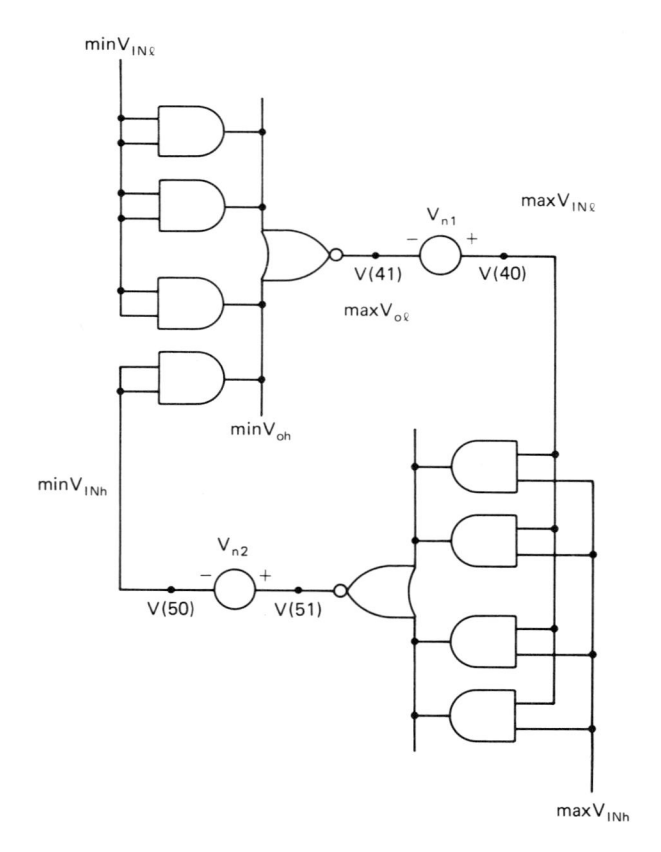

Figure 2.15 Test circuit for determining noise margins by means of computer simulation (from Triquint's GaAs Short Course Notes; by permission).

Sec. 2.5 Interconnection Lines 63

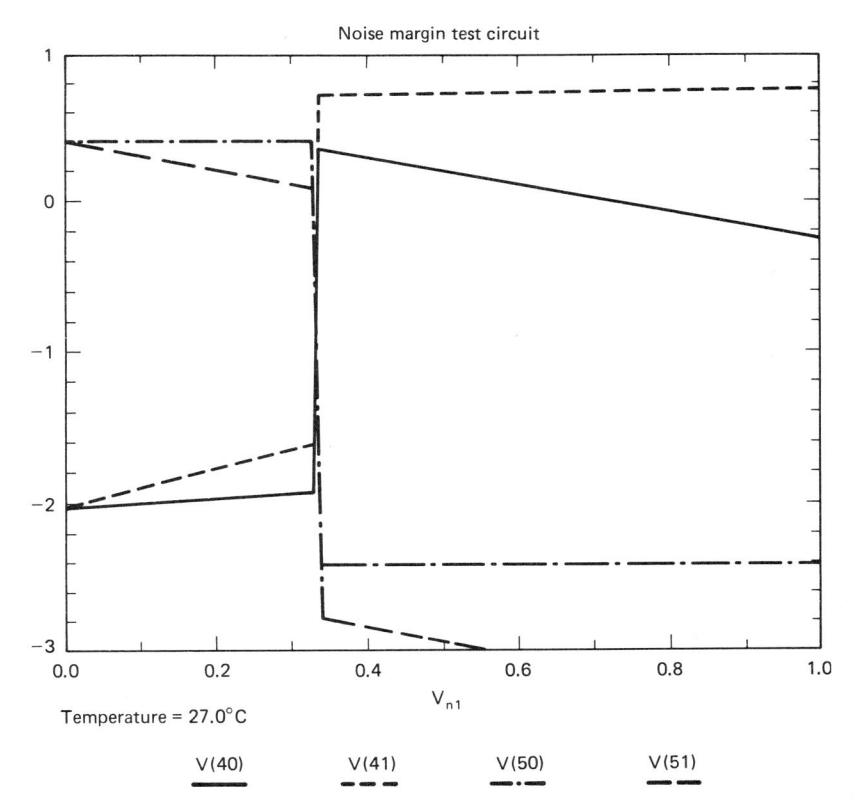

Figure 2.16 Computer simulation output of the noise margin test circuit (from Friquint's GaAs Short Course Notes; by permission).

2.5 INTERCONNECTION LINES

Interconnection lines play a major role in integrated circuits with regard to both chip area and dynamic behavior. In digital circuits (excluding memory) the required number of interconnection networks equals approximately the number of gates. Furthermore, the average interconnection line length increases with the complexity of the logic circuitry. Since the fraction of chip area occupied by the interconnection networks is proportional to the number of interconnection lines times their average length, their area requirements increase superlinearly with the number of logic gates.

The major role of interconnection networks is reflected, in addition to the required chip area, in the electrical properties of a logic circuit. The interconnection lines pass signals from one logic gate to the next and the rate of signal propagation is limited by delays occurring during the propagation through the interconnection. This delaying and at the same time band-limiting effect of

2.5.1 Electrical Properties of On-Chip Interconnection Lines

In a discussion of on-chip interconnection lines, we consider lines carrying signals between logic gates only. These are distinctly different from the power supply lines. In general, signal lines are driven by relatively high impedances which allow simple approximations to be made in determining their operation and effect on circuit behavior. The problems of the power supply lines are related mainly to switching noise, and their analysis requires numerical methods[13] which will not be discussed here.

The interconnection lines have distributed capacitance, inductance, and resistance. In high-speed systems interchip interconnection lines have to be terminated with a resistance equal to their characteristic impedance, Z_O, to avoid reflections. On-chip interconnections are unmatched. An equivalent-circuit representation of a signal line between two active elements is shown in Figure 2.17. We are interested in the line response in the frequency region below the cut-off frequency of the active devices. Skin effects at this frequency are not found to be important since R_O exceeds ωL_O,[14] where R_O is the line resistance per unit length, L_O is the line inductance per unit length, and ω is angular frequency. For small geometry technologies the relation of R_O and ωL_O is still valid because the reduction in line dimensions results in faster increase of R_O than the corresponding increase of L_O. The transfer function of the line in Figure 2.17 is

$$\frac{V_S}{V} = \cosh\gamma l + \frac{R_S}{Z_O}\sinh\gamma l \tag{2.33}$$

where

$$Z_O = \sqrt{\frac{R_O + j\omega L_O}{j\omega C_O}} \tag{2.34}$$

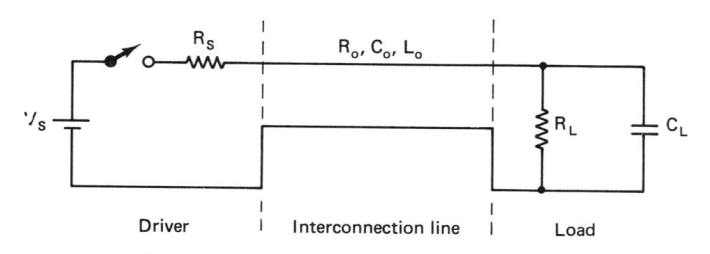

Figure 2.17 On-chip unterminated interconnection line.

Sec. 2.5   Interconnection Lines   65

and

$$\gamma = \sqrt{(R_O + j\omega L_O)j\omega C_O} = Z_O\, j\omega C_O \quad (2.35)$$

The terms l and C_O in the above expressions represent the line length and the line capacitance per unit length, respectively. If we neglect the L_O term, then the transient line response approaches the exponential charging of a capacitor $C = C_O \cdot l$ through the resistor R_S. Using this approximation, the line propagation delay can be calculated as:[15]

$$T_d = 0.7 R_S\, C_O\, l + 0.4 R_O\, C_O\, l^2 \quad (2.36)$$

The first term in Equation (2.36) considers the recharging of the line capacitance through the source resistance R_S. This fraction of T_d is proportional to line length. The second term in Equation (2.36) simulates the characteristics of RC lines. This fraction of T_d increases with the square of the line length.

The formula for T_d was derived on the assumption that $R_S \gg Z_O$. This assumption is valid in most cases and should be ensured by design, because if $R_S < Z_O$, the voltage at the receiving end (Figure 2.17) undergoes a series of oscillations at each transient. These oscillations cause noise margin degradation, and for sufficiently low R_S and long line length l, they will result in logic errors.

Figure 2.18 is a graphic representation of the line propagation delay as a function of line length for different line widths, w, and different source resistance, R_S, of the driving device. It is evident from this graph that long lines have a critical influence on the circuit dynamic behavior especially when the source resistance of the line driver device is large. Note however, that reduction of R_S will increase the average power per gate approximated by:

$$P_g = \frac{V_S^2}{2R_S} \quad (2.37)$$

assuming R_S is switched to ground by an active device and a duty factor of 50 percent.

The principal contributing factor in this line response is the capacitance per unit length, C_O, of the line. When it becomes possible to make this capacitance arbitrarily small, fast and low-power circuits will result. The extent of the reduction of this capacitance, however, is limited by the geometry of the lines and the fabrication process. Reducing the line width does not result in any significant reduction in line capacitance. This is because narrow lines no longer obey the laws of the plate capacitor. As horizontal dimensions begin to approach the vertical dimensions, mutual sidewall capacitances prevent, in general, C_O from decreasing below $\frac{0.1 pF}{mm}$ (fringe fields). Enlarging the separation, s, between adjacent lines, on the same level of interconnect, reduces the mutual capacitance but in general increases the average line length. Reducing the thickness, z, lowers the mutual capacitance but increases the resistance per unit length, R_O. The ratio $\frac{w}{h}$, where h is the distance from the ground plane (i.e., backside metal, usually)

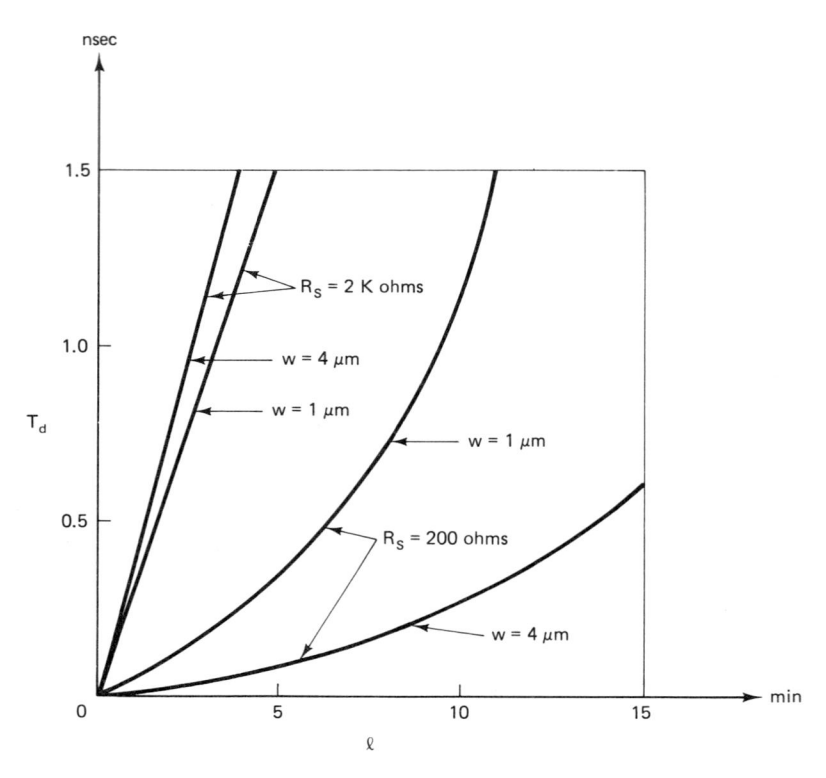

Figure 2.18 Propagation delay T sub d as a function of line length l, for different values of line width and source resistance.

for a given s, cannot become arbitrarily small because significant crosstalk will be introduced.

The resistance R_O, however, increases rapidly with the steady reduction in the size of interconnection lines. The resistance increases with the square of the size of the line structure because, for a given crosstalk boundary value, z and w have to be reduced by an equal amount. Figure 2.19 illustrates the line arrangement on a single layer of interconnect and the dimensions mentioned in the above discussion.

2.5.2 Proposed Solutions to Interconnection Line Problems

Line propagation delays and band-limiting occur irrespective of the semiconductor technology used for logic design. There are several ways, however, to reduce these effects on circuit performance and several approaches should be considered during the design process.

Sec. 2.5   Interconnection Lines   67

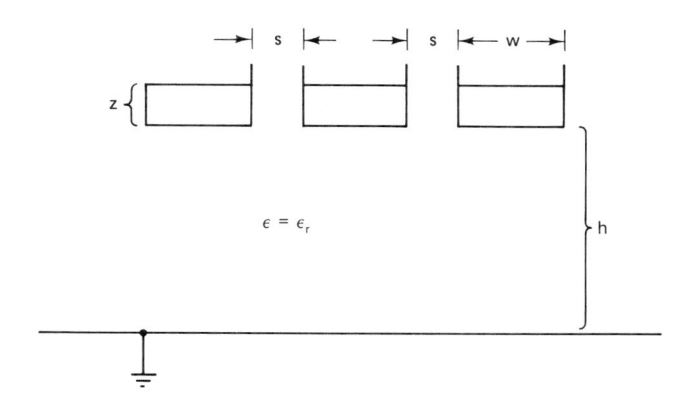

Figure 2.19 Line arrangement on a single layer of interconnect and parameters affecting capacitance.

The line capacitance can be reduced by reducing the dielectric constant. Although the dielectric constant is difficult to reduce below that of SiO_2 ($\varepsilon = 4$), there are some organic dielectric materials that have slightly lower values, and manufacturers are reporting reliable interconnections with air-bridges (i.e., $\varepsilon = 1$), which are described in Chapter 1 for a D-Mode GaAs process.

The effect of long lines on propagation delays can be alleviated by reducing the source resistance, R_S. Small source resistance results in a high switching current which should not be arbitrarily high due to the total power dissipation. However, for a given switching current I, reducing R_S implies a lower signal swing since $R_S = \dfrac{\Delta V_S}{I}$. Technologies allowing small signal swings are therefore suitable for realizing short propagation delays or large bandwidths for the interconnection lines. However, small signal swings must guarantee correct logic operation of the circuit under required noise margins.

A particularly effective approach for keeping propagation delays of interconnection lines within desired limits is the utilization of two or three interconnection layers. The utilization of multiple layers of interconnect allows higher packing density of the active circuitry and reduces the average line length.

Another option for reducing propagation delays and synchronizing signals of different length paths is to include active elements in the interconnection line. The repeated regeneration of the signal by the active elements can eliminate the l^2 term in Equation (2.36). The active elements, in this case simple buffers, introduce buffer propagation delays; however, a minimum propagation delay can be realized by a trade-off between the number of buffers used and the l^2 term of the propagation delay. The propagation delay of an interconnection line with length l, having N buffers at Δl intervals $\left[N = \dfrac{l}{\Delta l} \right]$ is

$$T_d = 0.7R_S \, C_O \, l + (0.4R_O \, C_O \, l\Delta l^2 + T_b)N \tag{2.38}$$

where T_b is the propagation delay of a buffer.

The minimum value for this expression is

$$T_d \, (\min) = l \left[0.7R_S \, C_O + \sqrt{1.6R_O \, C_O \, T_b} \right] \tag{2.39}$$

and can be achieved for

$$\Delta l = \sqrt{\frac{T_b}{0.4R_O \, C_O}} \tag{2.40}$$

The underlying assumption in the above expressions is that all N buffers have the same propagation delay T_b. Also, Equation (2.39) indicates that the effective propagation delay increases only linearly with the line length.

2.6 LOGIC GATE CONFIGURATIONS

GaAs logic gates are configured using transistors and diodes in D-Mode logic, while E-Mode logic usually uses only transistors. As in Si technologies, GaAs gates may implement a single logic function (i.e., NOR, NAND) or they may implement a complex logic function (i.e., AND-OR-INVERT). The choice of logic gates for implementing circuits depends on requirements of circuit speed, capacitive load driving, and power dissipation. For given technology and fabrication parameters, the designer may meet the circuit requirements with an appropriate choice of logic gates and device sizes.

D-Mode logic circuits are configured using depletion-mode transistors (normally-on) and Schottky diodes, which provide the required logic level shifting for proper operation. Implementation of these gates requires two power supplies in addition to ground.

Figure 2.20 illustrates a four-input NOR gate in BFL configuration. The source follower section provides the required level shifting and the current required to drive subsequent stages, while the front section of the gates implements the NOR function. Typical performance of a BFL gate fabricated with 1 μm technology and having 20 μm wide devices is reported[16] as 190 psec for propagation delay and 20 mw for power dissipation. These performance figures assume fan-out and fan-in ratios of three, plus 500 μm of interconnect implemented with an airbridge. Speed and power can be traded in this design by decreasing the width of the devices. Using the technology reported in Reference 16, a power dissipation of 5 mw per gate can be achieved when the propagation delay is increased to 600 psec. Fan-in for this gate is limited for practical purposes to four because of the contribution of the drain capacitance of the input transistors to C_{IN} and the area required by these devices.

An implementation of the same four-input NOR gate using the SDFL approach is illustrated in Figure 2.21. This approach uses diodes to perform the

Sec. 2.6　Logic Gate Configurations　　　　　　　　　　　　　　　　　　　　　　　69

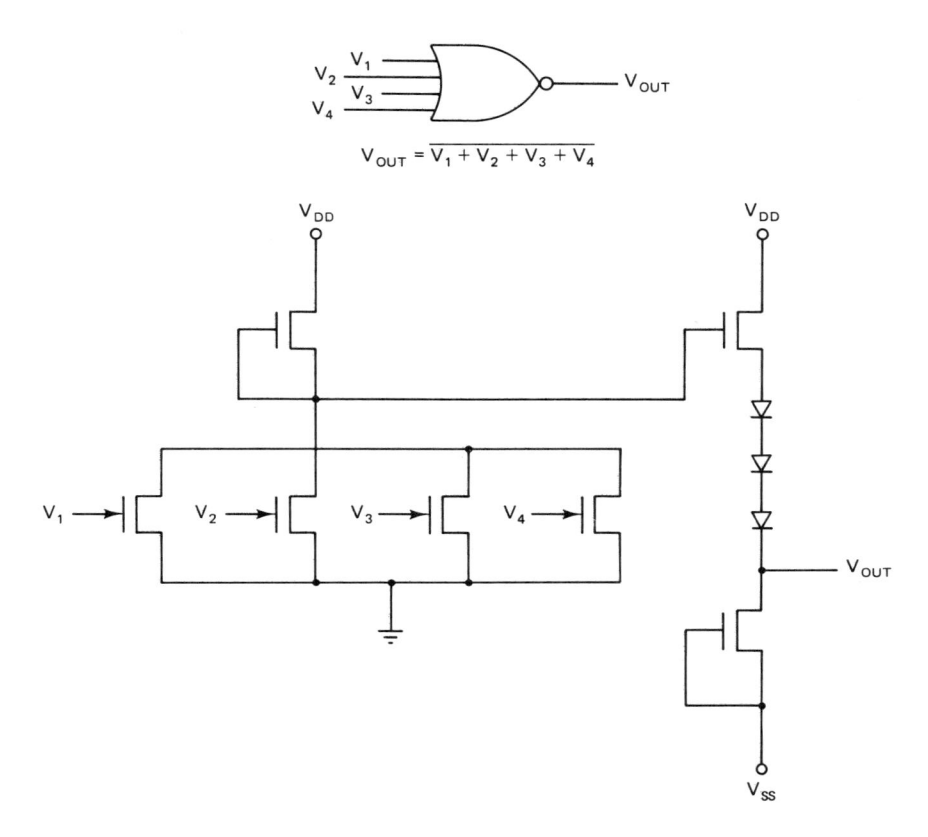

$$V_{OUT} = \overline{V_1 + V_2 + V_3 + V_4}$$

Figure 2.20　Four-input BFL NOR gate.

logical positive-OR function and the required voltage shift, while the transistors perform the logic inversion. This NOR gate occupies a smaller area and consumes less power than its BFL counterpart. This is because of the replacement of relatively large FETs with small diodes in the logic section, and the fact that the input diodes are not always forward-biased, which reduces power dissipation. Also, this SDFL gate structure allows for virtually unlimited fan-in. NOR gates with eight inputs have been described by Long et al.[17] The speed of this gate, however, is more sensitive to fan-out than the output of the BFL NOR gate. Thus its ability to drive large capacitive loads at high speed without buffering or without using very wide channel devices in the driving stage is limited. The use of input logic diodes has some process implications. These diodes require a lower carrier concentration and lower sheet resistance implant than the FET channel in order to optimize their reverse bias capacitance and series resistance. Therefore, fabrication of SDFL circuits requires two separate implant steps using localized implantation into selected areas of the substrate so that the input logic diodes can be formed.

A DCFL four-input NOR gate has the same configuration as an NMOS NOR gate, using both depletion and enhancement FETs. Figure 2.22 illustrates such a circuit. The arrows on the device input gates indicate an enhancement FET. This gate has the lowest design complexity (i.e., number of active devices),

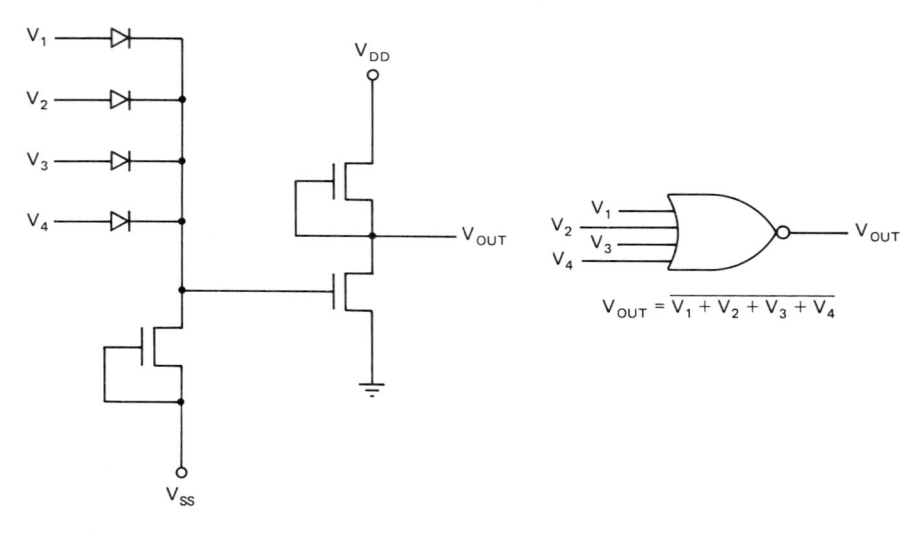

Figure 2.21 Four-input SDFL NOR gate.

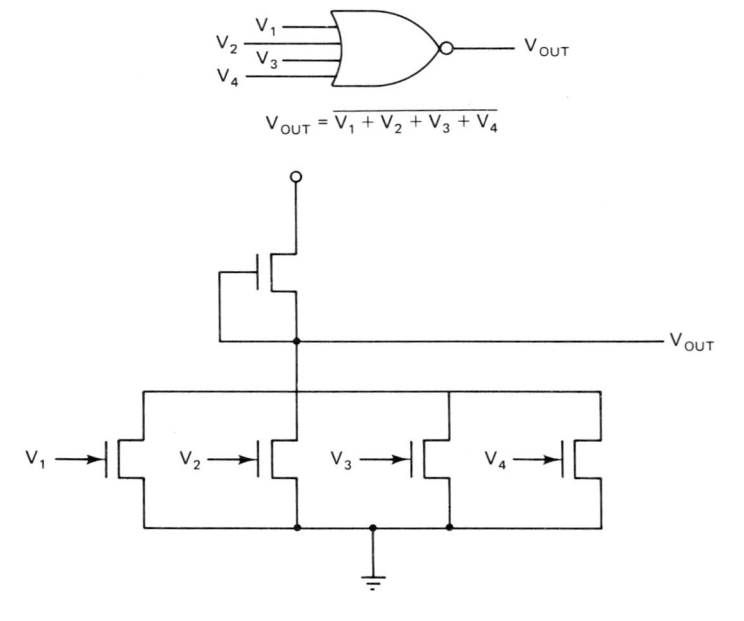

Figure 2.22 Four-input DCFL NOR gate.

Sec. 2.6 Logic Gate Configurations

area requirements, and power dissipation. Static power is reduced by keeping the current through the depletion device fairly low. This reduces the output logic "low" of the gate and increases the gate noise margin. However, low output current inversely affects the output rise time under heavy fan-out loading. From a dynamic point of view, the switching speed of the gate is reduced, as in NMOS, by the gate capacitance loading by a factor of approximately $\frac{1}{FO}$ (where FO is the fan-out). The DCFL gate is also sensitive to fan-in ratio. This can significantly change the threshold voltage for the logic gate and thus critically reduce the gate noise margin.

In all NOR gates implemented with different technologies, the gate width of the load device is made smaller than that of the input switching FET so that the high-gain transition region may be placed near the center of the logic swing.

The series connection of the input switching devices results in the implementation of the NAND function. The number of devices connected in series, however, is limited to two, as a rule of thumb (for all logic families), because the voltage drop in the series FETs results in threshold shift which adversely affects performance in terms of speed and noise margins. The two FETs in series are usually implemented as a dual gate device.[2] This device behaves electrically as two FETs in series; its layout, however, is very compact since the spacing between the two gates is on the order of the gate width of a single device. Figure 2.23 illustrates the layout of a single FET and that of a dual FET. Using the dual gate FET, NAND gates are configured in different logic families as shown in Figure 2.24. The implementation of NAND gates in the different logic families carries trade-offs similar to those in the NOR implementation discussed above.

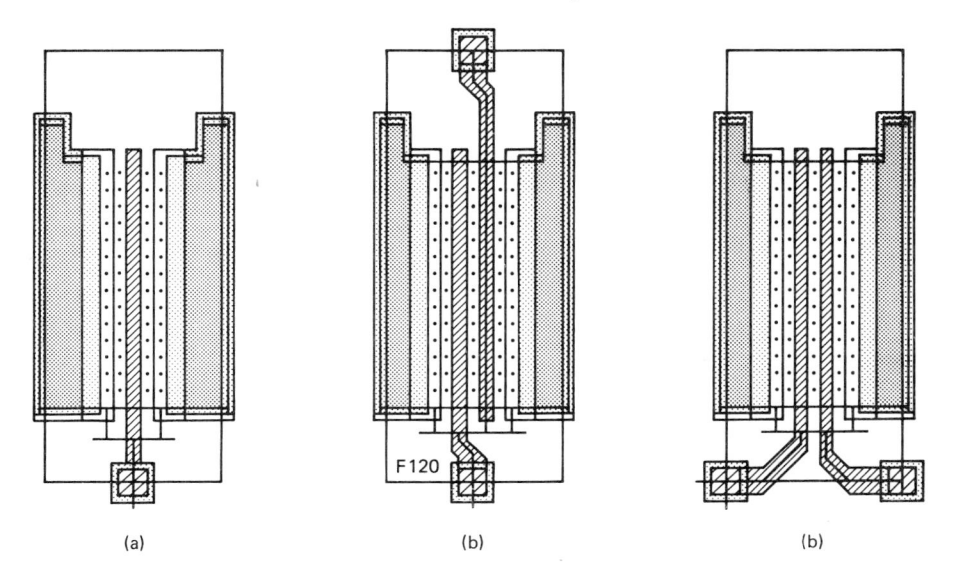

(a)   (b)   (b)

Figure 2.23 Layout of: (a) a single-gate FET, and (b) a dual-gate FET.

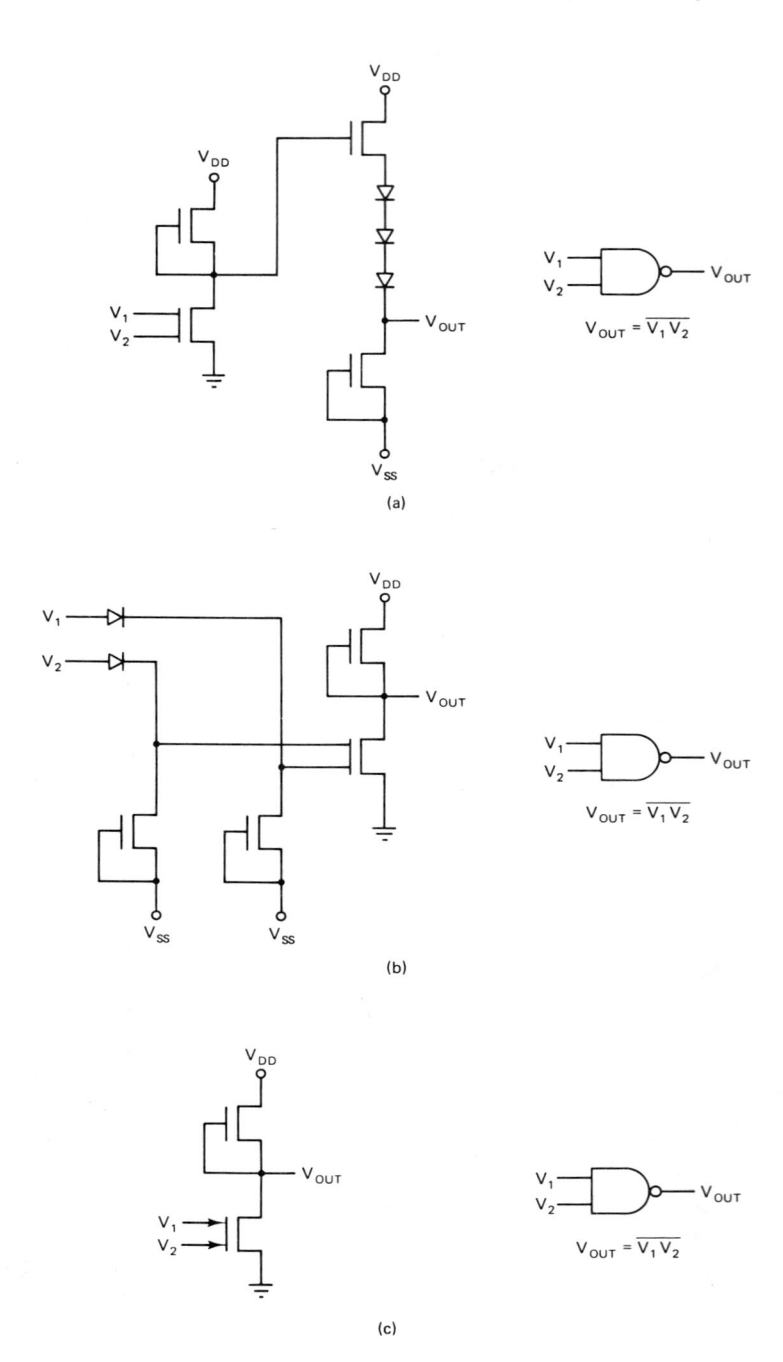

Figure 2.24 Two-input NAND configuration in: (a) BFL; (b) SDFL; and (c) CDFL.

Sec. 2.6   Logic Gate Configurations   73

However, one can observe from Figures 2.21 and 2.24 that the area efficiency of SDFL over BFL when implementing a NOR function is much greater than when implementing a NAND function because a WIRED-OR function may be performed by connecting the outputs of the input diodes together. This requires only a single additional diode and a single FET to implement the logic and level shift. It is advantageous, therefore, to utilize NOR gates as much as possible when implementing logic circuits in SDFL.

One of the most commonly used BFL logic gate configurations is the AND-OR-INVERT function which is implemented as a complex gate (i.e., more than one level of logic with a single gate delay). The AND/NOR gate uses the dual gate FET which results in compact layout. Figure 2.25 illustrates this gate configuration. As in the case of NOR gates, the number of FETs connected in parallel at the logic section is limited to four. If all the inputs to the logic

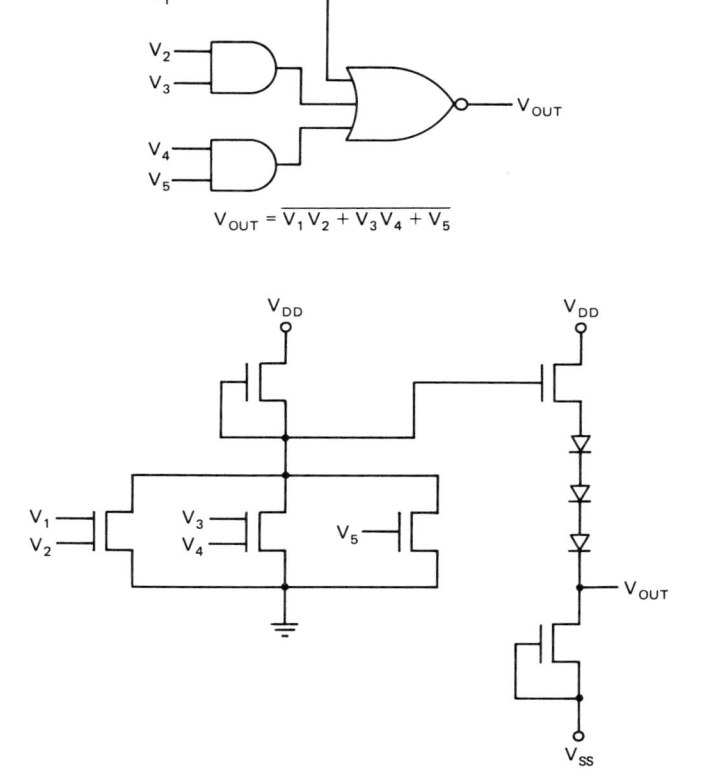

$$V_{OUT} = \overline{V_1 V_2 + V_3 V_4 + V_5}$$

Figure 2.25 BFL AND/NOR complex gate configuration.

section are inputs to dual gate FETs, then the logic function implemented is NAND/WIRED-AND. This is shown in Figure 2.26.

Complex gates implemented in SDFL take advantage of the efficient implementation of the OR function before the inversion. Therefore, commonly used complex gates in SDFL are of the OR/NAND type or the OR/NAND/WIRED AND type. Figure 2.27 illustrates the SDFL implementations of these logic functions.

DCFL complex gate implementation uses a similar approach to BFL. The difference is the utilization of enhancement devices for logic switching and the absence of the source-follower for level shifting.

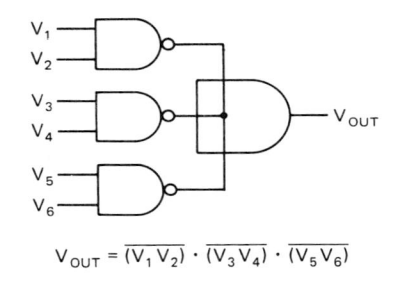

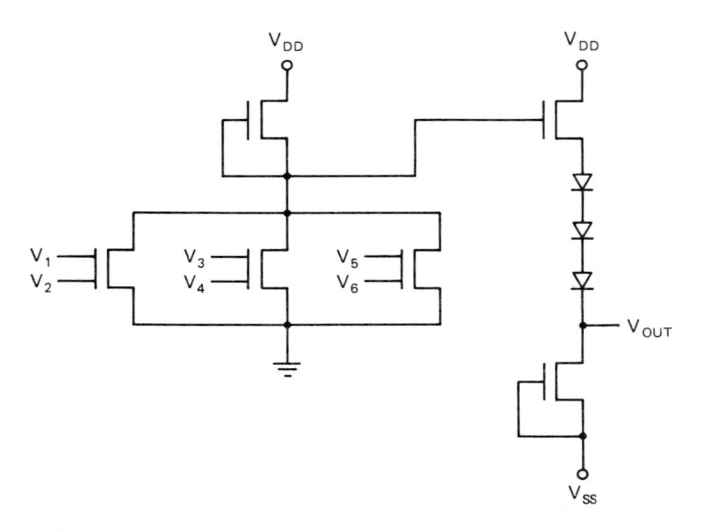

$$V_{OUT} = \overline{(V_1 V_2)} \cdot \overline{(V_3 V_4)} \cdot \overline{(V_5 V_6)}$$

Figure 2.26 BFL NAND/WIRED-AND complex gate configuration.

Sec. 2.6   Logic Gate Configurations

$$V_{OUT} = \overline{(V_1 + V_2)(V_3 + V_4)}$$

(a)

$$V_{OUT} = \overline{(V_1 + V_2)(V_3 + V_4)(V_5 + V_6)(V_7 + V_8)}$$

(b)

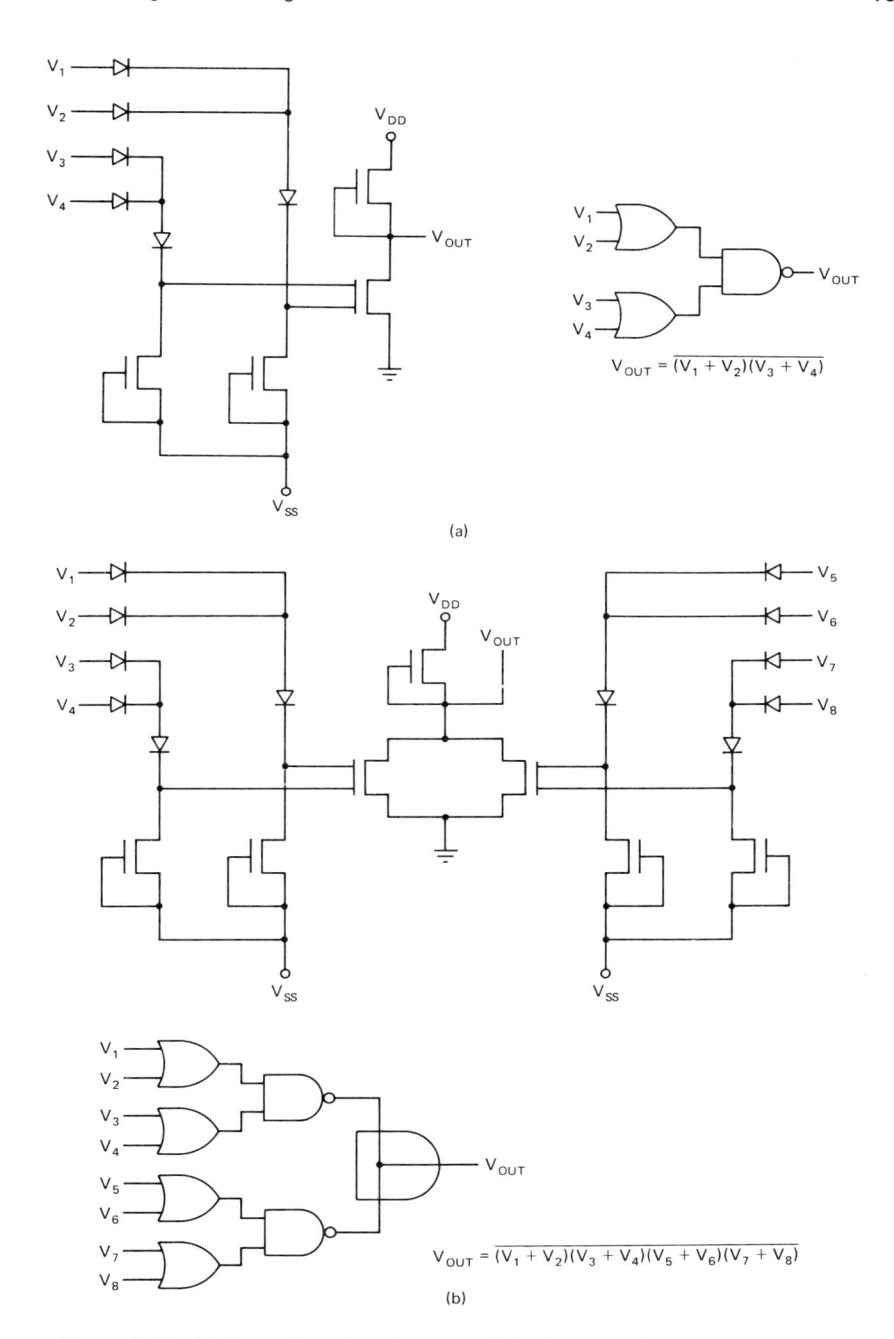

Figure 2.27 SDFL configuration of (a) or OR/NAND complex gate and (b) an OR/NAND/WIRED-AND complex gate.

2.7 LOGIC FAMILY SELECTION TRADE-OFFS FOR GATE IMPLEMENTATION

Several trade-offs should be made before selecting a logic family for implementing a GaAs circuit. The trade-offs should be made based on the circuit design requirements, the performance characteristics of each logic family, and the yield of the process used to implement a logic family.

BFL is most appropriate for implementing high-speed gates with high fan-out. The buffer transistor in the source-follower stage makes this logic family relatively insensitive to fan-out. SDFL is more sensitive to fan-out, and, for circuits with heavy output-node capacitive loading, the use of a push-pull output buffer may be necessary to meet the speed requirements. DCFL is the most sensitive logic family to fan-out.

For all logic families decreasing the width of the switching FETs affects the switching speed of a gate adversely, since transistor current decreases in proportion to W, while node capacitance contains a fixed width-independent offset value arising partially from lateral fringing capacitances.

The power dissipation per gate is the highest for BFL configurations because of the large output voltage swing and the fact that the diodes in the level shifting section are always forward-biased. Although SDFL gates have a similarly large output voltage swing, they use small power input diodes to implement the gate logic and these diodes are not always forward-biased. DCFL has the lowest power dissipation due to the low output current available through the depletion device and is the candidate technology for GaAs VLSI circuits.

Logic configurations in BFL are optimum when they can be expressed as AND/NOR functions, which are efficiently implemented in BFL. Using the basic AND/NOR complex gate will produce the most area-efficient layout. SDFL is more suitable for large fan-in NOR circuits since it replaces the FI switching transistors of BFL or CDFL by one switching transistor, one constant current transistor, and FI diodes. This results in compact layout since the diodes occupy a smaller area than the transistors and their use eliminates crossovers that cannot be avoided when using input transistors. DCFL is more area efficient compared to SDFL, mainly because of lack of level shifting requirements and, therefore, lack of second power supply routes. The level of integration, however, will be determined mainly from the power/gate requirements for the logic families rather than the area/gate requirements. Using the power/gate average for each family and then the average for a specified minimum power requirement for a circuit, the level of integration increases from BFL to SDFL to DCFL.

A very important factor in making trade-offs for selecting a logic family is the production yield of the circuits to be manufactured. If we ignore catastrophic failures during fabrication, such as shorts, missing contacts, defective implants, etc., then the primary cause of circuit rejection is the variation of operating points V_{oh} and V_{ol}. This variation results from lack of device threshold control. The tolerance of a circuit for variations of the input voltage is expressed by its noise margins. BFL and SDFL circuits have wide noise margins resulting from the large logic swing of the voltage at the output node of a gate, which is independent of the voltage gain $\frac{\partial V_o}{\partial V_{IN}}$. This is achieved by using the negative supply voltage V_{SS} in order to shift the positive voltage V_N of the node between the switching input device and the load FET by an amount V_K to a negative value $V_{OUT} = V_N - V_K < V_T < 0$. This, in turn, drives the input gate of the switching transistor of the next stage. The value of V_K is chosen so that the saturation current for the "low" output level is either zero (because V_{OUT} is below the negative threshold voltage) or is sufficiently small to provide a large noise margin. In DCFL, however, the "low" noise margin depends critically on the value of $(V_T - V_{ol})$, which is necessarily small because V_T has a positive value and V_{ol} is greater than zero. The small noise margins of DCFL can be achieved through strict control of device threshold uniformity during fabrication. This is a difficult task to perform, and it was the main cause for low yield of DCFL circuits in the early 80s compared to their BFL and SDFL counterparts. However, advancements in fabrication technologies made the fabrication of DCFL circuits possible in the late 80s, with comparable yield to D-mode circuits. Further increase in fabrication yield and process throughput will make DCFL the prominent GaAs technology because of its potential for large scale of integration.

PROJECTS

2.1. Design 2-input Exclusive-OR and 2-input Exclusive-NOR gates in BFL, UFL, SDFL, and DCFL logic families. Identify the implementation that requires the fewest number of transistors.

2.2. A foundry with a BFL process that can be modeled with the device models discussed in Section 1.5.4 is intended to be used for fabricating a 4-input NOR gate. The required propagation delay for this gate is 500 psec when it is driving an output load of 0.05pF. It is biased with $V_{DD} = 2V$ and $V_{SS} = -2V$. Design the NOR gate with appropriate device widths to meet the performance requirements. Select an appropriate input waveform and demonstrate that the gate operates as required. What is the expected power dissipation?

2.3. For the selected input waveform, what are the rise and fall times for the NOR gate? What would you do to equalize the rise and fall time? What are the noise margins?

2.4. Repeat Project 2.2 if the load is connected to the output of the gate through a 2 µm line with a line resistance of 200 ohms/mm, line capacitance of 0.1 pF/mm, and negligible line inductance. The requirement for the propagation delay at the load is again 500 psec.

2.5. A chip complexity is estimated as the number of 2-input NOR circuits with $FI = FO = 3$ and propagation delay of ≤ 1 nsec. Assuming that the minimum transistor size fabricated with the process assumed in Projects 2.2 and 2.3, is 2 µm × 1 µm, and the maximum power dissipation allowed for a chip is 1 W, what is the maximum number of transistors one can put on a chip using this process? Justify any approximations you may make.

2.6. Repeat Project 2.2 for a NAND gate. Compare the obtained device widths in the two cases and discuss any differences.

2.7. Provide the transistor level design in BFL, SDFL, DCFL, of the following Boolean expression

$$Q = D1\overline{D2}\overline{D3} + \overline{D1}D2D3$$

Which logic family implements this function with the least number of transistors?

2.8. If you have access to a GaAs foundry, obtain the process parameters that determine the appropriate device models and design the NOR and NAND gates as in Projects 2.2 and 2.6. Fabricate your designs and perform measurements on the prototypes. How close is the expected performance to the actual performance obtained through measurements? Discuss the observed differences and explain the reasons that may be responsible for these differences.

REFERENCES

1. R. C. Eden, "Comparison of GaAs Device Approaches for Ultrahigh-Speed VLSI," *Proceedings of the IEEE*, 70, no.1, January 1982.
2. R. L. Van Tuyl et al., "GaAs MESFET Logic with 4-GHz Clock Rate," *IEEE J. Solid State Circuits*, SC-12, p. 485, October 1977.
3. R. C. Eden, B. M. Welch, R. Zucca, "Lower Power GaAs Digital ICs Using Schottky Diode-FET Logic," *1978 Intl. Solid State Circuits Conf. Dig. Tech. Papers*, p. 68, February 1977.
4. M. R. Namordi, W. A. White, "A Low Power, Static GaAs MESFET Logic Gate," *Proceedings, GaAs IC Symposium*, p. 21, 1982.
5. M. J. Helix et al., "Improved Logic Gate with a Push-Pull Output for GaAs Digital ICs," *Proc. IEEE Intl. GaAs IC Symposium*, p. 108, 1982.
6. K. Lehovec and R. Zuleeg, "Analysis of GaAs FETs for Integrated Logic," *IEEE Trans. on Electron Devices*, Ed-27, No. 6, p. 1074, June 1980.
7. T. T. Vu et al., "The Performance of Source-Coupled FET Logic Circuits That Use GaAs MESFETs," *IEEE J. Solid State Circuits*, SC-23, no. 1, p. 267, February 1988.
8. R. L. Van Tuyl, C. A. Liechti, "High Speed Integrated Logic with GaAs MESFETs," *IEEE J. Solid State Circuits*, SC-9, no. 5, p. 269, October 1974.
9. A. Barna, C. A. Liechti, "Optimization of GaAs MESFET Logic Gates with Subnanosecond Propagation Delays," *IEEE J. Solid State Circuits*, SC-14, no. 4, August 1979.
10. W. Baechtrold et al., "Si and GaAs 0.5 µm-gate Schottky-barrier Field-Effect Transistors," *Electron. Lett.*, 9, p. 232, May 1973.
11. R. H. Dennard et al., "1 µm MOSFET VLSI Technology: Part II—Device Design and Characteristics for High Performance Logic Applications," *IEEE Trans. Electron Devices*, ED-26, p. 325, April 1979.
12. J. Lohstroh, "Static and Dynamic Noise Margins of Logic Circuits," *IEEE J. Solid State Circuits*, SC-14, p. 591, June 1979.

13. A. E. Ruehli, "Survey of Computer-Aided Electrical Analysis of Integrated Circuit Interconnections," *IBM J. of R-D*, 23, p. 626, 1979.
14. P. M. Solomon, "A Comparison of Semiconductor Devices for High Speed Logic," *Proceedings of the IEEE*, 70, no. 5, p. 489, May 1982.
15. W. Wilhelm, "Propagation Delays of Interconnect Lines in LSI Circuits," *Siemens Research and Development Reports*, BD-15, no. 2, 1986.
16. L. Pengue et al., "The Quick-Chip: A Depletion Mode Digital/Analog Array," *Proceedings Intl. GaAs IC Symposium*, p. 27, 1984.
17. S. I. Long et al., "Multi-Level Logic Gate Implementation in GaAs ICs Using Schottky Diode FET Logic," *IEEE Trans. Microwave Theory and Tech.*, MTT-28, p. 466, May 1980.

3
GaAs LOGIC CIRCUITS

This chapter discusses design and operation of basic digital GaAs circuits that can be implemented in different logic families. The design of these circuits takes advantage of the structures that are most efficiently implemented in each logic family. As a result, logic circuits to be implemented in BFL or DCFL use mostly AND-NOR gates, while circuits to be implemented in SDFL use OR-NAND gates.

Both combinational and sequential logic circuits are discussed in this chapter. These circuits can be used as building blocks in designing a monolithic integrated circuit. Several of these circuits are available in circuit libraries used by GaAs foundries. Some circuits available from GaAs foundries as standard products are also presented along with their performance and interface specifications. These off-the-shelf parts can be used for GaAs technology insertion into system design.

3.1 FLIP-FLOPS

Flip-flops designed in GaAs are clocked circuits with a single-phase or a two-phase clock. The utilization of a clock makes the change of the output synchronous to the clock and prevents erroneous states due to different delays at the paths that provide the input signals to the flip-flop. This is particularly important for GaAs circuits because, depending on the layout, the interconnect may introduce delays close to the propagation delay of the active elements.

Sec. 3.1   Flip-Flops

Flip-flops are used to store data temporarily before or after performing a function, and also as delay elements for signals that have to be synchronized, but they propagate on paths with different delays.

A two-phase clock flip-flop in master slave configuration is shown in Figure 3.1. Based on the discussion above, this circuit is best suited for implementation in BFL or DCFL. The data is sampled during clock (CLK) "high" and the output becomes valid during clock inverse ($\overline{CLK}$) "high." The feedback path, in both stages of the flip-flop, is always enabled while the data is enabled by the clock. Based on the transistor implementation of the AND-NOR gates, discussed in Chapter 2, the worst case data propagation delay from CLK to Q and also from CLK to $\overline{Q}$ is two gate delays. Clock skew is a problem with this structure because it can cause erroneous behavior if both stages are enabled at the same time. Enabling both stages at the same time will result in changing the output one clock phase prior to the intended change under fault-free operation. The effect of clock skew can be avoided by careful design of the inverter circuit that produces $\overline{CLK}$ from CLK. As a rule of thumb, this circuit should have half the propagation delay of the AND-NOR gate. If this rule is followed, then nonskewed clocks are established on nodes 1 and 2, before node 3 changes value based on the value of D. Therefore, the problem of transferring the data from D to Q in one clock phase is avoided. The underlying assumption of the above rule, however, is that the inverter providing $\overline{CLK}$ and the AND-NOR complex gate have the same threshold.

A single clock phase configuration of a master-slave flip-flop is illustrated in Figure 3.2. In the master section, the data is enabled by the clock while the feedback is always enabled. In the slave section the data is enabled by the feedback while the feedback is enabled by the clock. The worst case propagation delay from CLK to Q and also from CLK to $\overline{Q}$ is two gate delays. During clock "high" the data is sampled at the input of the master. At the same time, the clock in the slave disables the effect of the feedback on the sampled data and enables the slave to maintain its current state. When the clock becomes "low," the input of the master is disabled while the feedback of the slave is enabled, and the data, sampled during clock "high," can be transferred to the output. Since a single phase clock is used in the configuration, there is no clock-skew problem. However, the

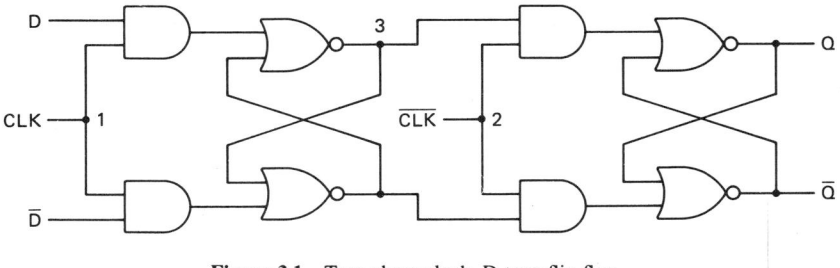

Figure 3.1   Two-phase clock, D-type flip-flop.

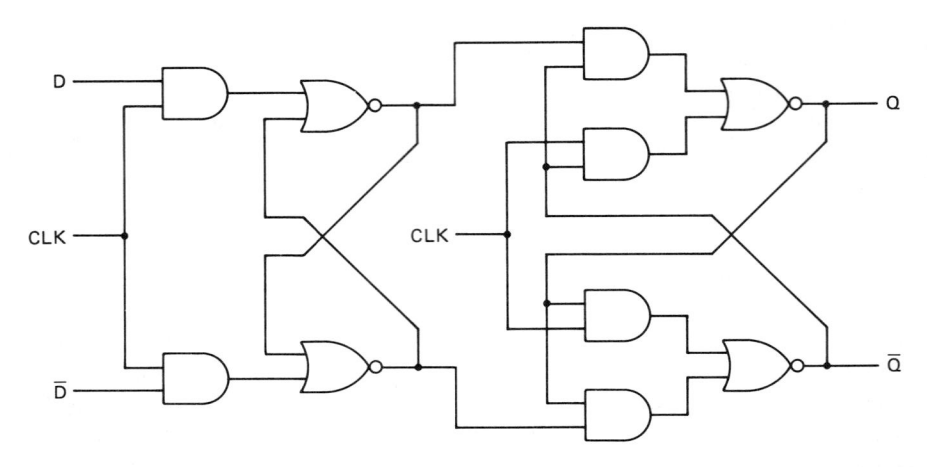

Figure 3.2 Single-phase clock, D-type flip-flop

trade-off is the larger number of logic gates, which implies larger area and power dissipation for this type of flip-flop.

Another design of a master-slave flip-flop using a two-phase clock is illustrated in Figure 3.3. While the two previously discussed designs require two propagation delays from CLK to Q, this design requires one gate propagation delay. This is due to the control of both the data and the feedback paths by the clock in contrast to the previous two designs. Therefore, this design is twice as fast in establishing a valid output than the previous two designs. However, several functional and implementation trade-offs must be considered in using this design. Its implementation requires a larger number of gates than the previous two designs and therefore occupies more area and dissipates more power than the previous

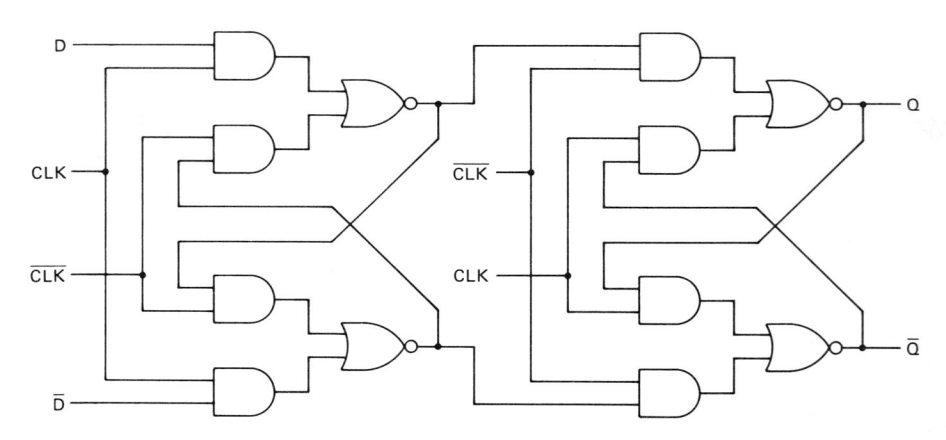

Figure 3.3 Two-phase clock, D-type flip-flop.

Sec. 3.1 Flip-Flops   83

designs. In addition, it is sensitive to clock skew. If both clocks are "high" then there is an ungated path from D to Q, causing input data to propagate to the output overriding the state of the slave earlier than normal operation assumes. This action can result in loss of data and introduction of logic faults in operations using data stored in the flip-flop. This situation will occur when the clock skew is greater than two times the propagation delay of the AND-NOR complex gate. When both clocks are "low" then both outputs become "high" independent of the input data. This action is caused by the AND gates which are enabled by the clock and drive the output NOR gates. With both clocks "low," the output of the AND gates is "low," forcing the output of the NOR gates to "high." This situation results in loss of data and can be the result of clock skew or anomalies of the clock. Special consideration should be given to the use of this flip-flop in case the clocks are asynchronous control signals and not free-running clocks. In this case, the design of the signals should guarantee avoidance of the conditions that cause the problems discussed above.

It should be noted that the first two flip-flop designs can be used for static data storage. This implies that although these are clocked circuits, the total absence of clock (i.e., all clock phases are "low") will not disturb the data stored. Both the two-phase and single-phase circuits will retain their state until new data is input by exercising the clock. This operating feature, however, is not possible with the third flip-flop design which will lose the data when the clock phases are both "low."

An approach to designing dynamic flip-flops using GaAs MESFETs as pass devices, just as in Si-MOSFET technologies, can result in significant reductions of circuit active area and power dissipation without compromising performance. Utilization of this approach to fabricate dynamic BFL circuits is reported in Reference 1.

GaAs MESFETs can be used as pass devices because of their symmetrical structure and their relatively low static input impedance and input capacitance. Their direct use with inverters that restore logic levels seems to be possible as long as the signals controlling the gate satisfy the following conditions:

$$V_{G(L)} \leq V_{(L)} + V_T \qquad (3.1)$$

$$V_{(H)} + V_T \leq V_{G(H)} \leq V_{(L)} + 0.7 \text{ V} \qquad (3.2)$$

where $V_{G(L)}$ and $V_{G(H)}$ are the low and high control voltages of the pass device gate, $V_{(L)}$ and $V_{(H)}$ are the logic low and high levels, and V_T is the MESFET threshold voltage. Equation (3.1) indicates the condition for the pass device to be off, while Equation (3.2) indicates the condition for the pass device to be on. For the N-off pass device, $V_T \approx -2\text{V}$, $V_{(L)} \approx -1.5\text{V}$, and $V_{(H)} \approx 0.5$ V. Consequently, for $V_{G(L)} \approx -4$ V, and $V_{G(H)} \approx -1$ V, the off and on states of the device are ensured. The circuit configuration and the timing diagram of a pass device driving an inverter are shown in Figure 3.4. As shown in Figure 3.4, when the input is "high" and the pass device is on, the "high" level is passed to node S with a dc

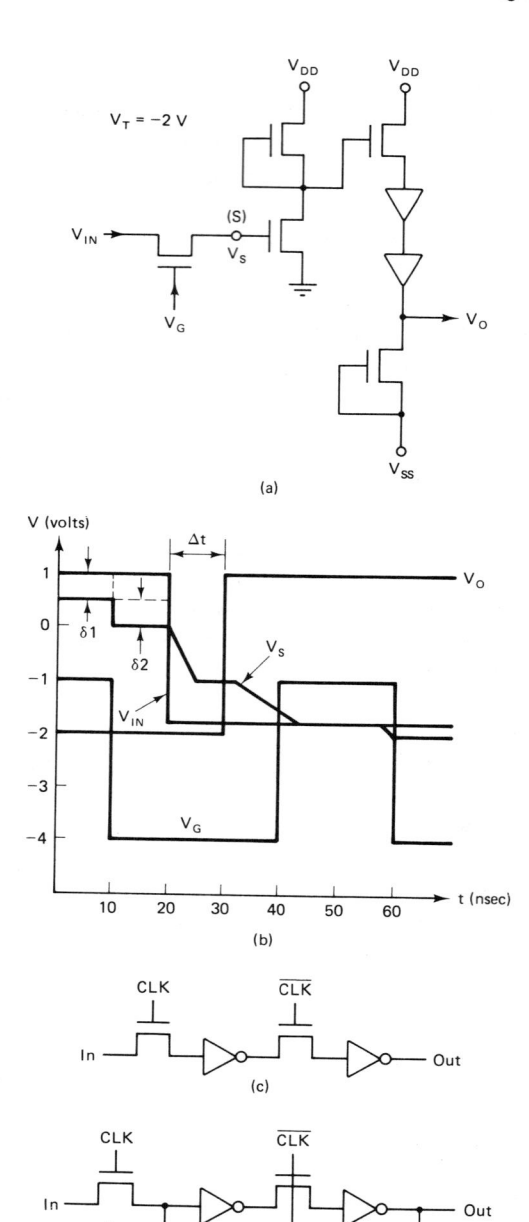

Figure 3.4 Pass device driving a BFL inverter; (a) circuit configuration and (b) timing diagram; (c) dynamic and (d) semidynamic implementation of a flip-flop based on the pass device approach.

Sec. 3.1   Flip-Flops   85

degradation $\delta 1$. The degradation occurs because the pass device is almost pinched off and $\delta 1 = V_{(H)} R_G / (R_I + R_G)$, where R_G is the on-resistance of the pass device and R_I is the input resistance of the inverter input MESFET. When the pass device switches off, an additional degradation occurs at node S. This degradation $\delta 2$ is capacity-induced noise and is given by:[1]

$$\delta 2 = V_G \frac{C_G}{C_G + C_{IN}} \qquad (3.3)$$

where V_G is the voltage swing on the gate of the pass device, C_G is the capacitance of the pass device between the device and the inverter input MESFET, and C_{IN} is the inverter input MESFET capacitance. As soon as the pass device turns off and the input becomes "low," C_{IN} will discharge across the pass device off resistance. When the pass device is in saturation, the current across the device can be considered constant, and the discharge is linear in time up to the point where $V_o \sim -1$ V (Figure 3.4b). When the logic threshold is reached, after a discharge time Δt, the logic stage of the inverter changes state, driving the source follower to $V_o = V_{(H)}$. When V_{IN} is switched to V_H while the pass device is off, C_{IN} is charged up the logic threshold level within a time interval $\Delta t'$ such that $\Delta t' \ll \Delta t$. Therefore, $\delta 1$, $\delta 2$, and Δt are the factors determining the operating limitations of the pass device. The minimum operating frequency for the structure of Figure 3.4 is determined by the discharge time of the inverting stage. For the 40-μm wide MESFETS reported in Reference 1 this time is 2 nsec and consequently the minimum frequency for the pass device gate control signal is 500 MHz. The maximum frequency, on the other hand, depends on the propagation delay (i.e., $t_{pd} = T/2$) of the pass device. For an 8-μm wide pass device, the propagation delay was reported at about 50 psec[1] and therefore the maximum frequency of operation can be 10 GHz.

A dynamic flip-flop configured with pass devices and inverters and clocked with a two-phase clock is shown in Figure 3.4c. This design represents a significant reduction of active components as compared to flip-flop designs in Figures 3.1, 3.2, and 3.3. However, since the discharge time of a pass device is very short, proper operation of this circuit necessitates a very high frequency refresh cycle. A semidynamic approach that preserves the high value of maximum operating frequency without the limitation for minimum refresh cycle is shown in Figure 3.4d. Such a design represents a 20 percent reduction in transistor count when compared with the flip-flop of Figure 3.3, which implies also reduced power dissipation. Operation of such a dynamic circuit with 5 GHz clock rate was reported in Reference 1.

A close examination of the flip-flop designs presented shows that internal nodes do not have a fan-out greater than two. This observation may be used to reduce area and power dissipation by using unbuffered FET logic, since the load is well within the capability of UBFL without significant speed degradation. If UBFL is used, then there is no switching transistor in the source follower, just the voltage level shifting diodes. Figure 3.5 illustrates the design of the flip-flop in

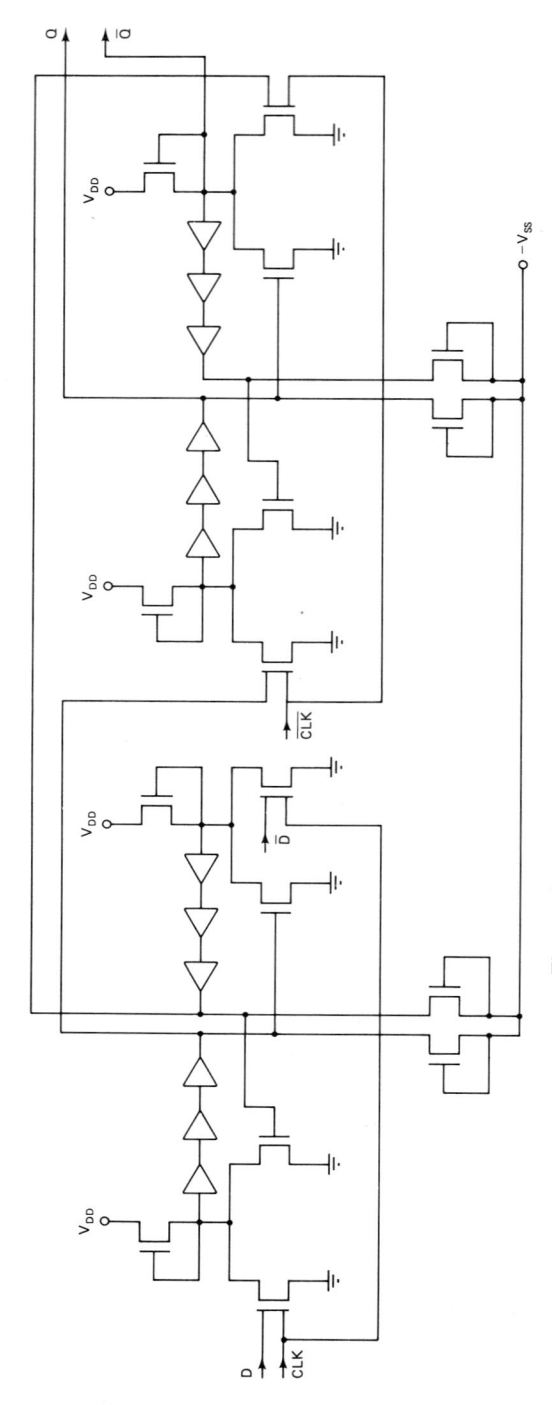

Figure 3.5 UBFL master-slave flip-flop with two-phase clock.

Figure 3.1 using the UBFL approach. The flip-flops in Figures 3.2 and 3.3 can be designed in a similar manner.

If the flip-flops discussed were implemented in DCFL, then the selection of gates would be the same as in BFL (Figures 3.1, 3.2, 3.3). However, if the technology of choice is SDFL, then a combination of OR-NAND gates would result in the most area-efficient implementation. As an example, the logic design of the flip-flop in Figure 3.3, for SDFL implementation, is shown in Figure 3.6. Although both designs use the same number of gates, the choice of logic gates is based on the implementation efficiency of a logic gate for a given technology.

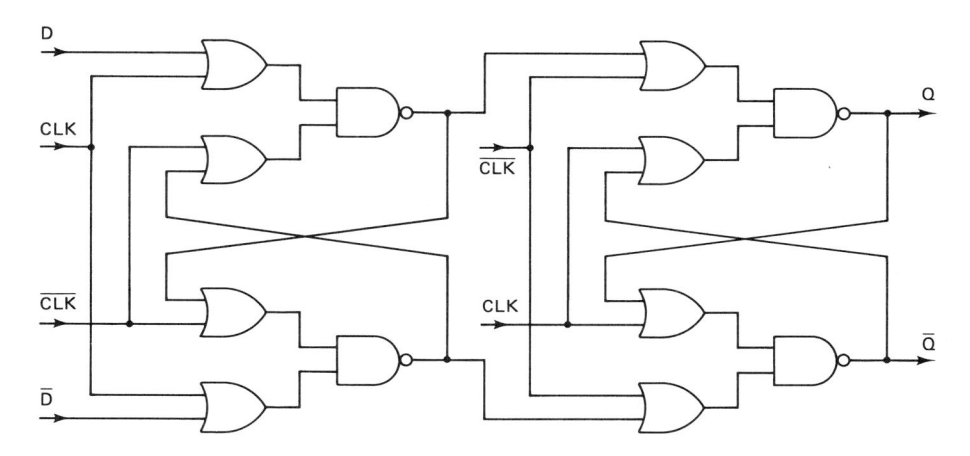

Figure 3.6 SDFL master-slave flip-flop with two-phase clock.

3.2 SHIFT REGISTERS

Shift registers are simple memory circuits that can accept, store, and read out data for arithmetic or logic operations. In addition, shift registers may serve as a convenient means of transferring serial data to parallel data and vice versa. The implementation of shift registers is based on flip-flop arrangements with proper control and clocking schemes.

Any of the flip-flop designs discussed in Section 3.1 can be used to form a shift register. For illustration of different shift register designs, however, we will use the flip-flop design of Figure 3.1. The number of flip-flops required to design a shift register depends on the number of bits that need to be stored, or shifted, or on the number of clock periods that a binary sequence has to be delayed. A flip-flop, called a shift register stage, can store a single bit or provide delay of a single clock period. An example of a 4-bit serial in serial/parallel out shift register is shown in Figure 3.7a. More or fewer bits can be accommodated by adding or deleting flip-flops. As Figure 3.7 illustrates, the interconnections between flip-flops are such that the logic level at an input terminal is determined by the state of

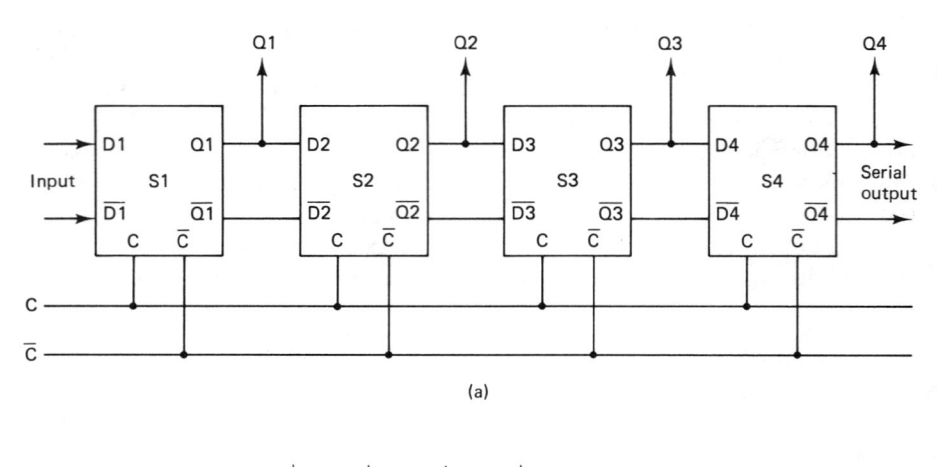

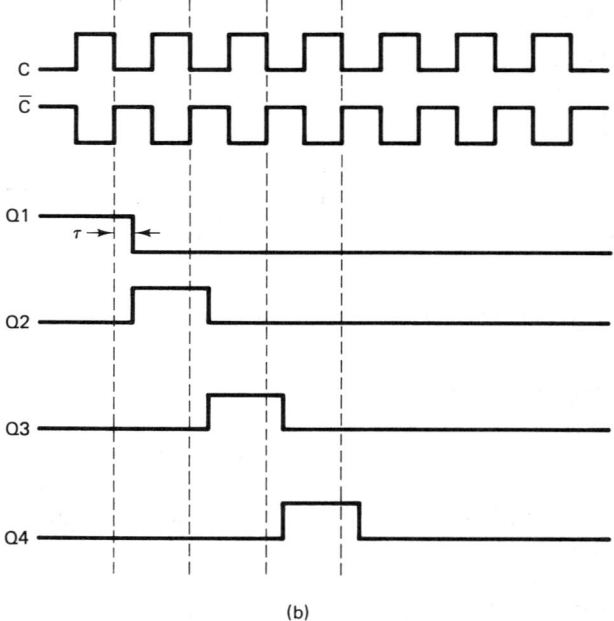

Figure 3.7 Four-bit serial-in, serial/parallel-out shift register: (a) circuit diagram and (b) timing diagram of operation.

the preceding flip-flop and all flip-flops are clocked in parallel. These are the distinguishing features of a shift register circuit.

Data enters at the input terminal of the 4-bit shift register and can be received serially through the output terminal, or in parallel through the intermediate flip-flop outputs. Data transfers from one stage to the other every clock

Sec. 3.2　Shift Registers

period. The bit registered in the last flip-flop (S4) is overwritten by the bit in Stage 3 (S3) while Stage 1 (S1) goes to the state determined by its input D1. Data is sampled on one phase of the clock (C) and the output is valid on the second phase of the clock ($\overline{C}$). A timing diagram of the shift register operation is shown in Figure 3.7b. The time difference τ between the rising edge of the clock phase $\overline{C}$ and the change of a flip-flop output Qi equals the propagation delay through two gates that the signal has to travel to establish a valid output after the slave has been enabled by $\overline{C}$.

The operation of the shift register for an arbitrary input sequence of 11010 is illustrated in Figure 3.8, assuming the initial states are $Q1 = Q2 = Q3 = Q4 = 0$. As shown in this figure, input data is sampled during C, and the output is valid shortly after $\overline{C}$ becomes "high." This is a level enabled operation and certain design considerations should be taken into account to guarantee error-free operation of the circuit. Since input data is sampled during C, the input data should be valid before the falling edge of C. In other words, the input data should not change after $C_t - \tau$ sec from the rising edge of C. This requirement is due to the need for time τ, equal to two gate propagation delays, to establish valid data in the master flip-flop of each shift register stage. Although this is a boundary value of the time window in which the input signal is allowed to change, a good design practice should provide for a stable input at the rising edge of C. The output from each stage is valid shortly after the rising edge of $\overline{C}$ and remains valid for a clock period.

The clock distribution to all shift register stages is of critical importance. All stages should receive the two clock phases without skew because clock skew

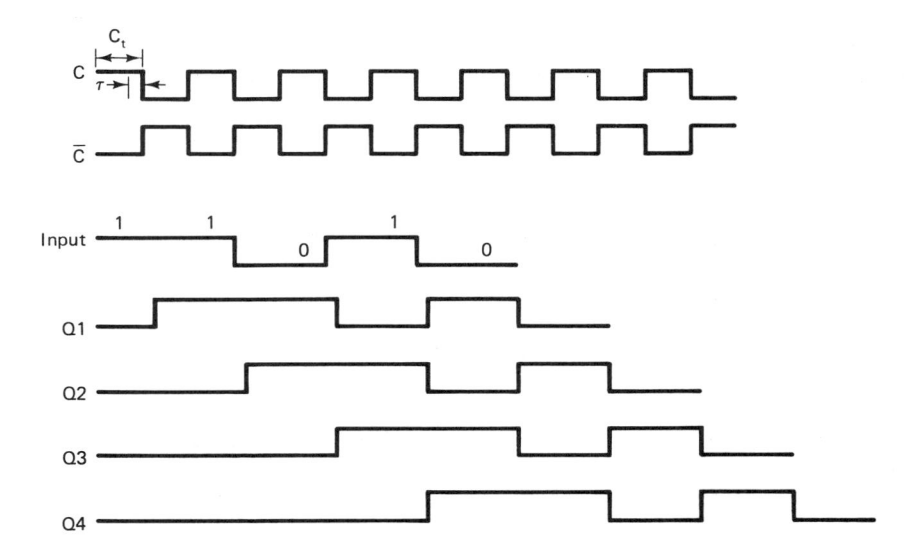

Figure 3.8 Shift register operation for the 11010 input sequence.

can cause logic errors as discussed in Section 3.1. Since GaAs circuits can operate at very high frequencies, the clock generation circuit should be designed so that unskewed clock phases are guaranteed in these frequencies for a given number of shift register stages. The implication of many shift register stages enabled with the same clock is that the clock generation circuit has to drive relatively large capacitive load, propagate the clock signal on uneven length paths (between the first and the last stage) and yet maintain unskewed clock phases.

The shift register circuit discussed above has serial data input and serial or parallel output. However, many applications often require the serialization of parallel data. In this case, a parallel-in serial-out shift register has to be used. A parallel-in serial-out shift register is designed similarly to the shift register in Figure 3.7. In this case, however, the individual flip-flops have some additional capabilities, namely SET and CLEAR. A modification of the flip-flop in Figure 3.1 which includes these additional capabilities is shown in Figure 3.9. The signals SET and CLEAR are always complementary when valid. In other words, when a SET operation is enabled with SET being "high," CLEAR remains "low" and vice versa. Both SET and CLEAR operations are asynchronous. When SET goes "high," the output of gate-1 becomes 0 and so does the output of gate-6, which forces the output of gate-3 to be 0 and therefore, Q goes "high" since CLEAR is "low" at this time. Based on the description of the SET operation, it is evident that this operation can be used to store a 1 in the slave without shifting the datum through the serial input. If the CLEAR signal goes "high," then the output Q becomes 0, the output of gate-5 goes to 0, forcing the output of gate-4 to 0 and thus, $\overline{Q}$ becomes 1 since SET is "low" at this time. Therefore, CLEAR can be used to store a 0 in the slave. Both operations are extremely useful because they allow the establishment of a state in the flip-flop asynchronously. This property is used for designing parallel-in serial-out shift registers. In many

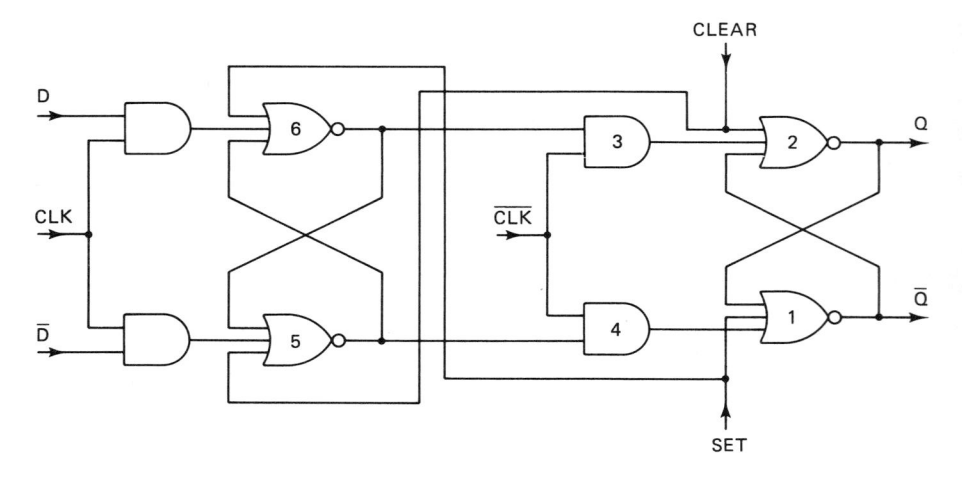

Figure 3.9 Two-phase clock, master-slave flip-flop with SET and CLEAR.

Sec. 3.2   Shift Registers   91

shift register designs, CLEAR is used to establish a known state without need for data shifting through the shift register stages, a desirable feature for testing purposes.

A parallel-in serial-out shift register design is shown in Figure 3.10. The shift register stages are based on the flip-flop design in Figure 3.9. In most respects, this circuit is identical to the serial-in serial/parallel-out shift register shown in figure 3.7. The interconnection of the stages and the clocking scheme are the same for both designs. The essential difference between the two designs is the added capability of SET and CLEAR, to the stages of the circuit in Figure 3.10, which provides for asynchronous loading of the stages from a set of "preset" inputs S_1 through S_4. Whenever the LOAD EN control is set to logic 0, the NOR gates deliver complementary logic levels from the S inputs to the SET and CLEAR inputs of each stage. This action overrides the normal synchronous stage action of a serial in shift register and forces the logic levels at the S inputs to appear at the corresponding data outputs as long as LOAD EN is held to logic 0. This is a parallel loading operation often referred to as *jam loading* because it stops the serial shifting effect altogether when it is asynchronously enabled.

When the LOAD EN control signal is set to logic 1, the NOR gates are effectively turned off and their outputs forced to 0, regardless of the logic levels appearing at the S inputs. This action returns the shift register stages to the serial

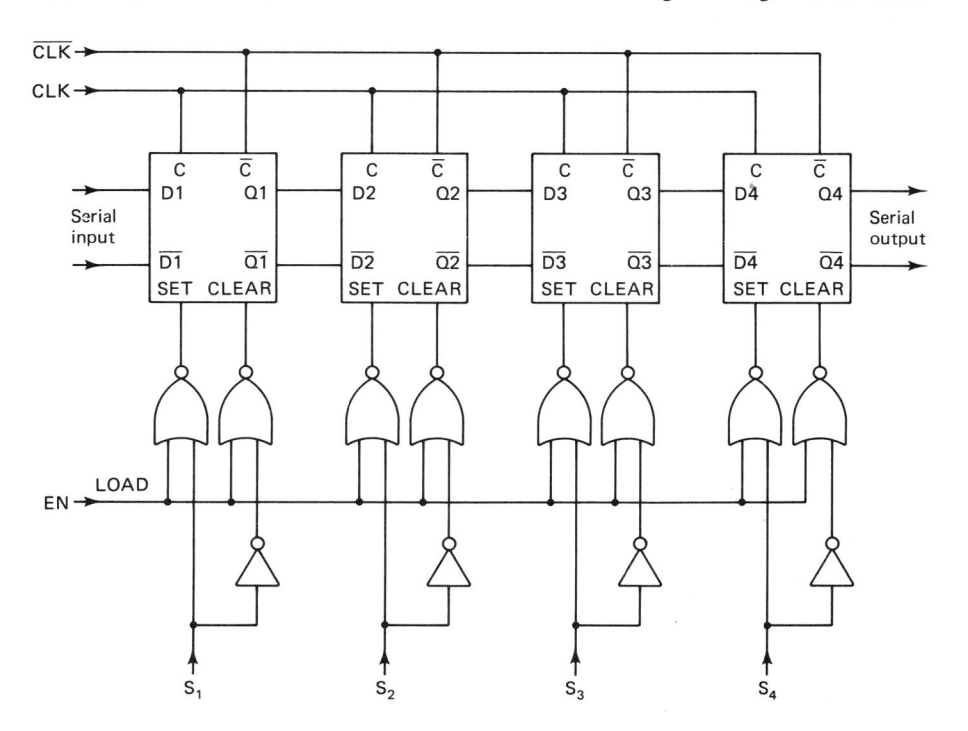

Figure 3.10 Four-bit parallel-in, serial-out shift register.

shifting mode and the 4-bit data is shifted to the right followed by whatever logic level is valid at the DI and $\overline{DI}$ inputs at the time. Data can be loaded in parallel with one LOAD EN pulse and then shifted out serially using the two phase clock. This register design, in effect, can work as a parallel-to-serial data converter which is particularly useful for computational structures where arithmetic operations are performed bit-serially but input/output operations are performed in a bit-parallel fashion. An implementation concern of this design is the loading on the LOAD EN signal line. The logic circuit that produces LOAD EN has a fan-out of 8 which may require the LOAD EN pulse to be "wider" than a clock phase.

The serial shifting effects described thus far all have been in the direction from the least-significant to the most-significant bit position. These are examples of right-shift registers. A left-shift register, by contrast, moves the data in the opposite direction. That is, data is entered serially, most-significant bit first, and proceeds through the circuit toward the least-significant bit position. This shifting discipline is also valid when data is loaded into the shift register in parallel. Figure 3.11 illustrates a modified design of the shift register in Figure 3.7, which operates as a shift-left shift register. Whether a serial shift operation is considered right-or-left shift depends on the position of the least-significant bit in the data shifting sequence, and not on how the circuit is drawn.

A left-shift register can be used to store input and output data in bit-serial arithmetic operations, such as division, where the most-significant bit is required first by the computation.

The capabilities of shifting data to the right and to the left can be incorporated into a single shift register design which is called a shift-right, shift-left register. Figure 3.12 illustrates the design of this shift register. The shift-right operation is enabled when SRE is at logic 1. In this configuration data input is applied at SR and is shifted toward the output $SROUT$. During this operation, the logic value on the SL input has no effect since the shift-left operation is disabled.

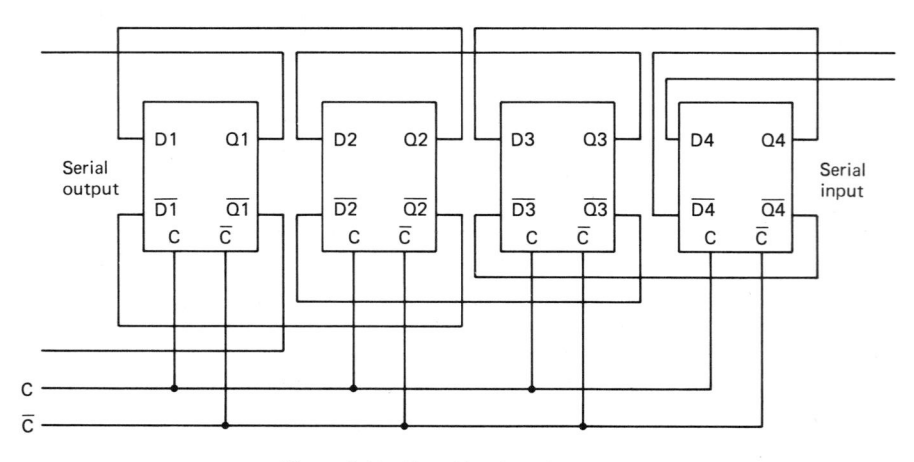

Figure 3.11 Four-bit left-shift register.

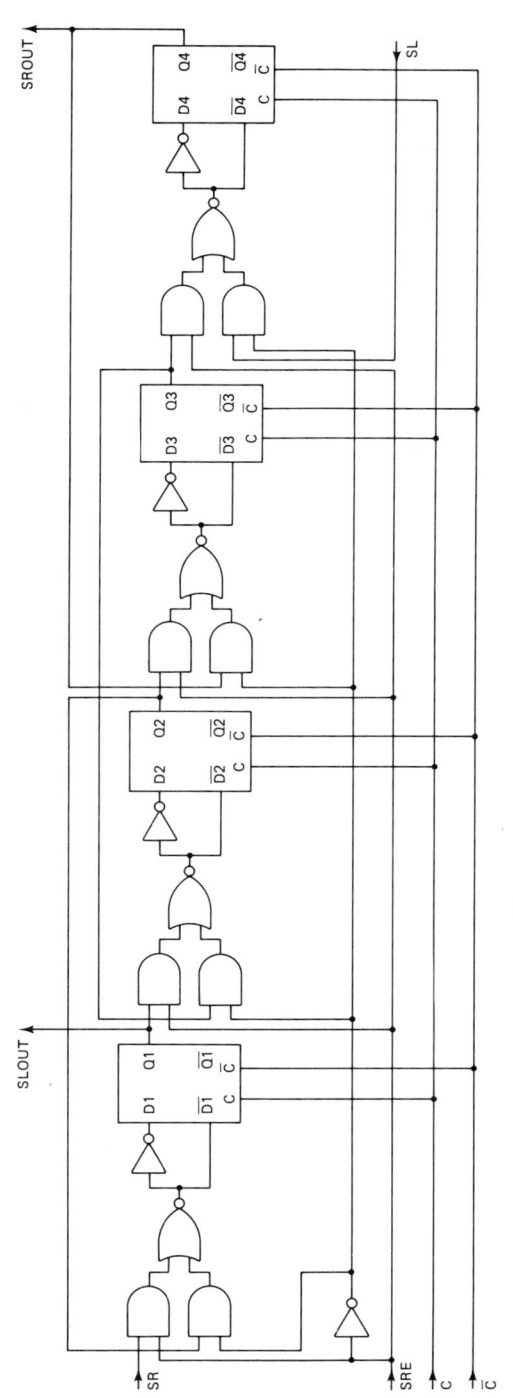

Figure 3.12 Four-bit shift-right, shift-left register.

93

The shift-left operation is enabled when *SRE* is at logic 0. In this case, data enters at the *SL* input and shifts towards the *SLOUT* output. The shift-right operation is disabled during this time and logic values of the *SR* input have no effect. The shifting in both directions is controlled by the two-phase clock.

Notice that the control logic between the shift register stages is performed by the AND-NOR function, which is chosen because of its efficient implementation, as a complex gate, in *BFL*. This is in agreement with the implied circuit of Figure 3.1 for the shift register stages. Right-shift registers implemented in *BFL* are reported operating reliably for clock frequencies ranging from 1 KHz to 2.5 GHz.[2]

All the shift register designs described up to this point are based on the basic flip-flop design of Figure 3.1. Configurations can also be obtained if the shift register stages are implemented using the flip-flop designs shown in Figures 3.2 and 3.3. Area and power requirements can be reduced, however, if shift registers are designed using pass devices as shown in Figure 3.4. A dynamic 4-bit shift register design utilizing pass devices is illustrated in Figure 3.13. As with the flip-flop of Figure 3.4, this shift register requires a minimum clock frequency, given by the discharge time of the inverting stage, to operate properly. Operation at maximum clock frequency of 5 GHz has been reported for this dynamic shift register design.[1]

To alleviate the minimum frequency limitation and yet preserve high maximum frequency operation, one may use a semidynamic approach where the data is latched in every shift register stage during either phase of the clock. The design of this semidynamic shift register is shown in Figure 3.14. In this design two pass devices and two inverters implement one stage of the flip-flop used for static shift register designs. The flip-flop with the least number of gates in a static configuration is the one shown in Figure 3.1, and for one stage (i.e., either master or slave), the design requires two AND-NOR complex gates.

Pass devices can also be used to design more efficiently parallel-in, serial/parallel-out shift registers. This design approach is taken in Reference 3 for designing a dynamic shift register with parallel load capability. Although the idea of using pass devices results in structures similar to those in Reference 1, the placement of the pass devices in the shift register design reported in Reference 3 is different. Figure 3.15a illustrates a 4-bit shift register using pass devices and having parallel load capability, while Figure 3.15b illustrates the circuit design of a single stage. As shown, the stage consists of two serially connected BFL inverters. The pass devices are placed within the inverter circuit for controlling data transmission between the two branches of the logic inverter. Data ripple through the stage is avoided by ensuring that the two phases of the clock are nonoverlapping. Parallel data load and parallel output are provided by the other pass devices shown in Figure 3.15b. The control signal for parallel load is ENL, which enables the parallel loading of data when it is in logic 1. Parallel output occurs synchronously with $\bar{C}$ and data is available when $\bar{C}$ is "high." A shift register based on this design was operationally demonstrated with clock frequencies

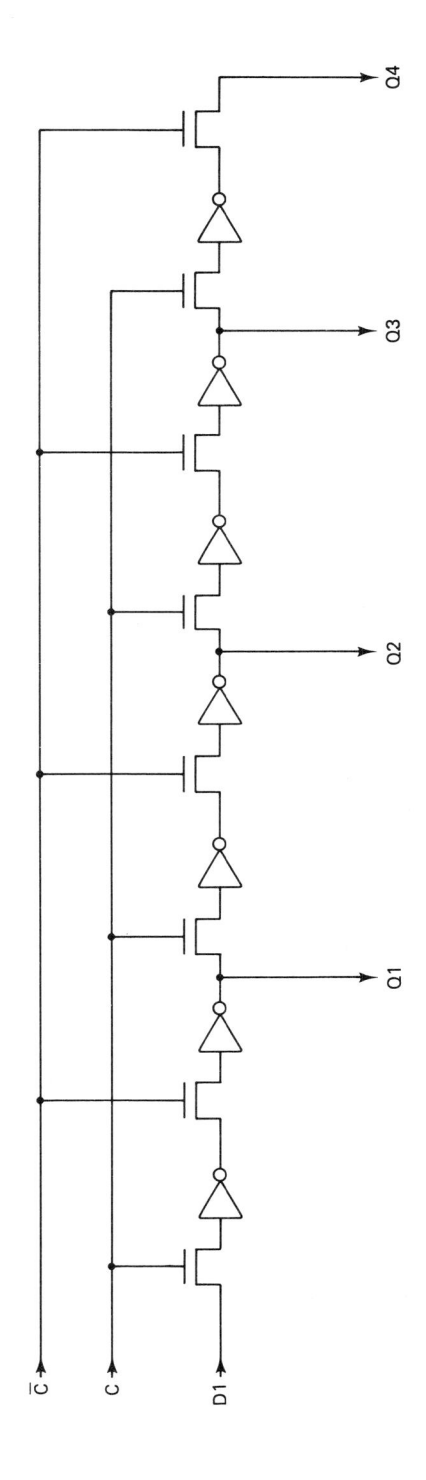

Figure 3.13 Dynamic 4-bit shift register using pass devices.

Figure 3.14 Semidynamic 4-bit shift register using pass devices.

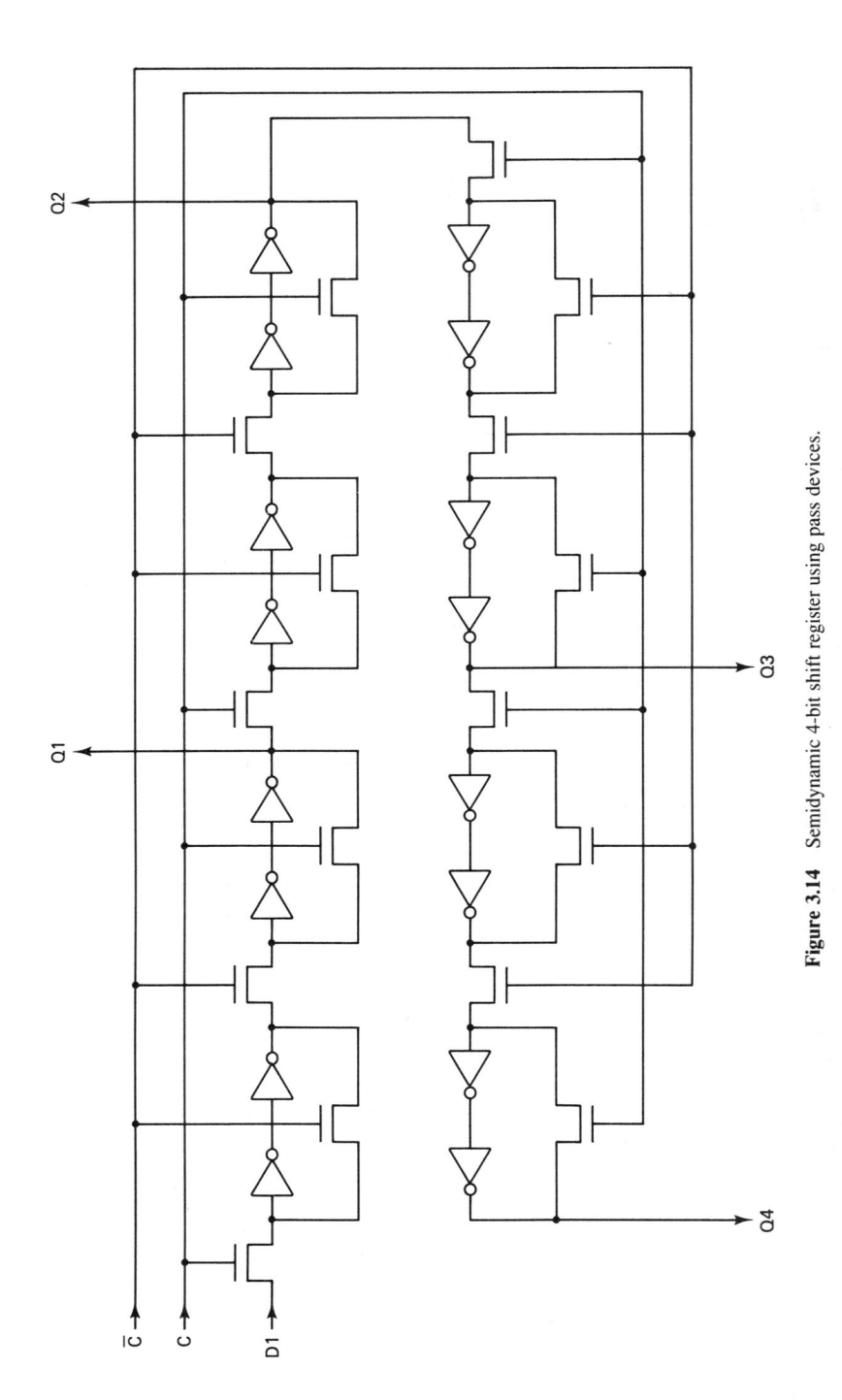

Sec. 3.2　Shift Registers

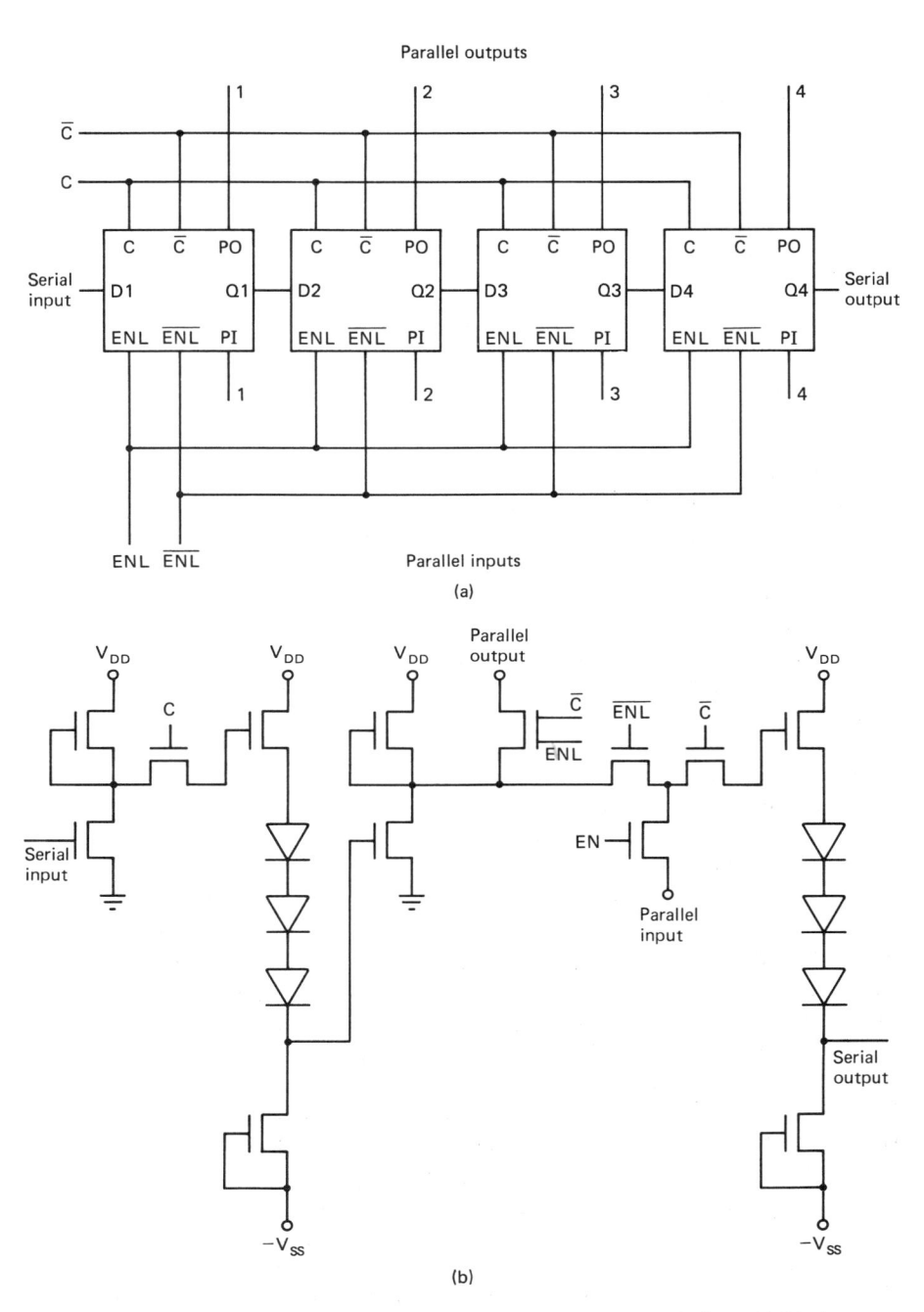

Figure 3.15 Four-bit shift register with parallel load using pass devices; (a) block diagram and (b) circuit design of a single stage.

ranging from below 1 MHz up to 2.2 GHz.[3] In this circuit, the absolute value of the pinch-off voltage of the pass devices was designed to be greater than the pinch-off voltage of the FETs in the logic gates. This action ensures that the whole logic swing is passed by the pass devices, and the time constants near the end of the logic transitions are also reduced.

3.3 COUNTERS

Counters are among the most commonly used circuits in the control section of digital designs. It is hard, indeed, to envision a digital system which does not contain one or more counters. The design of GaAs counters is especially important because, due to their high performance, they allow for direct frequency division in microwave synthesizers. Until recently, synthesis of local oscillators and carrier signals has relied upon heterodyning down a high-frequency oscillator for low-speed processing, or synthesizing at low frequency and then multiplying up.[4] These operations require the use of bulky analog mixers, amplifier and filter combinations, and frequency multiplier chains. The use of GaAs digital counters eliminates the need for such circuits and at the same time can result in better phase-noise performance and increased frequency agility.

The study of counter design principles is based on the operation of a frequency divide-by-two circuit. Frequency dividers can be split into two major groups according to their clocking discipline—single-phase clock dividers and complementary (two-phase) clock dividers. A single-phase clock divider is shown in Figure 3.16. This circuit is almost identical to the flip-flop in Figure 3.2, the only difference being that the Q output is fed back to the $\bar{D}$ input and the $\bar{Q}$ output to the D input. This feedback connection forces the flip-flop to operate in its toggle mode where the Q output changes every time the clock switches from logic 1 to logic 0. This operation is illustrated by the timing diagram shown in Figure 3.17. Some observations can be made about the operational characteristics of this frequency divider design. The output has two states, 0 and 1, advancing from state to state with each cycle of the input clock waveform. In other words, the output waveform is like the input waveform with half the frequency of the input waveform (Figure 3.17), thus the name frequency divider by two. As was discussed for the flip-flop in Figure 3.2, this circuit has excellent immunity to data loss and/or race conditions. The NOR-based latch has two basic properties: A stable output "low" is present before an output "high" and the input "high" is the forcing function of the latch. When a stable output "high" and a nearly stable output "low" are present from the master, the slave can be clocked. Consequently, this divider can divide when operated at a maximum clock frequency of $1/3t_{pd}$ where t_{pd} is the gate propagation delay. The outputs of the circuit are skewed (Figure 3.17) by t_{pd} because after the output is enabled by the clock, the new state in the slave is established, propagating through the feedback path.

Sec. 3.3　Counters　　　　　　　　　　　　　　　　　　　　　　　　　　　　　　　99

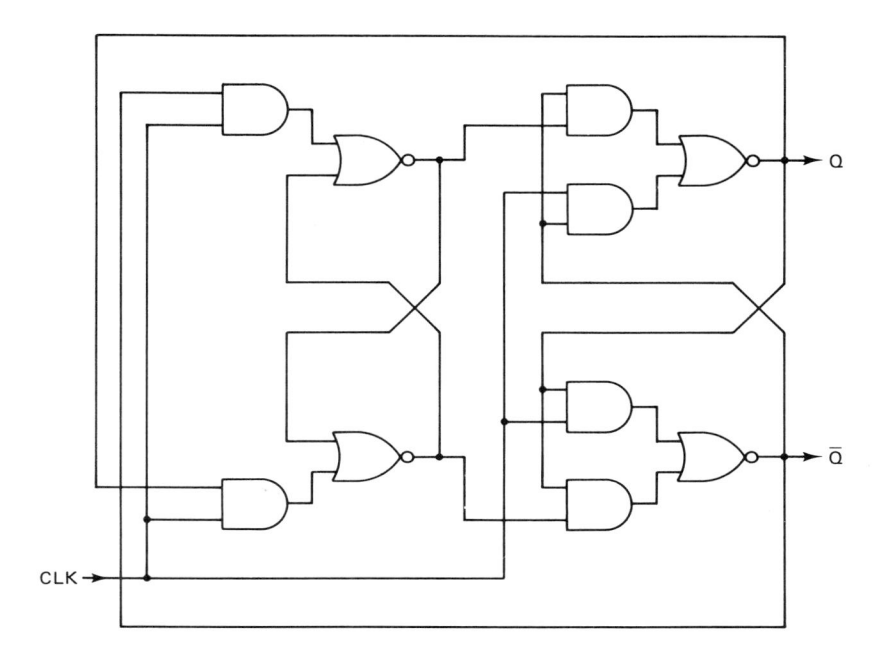

Figure 3.16　Single-phase clock frequency divider-by-two.

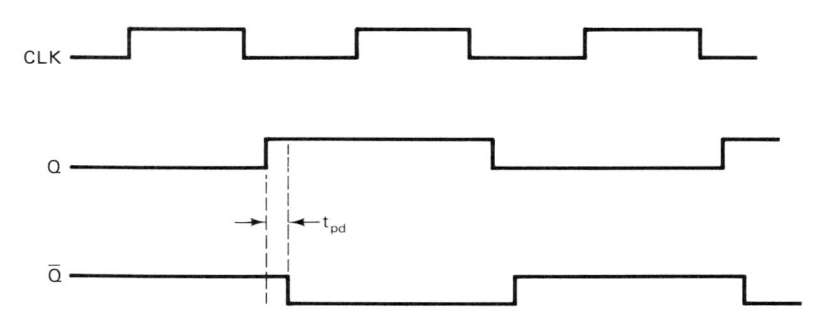

Figure 3.17　Timing diagram of the operation of the single-phase clock frequency divider.

A complementary clock frequency divider is shown in Figure 3.18. This circuit is essentially the flip-flop in Figure 3.3 with its Q output connected to its $\bar{D}$ input and its $\bar{Q}$ output connected to its D input. A timing diagram illustrating the operation of this frequency divider is shown in Figure 3.19. As in the single-phase clock divider case, the output waveform of this circuit is like the input waveform with twice as long a period. The output state advances on the "high" to "low" transition of the input $\overline{CLK}$ signal. Since the feedback is controlled by a clock phase, the establishment of a state during CLK or $\overline{CLK}$ does not depend on

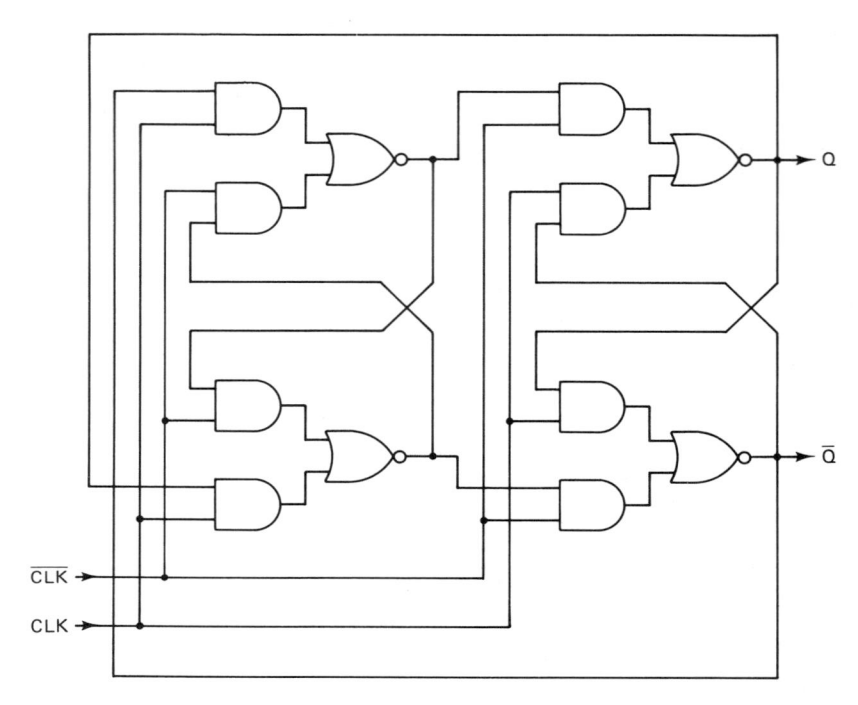

Figure 3.18 Complementary clock frequency divider-by-two.

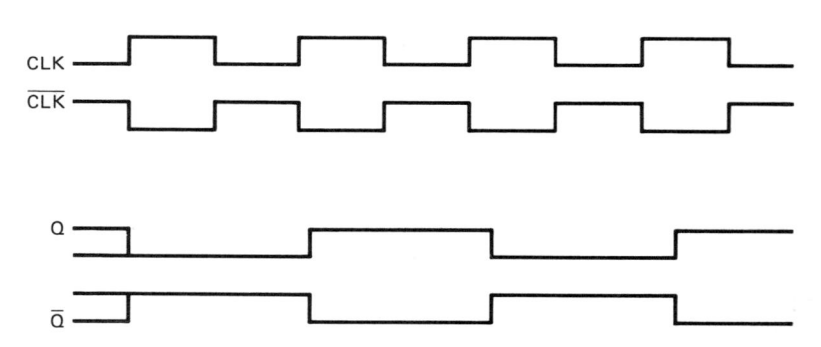

Figure 3.19 Timing diagram of the operation of the complementary clock frequency divider-by-two.

propagation through the feedback path; therefore, this divider can be operated with a clock rate of up to $1/2t_{pd}$ —the highest possible for divided-by-two circuits. In addition, the outputs of these circuits have minimal skew.

However, this type of frequency divider is subject to data loss when both phases of the clock are in the same logic state. For example, when both CLK and $\overline{\text{CLK}}$ are "low," all the AND gate outputs are forced to "low" which results in

Sec. 3.3   Counters   101

both Q and $\bar{Q}$ being "high," thus overwriting the previous state of the flip-flop. On the other hand, when $\overline{\text{CLK}}$ and CLK are both "high," both the master and the slave are enabled and thus the flip-flop is transparent and it will self-oscillate at a rate of $1/8t_{pd}$. Because of its critical sensitivity to the overlapping of the clock phases or to time skew between these phases, proper operation of this divider requires phase-matched complementary clocks. Providing usable clock signals for optimum performance of this divider will be, in general, a serious difficulty.

A complementary clock frequency divider also can be implemented using the dynamic circuit design approach. In this case, an inverter and two buffer stages combined with two pass devices suffice to perform the divide by two function. Figure 3.20 illustrates such a design. The same observations for the complementary clock phases can be made for this design as for the design discussed above. In both designs the maximum clock rate can be $1/2t_{pd}$, but the propagation delay time of the inverter in the dynamic divider can be about two times shorter than the four input AND-NOR gate with a fan-out of two in the static divider, thus resulting in higher performance. This circuit, however similar to the dynamic flip-flops discussed in Section 3.1, cannot operate below a minimum

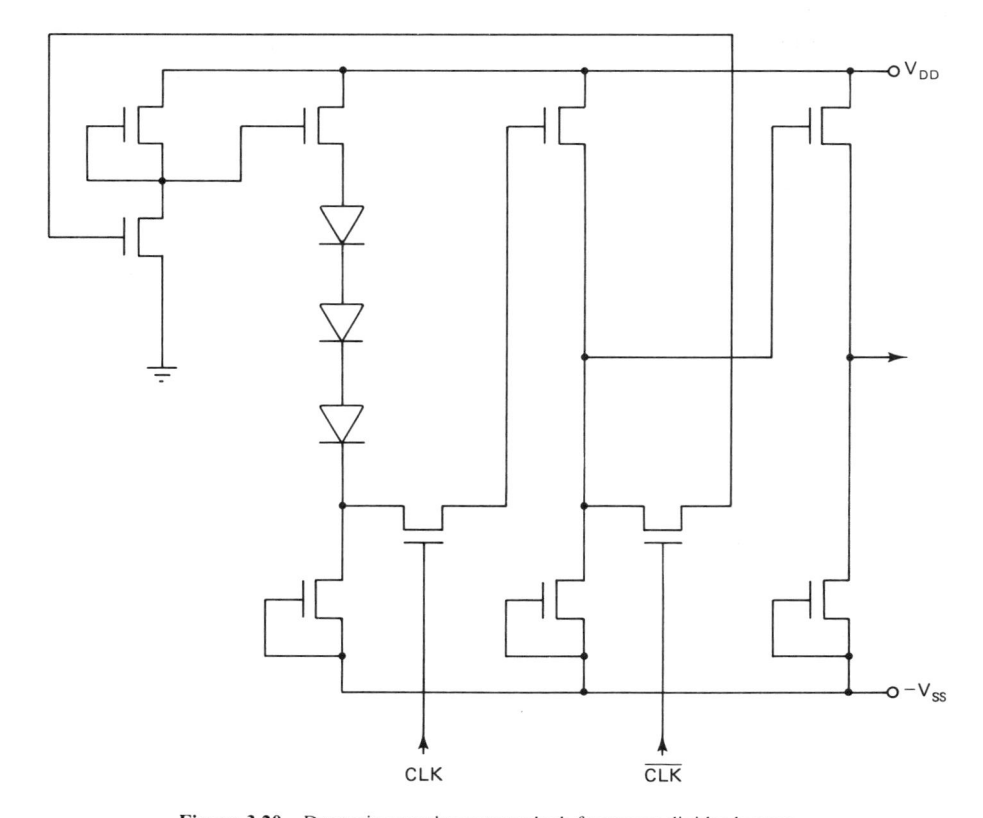

Figure 3.20 Dynamic complementary clock frequency divider-by-two.

frequency. The minimum operating frequency in this case is given by the discharge time of the noninverting stage. For 40-μm wide MESFETs this time is reported to be about 2 nsec.[5] Therefore, the minimum clock rate for this divider should be approximately 500 MHz.

3.3.1 Ripple Counters

Ripple counters are implemented as flip-flop arrays with different interconnection schemes. Since a flip-flop has two states, an array of N flip-flops has 2^N states, where a state of the array is defined as a particular combination of states of the individual flip-flops. A ripple counter is an array of flip-flops in which the flip-flops are interconnected in a manner that allows the array to advance from state to state with each cycle of an input sequence. If, starting at some initial state, the counter returns to that initial state after K cycles of the input sequence, the counter is classified a base-K counter or a modulo-K counter. In other words, the modulus of a counter is the number of different output states it has. Therefore, if a counter is designed as an array of N flip-flops, and if the counter advances through every possible state (this mode of operation is not always necessary, nor is it desirable for a class of applications), the counter is of modulo 2^N.

One of the oldest ripple counter designs is the ring counter. This counter is simply a flip-flop array where the output of the last flip-flop is fed back to the input of the array and all array elements are advanced with the input sequence. Figure 3.21a illustrates a 4-bit ring counter. Proper operation of this counter

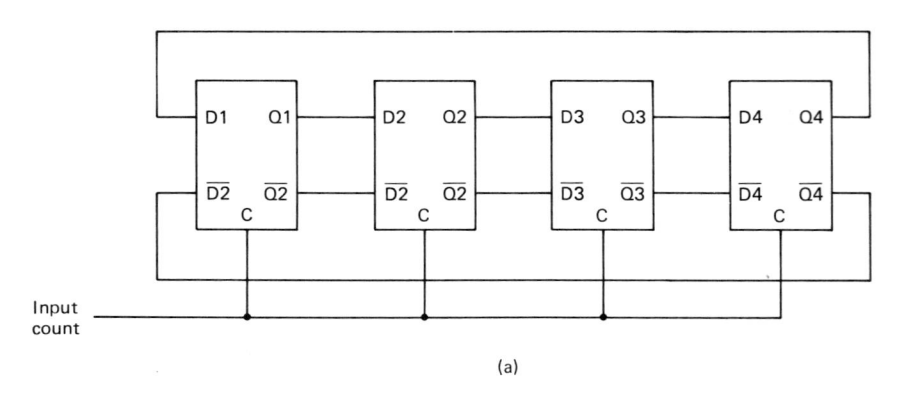

(a)

Count sequence	Outputs			
	Q1	Q2	Q3	Q4
0	1	0	0	0
1	0	1	0	0
2	0	0	1	0
3	0	0	0	1
4	1	0	0	0

(b)

Figure 3.21 Four-bit ring counter: (a) block diagram and (b) counter sequence.

Sec. 3.3 Counters

requires that one of the Q outputs is initially at logic 1, all others being at logic 0. The ring counter has a maximum count of N, where N is the number of flip-flops in the array, as shown in Figure 3.21b. A simple modification of the ring counter design can result in a maximum count of $2N$. The counter achieving this count is called a twisted-ring or Johnson counter and its block diagram is illustrated in Figure 3.22a. This counter can be implemented with flip-flops like the ones shown in Figure 3.2, all advancing simultaneously with the input counting sequence. Correct operation of the counter requires the flip-flops to have a reset function that can be implemented as shown in Figure 3.9. In case the output of the Johnson counter is used to indicate the binary count of the input sequence, then the counter states need to be decoded because the counter output code is non-standard, as shown in Figure 3.22b. The decoding of the output code can be implemented using two-input NOR gates; this is preferable when using BFL or DCFL technologies.

A commonly used ripple counter in digital design is the binary count-up counter. A 4-bit configuration of this counter is shown in Figure 3.23a. This counter is designed using single-phase clock flip-flops (Figure 3.2) connected in their toggling mode. The input signal whose cycles are to be counted is applied

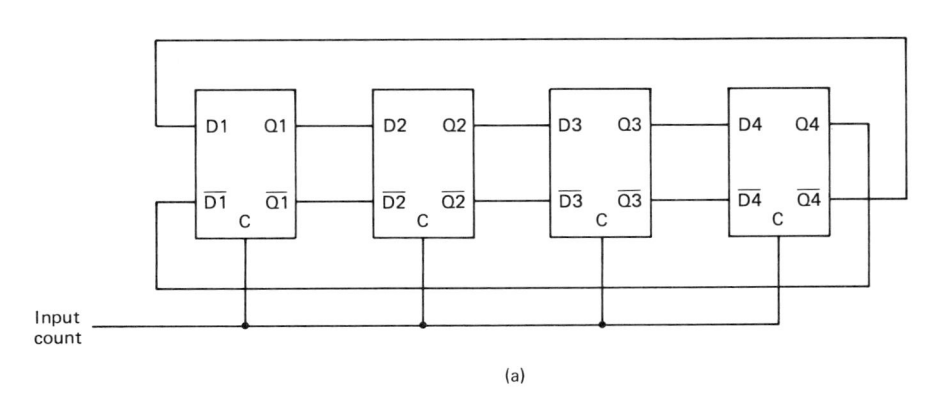

(a)

Count number	Q1	Q2	Q3	Q4
0	0	0	0	0
1	1	0	0	0
2	1	1	0	0
3	1	1	1	0
4	1	1	1	1
5	0	1	1	1
6	0	0	1	1
7	0	0	0	1
8	0	0	0	0

(b)

Figure 3.22 Four-bit Johnson counter: (a) block diagram and (b) counting sequence.

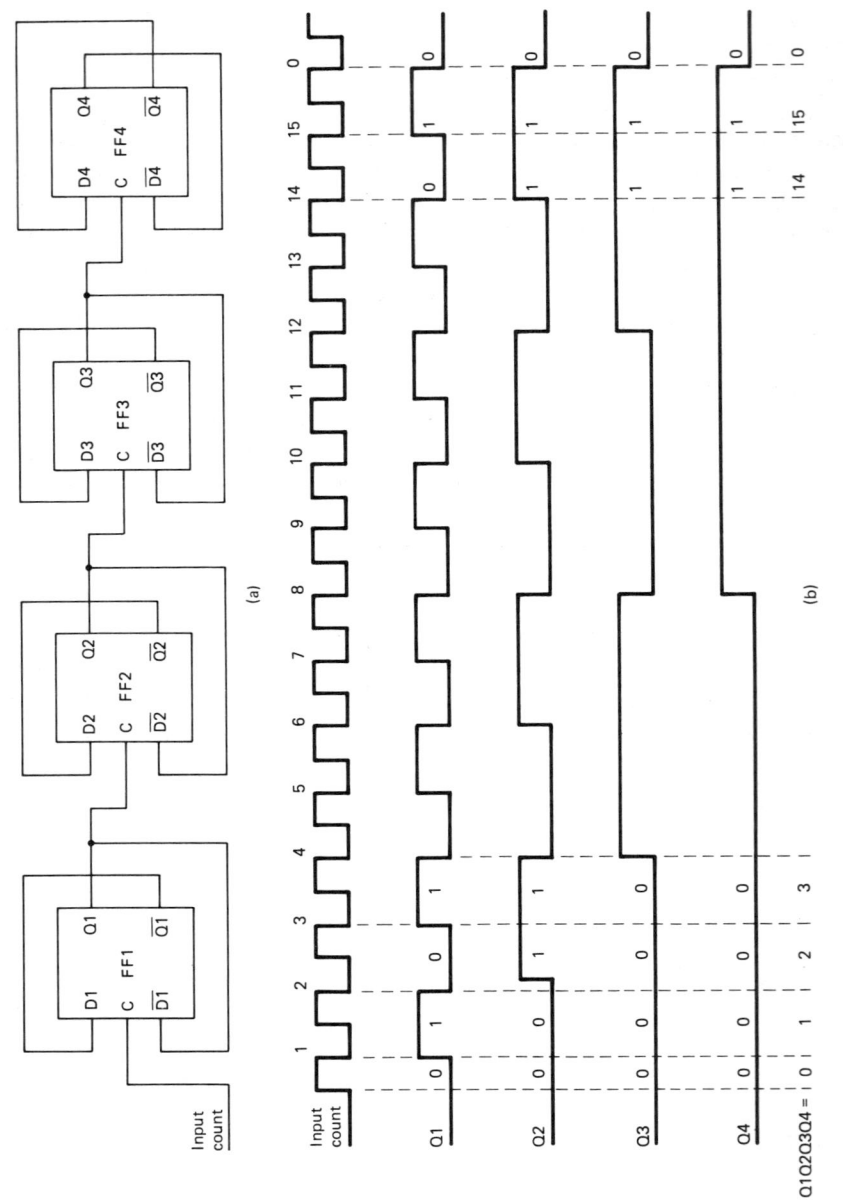

Figure 3.23 Modulo-16 binary-up counter: (a) block diagram and (b) timing diagram of counting sequence.

to the clock of flip-flop one ($FF1$) while the other flip-flops are clocked with the output of the preceding flip-flop. If the counter is initialized arbitrarily at the $Q1 = Q2 = Q3 = Q4 = 0$ state, then the input sequence will take it through all its $2^4 = 16$ possible states before it returns to the initial all-zero state. The counter is called an up-counter because each state advancement is indicated by an increase of 1. This is illustrated by the timing diagram and the indication of output states in each count shown in Figure 3.23b. This counter is also referred to as an asynchronous counter because there is no common clocking signal applied to all flip-flops to induce transitions to each flip-flop simultaneously. Although Figure 3.23b shows a periodic input sequence driving the counter, in general the input sequence need not be periodic. This counter design minimum load on the clock input since only $FF1$ requires this input. Also, the output of each flip-flop has to drive a relatively small load (fan-out = 2) in providing the clock for the following stage. Generally, this clocking scheme yields the highest counting rates for ripple counter designs. The timing diagram in Figure 3.23b illustrates an ideal response of this counter. A real circuit will experience some time propagation delay during state transitions, and during this time the counter outputs will overlap on nonvalid states. This is due to the ripple of the clock signal and is illustrated in Figure 3.24. This performance characteristic of this counter is a disadvantage for its utilization as a timing element. Note that if the counter outputs are to be decoded, during the output overlap a decoding error (glitch) will occur. The decoding error increases with the number of flip-flops in the counter (Figure 3.24).

If the design of the counter is modified so that the clock of a flip-flop is provided by the $\overline{Q}$ output of the preceding flip-flop, then the resulting design is a down-counter commencing with the maximum count, 15, and decreasing by 1 every time the counter state advances. Although this type of a down-counter is functionally sound, its use is very limited. Instead, a down-counter is required to have all-zero output at initialization, assume the maximum modulo (15 in this case), and decrease subsequently.

Both up- and down-counters should have the capability to CLEAR so initialization can be established. This utility can be incorporated into the flip-flop design, as was illustrated in Figure 3.9.

3.3.2 Synchronous Counters

A counter design that minimizes the state decoding error due to ripple propagation delay is that of a synchronous counter. In a synchronous counter all flip-flops are driven by the same clock waveform and, therefore, change of state during any particular count occurs almost simultaneously. However, even in this case decoding errors can occur due to possible differences in propagation delay of the flip-flops, clock distribution, and nonequal rise and fall times of the flip-flop outputs. A decoding error in this case persists only as long as it takes for the flip-flop to change state plus the maximum time difference in propagation time between flip-flops. What is important, however, in contrast to asynchronous

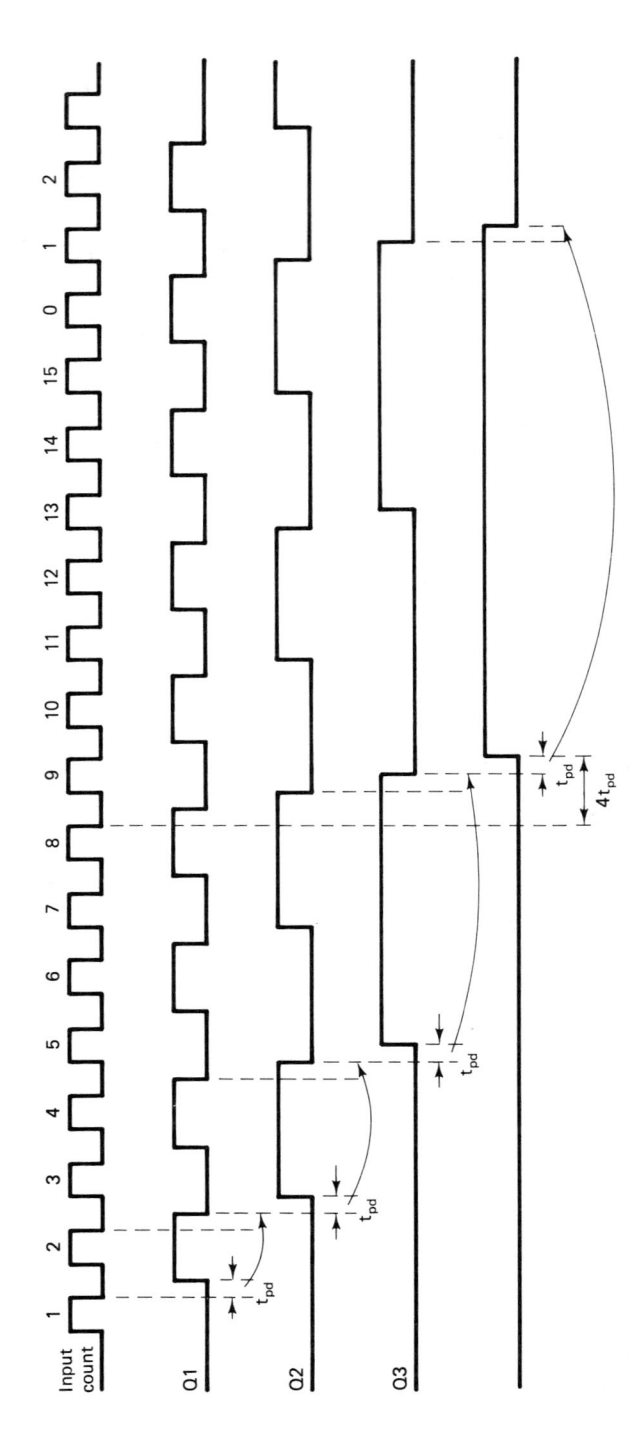

Figure 3.24 Output timing of a mod-16 up-counter, including ripple propagation delay.

counter designs, is that the decoding error interval does not depend on the number of flip-flops in the counter. Figure 3.25 illustrates a 4-bit synchronous binary counter.

Flip-flops in this counter are assumed to be single-phase clock flip-flops like the one in Figure 3.2. The first flip-flop is connected in its toggling mode while succeeding flip-flops determine their state based on information concerning the state of preceding flip-flops. State information is carried to succeeding flip-flops through a series of NOR gates and hence, this counter design is referred to as a ripple or serial carry synchronous counter.

Based on the design in Figure 3.25, one can observe that $FF2$ responds to clock transition with a change of state only if $\overline{Q1} = 0$. If $\overline{Q1} = 1$, then $D2$ and $\overline{D2}$ are both 0, and, based on the flip-flop design of Figure 3.2, the present state is preserved. Similarly, $FF3$ responds only when both $\overline{Q1}$ and $\overline{Q2}$ are 0, and $FF4$ responds only when $\overline{Q1} = \overline{Q2} = \overline{Q3} = 0$. On this basis, it can be verified that the above circuit has the same counting sequence as the asynchronous ripple counter shown in Figure 3.23. The fact that information from preceding flip-flops has to propagate through the series of NOR gates in order for a flip-flop to establish its state is a design consideration. The reason for this consideration is that if the duration of the single clock phase is shorter than the time required for the logic levels to propagate on the longest path (from $FF1$ to $FF4$ in the case of the circuit in Figure 3.25), then the counter will operate erroneously when changing of states requires information propagated through this path. Since a modulo-N counter requires $N - 2$ NOR gates, it will count correctly as long as the duration of the single-phase clock is greater than the propagation delay time $(N - 2)t_{pd}$ of the NOR gates, plus the propagation delay time of $FF1$, and the propagation delay of the interconnect.

In comparing the synchronous with the asynchronous counter we can observe that it the asynchronous counter has no limitation on the counting speed other than the inherent speed of the individual flip-flops. However, the speed of reading the output should be such that it allows for the settling of the decode errors. For the synchronous counter, on the other hand, exceeding some counting speed causes erroneous operation where no problem is encountered in correct decoding. In the asynchronous counter, the loading on the clock (input count) is minimal, whereas in the synchronous counter the loading on the clock is proportional to the number of flip-flops in the counter.

An improvement in the performance of the synchronous serial-carry counter is achieved in the synchronous parallel counter shown in Figure 3.26. This design avoids the need for rippling through the series of NOR gates by bringing the logic levels required directly to the appropriate flip-flop inputs. As can be verified, the counting sequence of this counter is identical to that of the serial-carry synchronous counter and the asynchronous counter. Since changing of states does not depend on logic levels rippling through a series of gates, this counter can count (error-free) faster than the serial-carry synchronous counter. There are several design trade-offs, however, that must be considered when using such a counter.

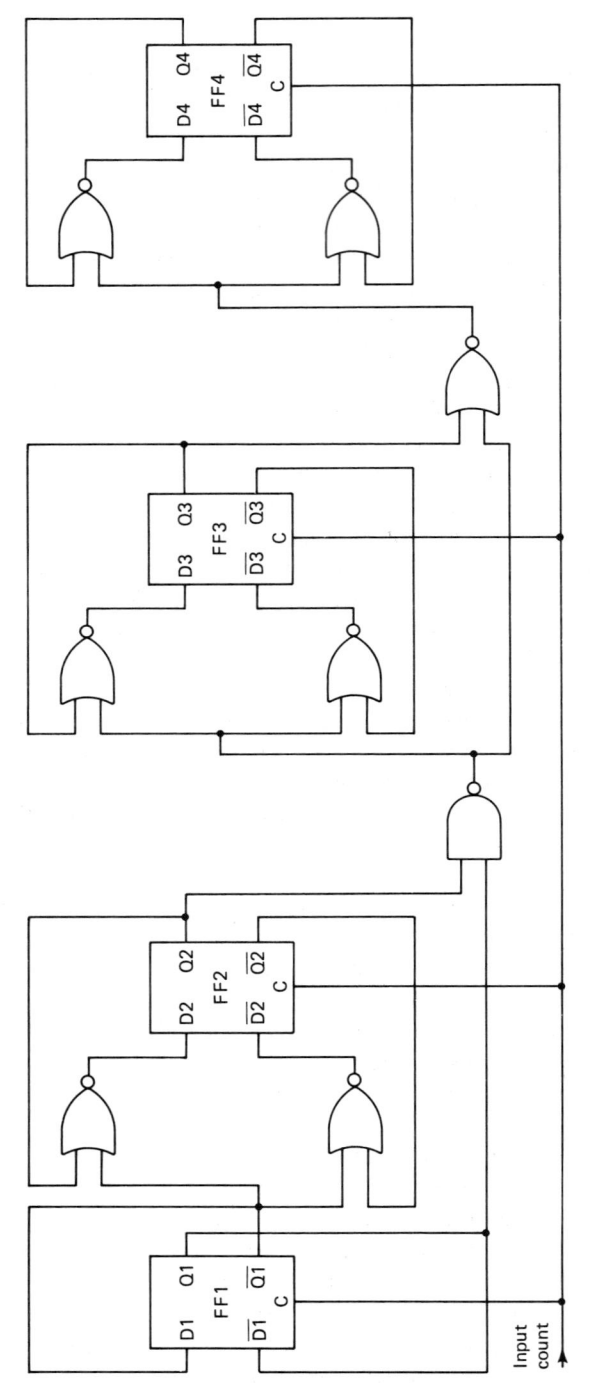

Figure 3.25 Four-bit synchronous binary counter (serial carry).

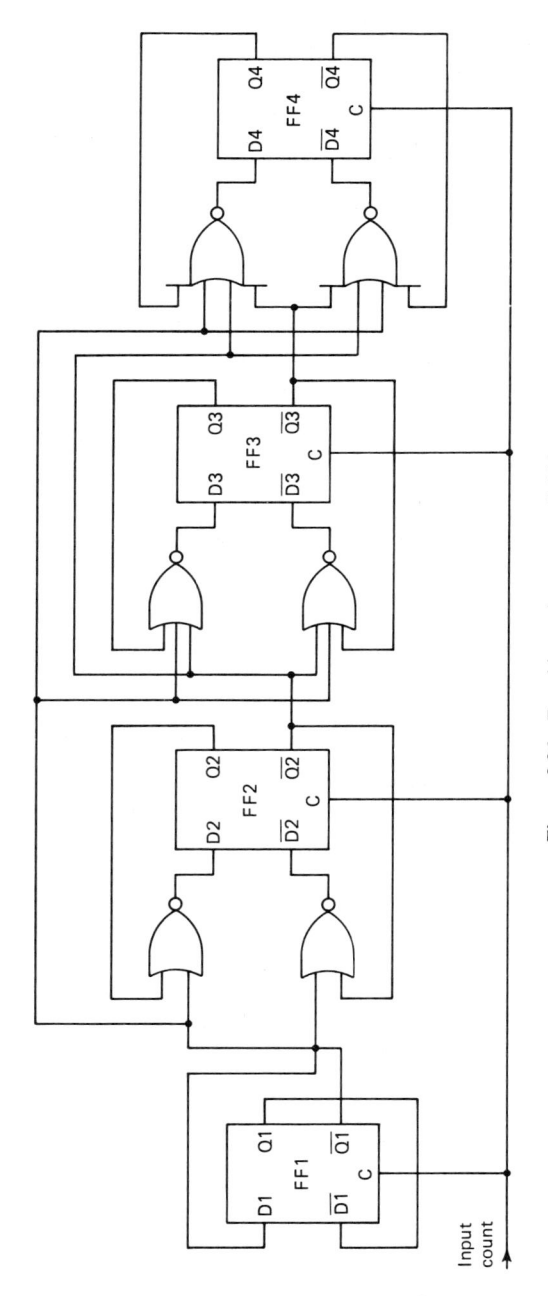

Figure 3.26 Four-bit synchronous parallel binary counter.

As Figure 3.26 illustrates, the fan-in of each succeeding NOR gate, at the inputs of a flip-flop, increases by one. Increasing fan-in increases the gate propagation delay unless wider MESFETs are used in gates with succeedingly higher fan-in. Furthermore, each additional NOR gate increases by one the fan-out each flip-flop is required to drive. Each additional load on a flip-flop output decreases the speed of the flip-flop itself, assuming all flip-flops are designed with the same size MESFETs. Therefore, *FF1* has the maximum fan-out and the longest propagation delay.

Synchronous counters are used for timing devices, because there is not a problem in correctly decoding their output—unlike asynchronous counters. For general applications, however, design trade-offs should be carefully considered before selecting a synchronous over an asynchronous counter or vice versa.

All the counter designs discussed so far are composed of flip-flops like the one shown in Figure 3.2 with an added CLEAR function. An additional feature, desirable in many applications, is the asynchronous preset function. The design of the flip-flop in Figure 3.2, modified to perform both the CLEAR and PRESET functions, is shown in Figure 3.27.

A presettable counter gives the user the option of asynchronously preloading

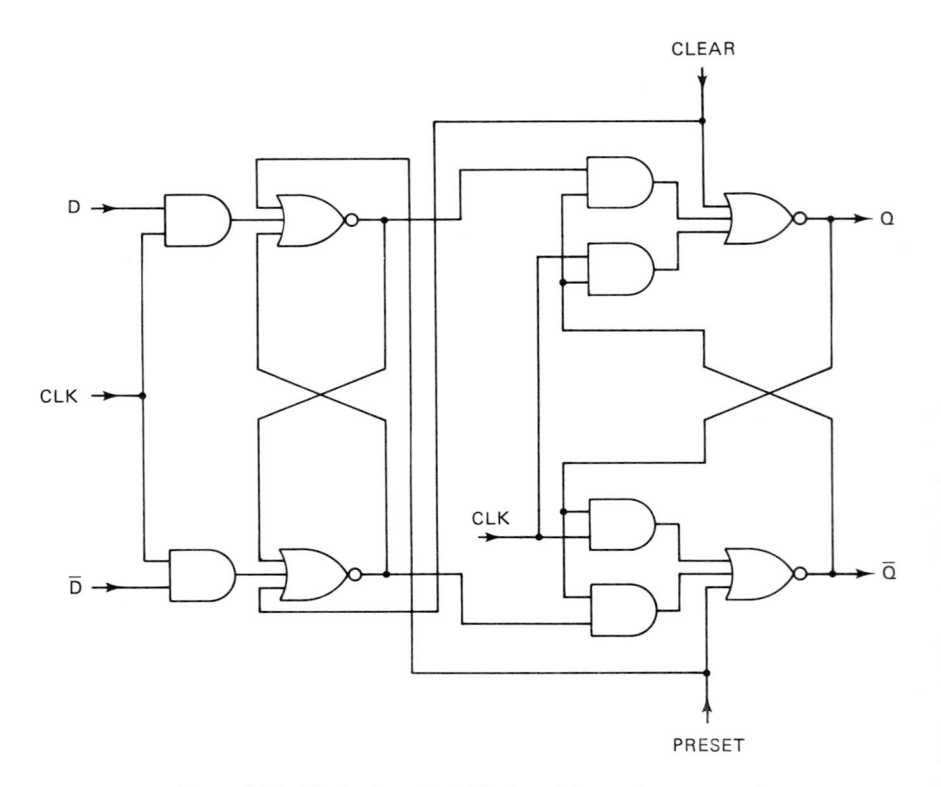

Figure 3.27 Single-phase clock flip-flop with asynchronous preset.

the counter's output with any desired number and then letting the count resume from the preset number. Based on the design of the flip-flop in Figure 3.27, the Q output of the flip-flop can be asynchronously set to 1 or 0 by setting CLEAR = 0 and PRESET = 1 or CLEAR = 1 and PRESET = 0, respectively. PRESET and CLEAR are both set to 0 during counting and have no effect on the operation of the counter. Therefore, upon exercising these signals the counter will assume the intended state to preset, and it will retain this state until both signals go to 0, when it will resume counting on the next clock. Use of PRESET and CLEAR overrides the effect of the clock and presets the counter output to the intended state irrespective of what the Q and $\bar{Q}$ outputs might be showing at the time.

The asynchronous preset function is mostly used for asynchronous control circuits, where, based on certain data dependent conditions, a specific state has to be assumed by the circuit.

3.3.3 Nonbinary Counters

All the counter designs presented up to this point have a modulo which is a power of 2. Therefore, based on these designs a counter with N flip-flops is a mod-2^N counter, which implies that starting at some initial state the counter cycles through all its possible 2^N states before returning to its initial state. A nonbinary counter is defined as a counter whose modulus is not a power of 2. For example, a decade or mod-10 counter is of special interest.

In general, the design of a mod-K counter ($K \neq 2^j$, $\forall j \varepsilon Z$) can be based on N flip-flops where N is the smallest integer for which $2^N < K$. Therefore, in order to implement a mod-K counter we can use a mod-2^N counter and provide the means to omit some of the possible states of this counter. More precisely, for a mod-K counter one has to omit $2^N - K$ states. There are not restrictions, in general, as to which states are to be omitted; this implies that there are many different ways of designing a mod-K counter. In most cases a decoder is used between a counter and the circuit using the counter output. The decision about which states are to be omitted, in designing a mod-K counter, is often aimed at simplifying the decoder design.

One of the most popular nonbinary counter designs is that of a decade or decimal counter. This counter generates the ten numbers required for ordinary decade (decimal 0 through 9) counting. Figure 3.28 illustrates the logic diagram, timing diagram, and counting sequence of a basic decade counter. This counter is designed with flip-flops having the PRESET and CLEAR functions of the flip-flop shown in Figure 3.27. If the count starts at binary 0000, it advances normally through binary 1001, which is decimal 9. The next clock pulse sets $\overline{Q2}$ to 0 and since $\overline{Q4} = 0$ at this time, the output of the NOR gate becomes 1 which activates the CLEAR function that forces the counter to 0000 state. The PRESET signal is 0 during this operation. As the timing diagram and the counting sequence indicate, the output of this counter is the same as the output of the 4-bit ripple counter up to the 1001 count. Since this output is the binary equivalent of the ten decimal

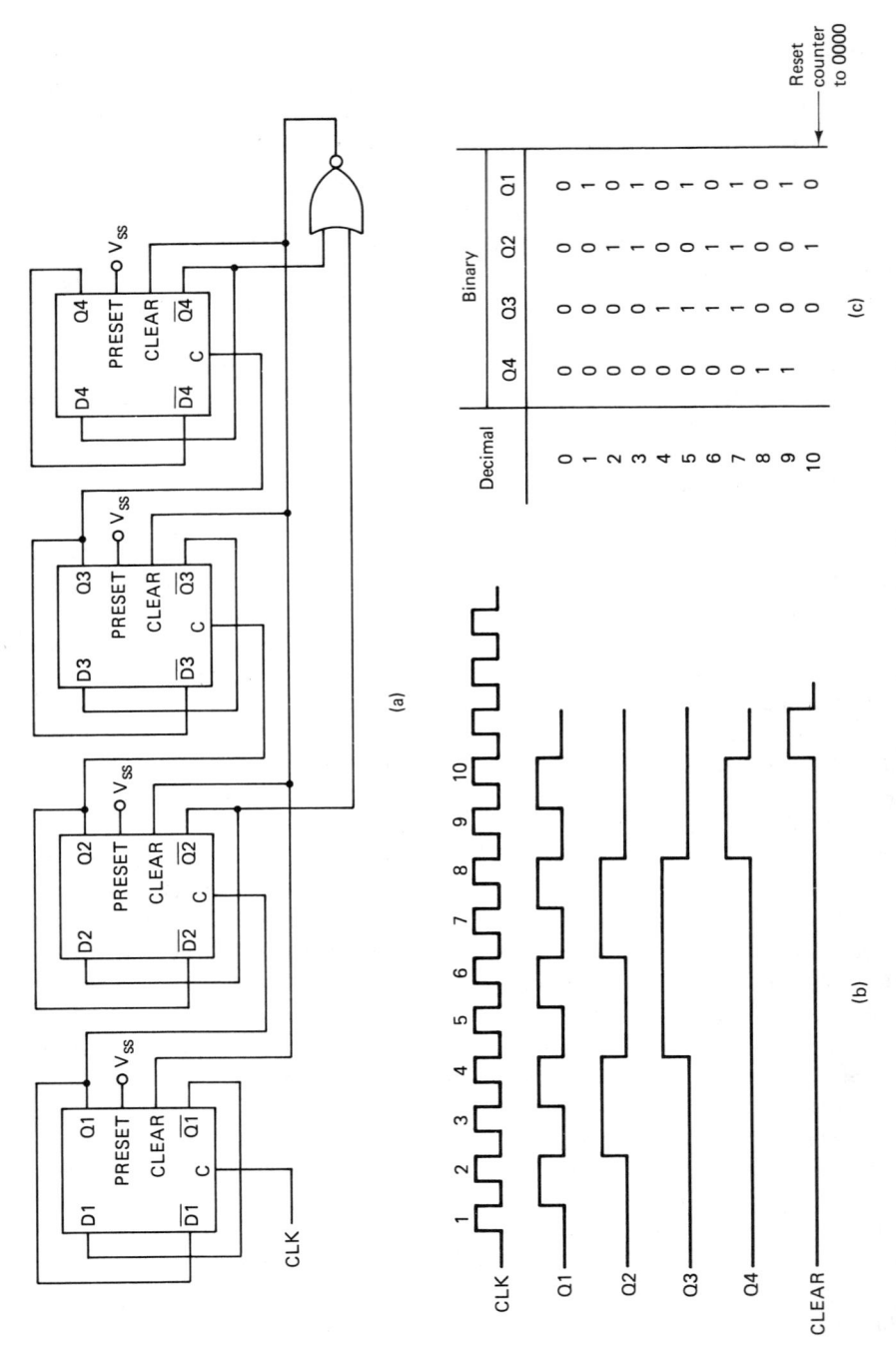

Figure 3.28 Basic decade counter: (a) logic diagram; (b) timing diagram; and (c) counting sequence.

112

numbers, this counter is commonly known as a Binary-Coded Decimal (BCD) counter.

In principle, the counter goes to state 1010 (decimal 10) only long enough to reset to the all-zero state. Special design consideration should be given to the NOR gate that generates the CLEAR signal. As is shown in Figure 3.28, this gate has a fan-out of eight. The 1010 output will persist as long as it takes for the NOR gate to propagate its logic level to all four flip-flops. Since the 1010 output is not a valid output for this counter, there must be a provision for strobing the counter output when it is valid. That will guarantee proper operation of a circuit that may be driven by the output of a BCD counter. One way to reduce the load that the NOR gate has to drive is to use two NOR gates in parallel, each driving two flip-flops or, equivalently, each having a fan-out of four.

Basic BCD counter modules can be cascaded to generate more than one decade of decimal numbers. Cascading basic modules can be done simply by using the $Q4$ output of one module as the clock sequence for the next module. Cascading of two basic modules is illustrated in Figure 3.29. This circuit generates two decades of decimal numbers, namely 00 through 99. The counter section providing the Most Significant Digit (MSD) is advanced when the $Q4$ bit of the Least Significant Digit (LSD) section switches from logic 1 to logic 0, which is the time of completing counting one decade. Therefore, the MSD section advances every time the LSD section completes the counting of one decade. Of course, one can cascade in this fashion any number of BCD counters to count any number of decades.

The $Q4$ of a basic BCD counter has a modulus of 10. This makes a BCD counter attractive for applications requiring frequency division by 10. In case BCD modules are cascaded, then the overall frequency division of a BCD counter chain will be 10^n, where n is the number of basic BCD counter modules in the chain. Such counter chains are often used in equipment with precision timing requirements where the output of a crystal-controlled oscillator needs to be divided down to intervals of multiples of 10.

As a second example of a fixed-module nonbinary counter design, we consider the mod-5 counter. The counter design uses flip-flops without PRESET and CLEAR functions (Figure 3.2). Three flip-flops are required since the smallest integer for which $2^N > 5$ is $N = 3$. Figure 3.30 illustrates the mod-5 logic diagram, timing diagram, and counting sequence. As also can be seen, this is a

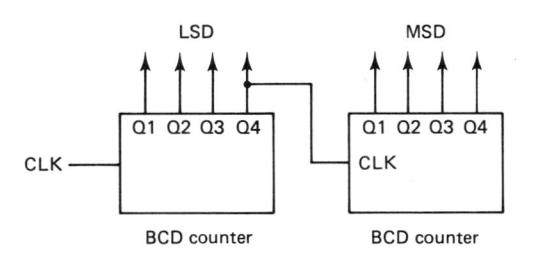

Figure 3.29 Two-decade BCD counter configured by cascading two basic BCD counters.

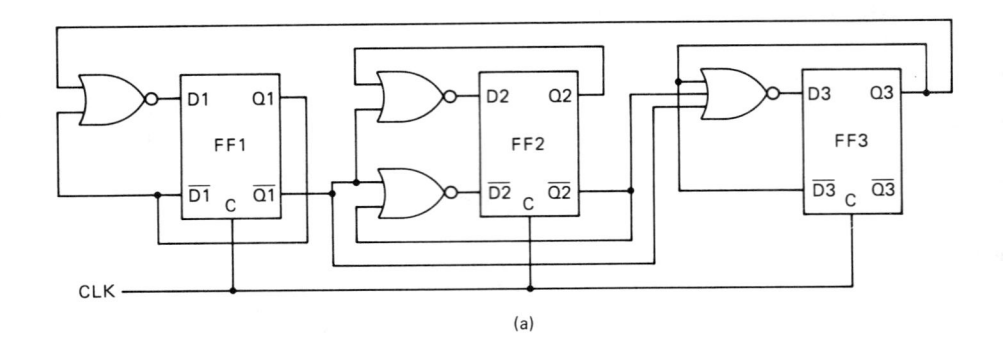

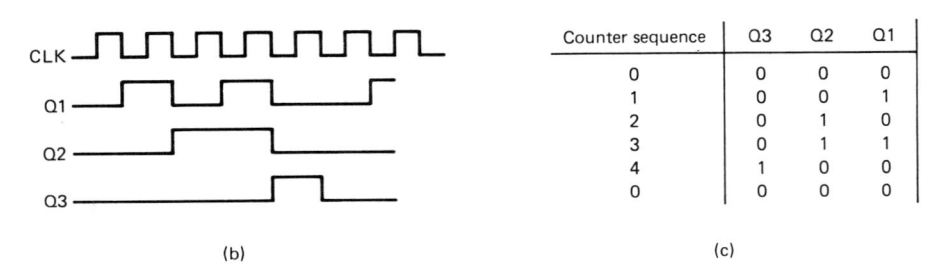

Figure 3.30 Mod-5 counter: (a) logic diagram; (b) timing diagram; and (c) counting sequence.

Counter sequence	Q3	Q2	Q1
0	0	0	0
1	0	0	1
2	0	1	0
3	0	1	1
4	1	0	0
0	0	0	0

synchronous counter in contrast to the BCD counter design discussed above. The state assignment of the counter is arbitrary and can be modified by modifying the circuit design. Since there is no CLEAR function available, the counter avoids the unused states by proper feedback interconnections and proper combination of the outputs of flip-flops 1 and 2, used as input to flip-flop 3. However, it is assumed that the initial state of the counter is always a valid state.

3.3.4 Combinations of Counter Modules

Nonbinary counter modules can be combined to form nonbinary counters whose modulus is the product of the modulus of the combined modules. Thus, a mod-2 and mod-3 counter in cascade constitute a mod-6 counter. Two mod-3 counters yield a mod-9 counter, while a decade counter can be constructed by combining a mod-2 and a mod-5 counter. The individual counter modules can be combined using a ripple connection, or the counters can be operated in synchronism with one another, independently of the clocking discipline of the individual modules. Furthermore, the order of individual modules in a counter chain can be permuted without affecting the module of the counter chain. However, such permutation will cause a substantial difference in the code in which the counter state is to be read.

3.3.5 Variable Modulus Counters

Although fixed-module counters have some important functions in different types of digital circuits, it is often desirable to be able to change the modulus of a counter when certain conditions are met during the operation of a digital circuit. This can be achieved by using a variable modulus counter.

There are several approaches to designing variable modulus counters that provide the option of selecting the counting range or amount of frequency division in the simplest way possible. An example of a synchronous variable modulus counter design based on the single-phase clock flip-flop of Figure 3.2, is shown in Figure 3.31. This counter assumes a different modulus depending on the control signals SEL and $\overline{SEL}$ which enable different connections on the feedback path. When SEL is 0 and thus, $\overline{SEL} = 1$, the feedback path is only sensitive to $Q2$ and $\overline{Q2}$. Since these signals are driving the 2-AND-NOR complex gates, they effectively become $\overline{Q2}$ and $Q2$ driving the $D1$ and $\overline{D1}$ inputs, respectively. It can be verified that with these feedback connections in effect, the counter in Figure 3.31 functions as a mod-5 counter with the counting sequence shown in Figure 3.31b. When SEL = 1 and thus $\overline{SEL} = 0$, the feedback path becomes sensitive to the outputs of all flip-flops which are combined by the 2-AND-NOR gates to provide the signals that drive the $D1$ and $\overline{D1}$ inputs. It can be verified that the counter is now operating as a mod-6 counter and its counting sequence is the sequence shown in Figure 3.31c.

Some observations can be made about this counter design. The feedback path limits the toggle rate of the counter, since any change of state has to propagate through the feedback path. If counting at high clock rates is required, the design of such a counter should keep the logic controlling the feedback path to a minimum. The output duty cycle of the counter is variable, and it depends on the feedback connections enabled in each configuration. There is no direct way to initialize this counter; for correct operation it is assumed that its initial state is a valid state or a state that can eventually lead to a valid state.

Another approach to designing a variable modulus counter is to combine the features of a binary down-counter with those of a presettable counter. Figure 3.32 illustrates the design of a 4-bit presettable binary down-counter. The counter counts down until it reaches 1111 (right after 0000), and at that time the feedback enables the PRESET inputs $P1$ through $P4$ which preset the counter to any given value, and the count-down operation resumes from that value. The PRESET function is controlled by the logic on the feedback implemented as a 2-AND-NOR complex gate. The counter can be preset to any value—counting begins from that value and then decrements from that value toward 0000. Eventually, the value 0000 is reached, and on the next trigger the counter attempts to go to 1111, which subsequently triggers the feedback action—thus enabling the counter to be preset again to the previous count value or to a new count value. As is shown in Figure 3.31b, this circuit can be set anywhere between 0000 and 1110 which implies that the modulus of this counter can be anywhere between 1 and 15.

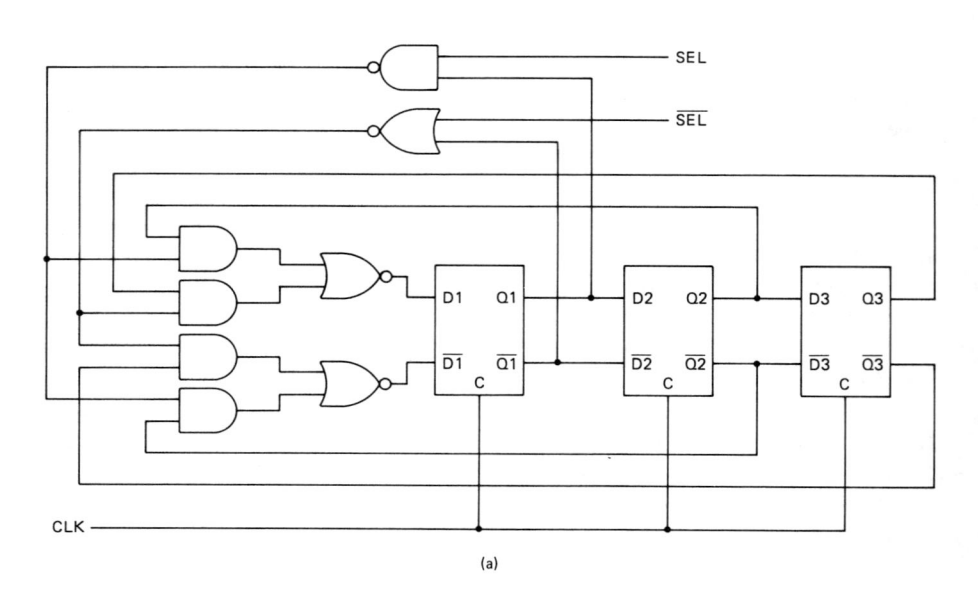

Figure 3.31 Synchronous variable modulus counter: (a) logic diagram; (b) mod-5 counting sequence; (c) mod-6 counting sequence.

(b) mod-5 counting sequence:

Counter sequence	Q3	Q2	Q1	SEL	$\overline{SEL}$
0	0	0	0	0	1
1	0	0	1	0	1
3	0	1	1	0	1
6	1	1	0	0	1
4	1	0	0	0	1
0	0	0	0	0	1

(c) mod-6 counting sequence:

Counter sequence	Q3	Q2	Q1	SEL	$\overline{SEL}$
0	0	0	0	1	0
1	0	0	1	1	0
3	0	1	1	1	0
7	1	1	1	1	0
6	1	1	0	1	0
4	1	0	0	1	0
0	0	0	0	1	0

Special design considerations should be given to the 2-AND-NOR complex gate since it has a fan-out of eight and the rise time of the feedback circuit is required to be fast. The flip-flops in this design are the same as the flip-flop shown in Figure 3.25.

Sec. 3.3 Counters

(a)

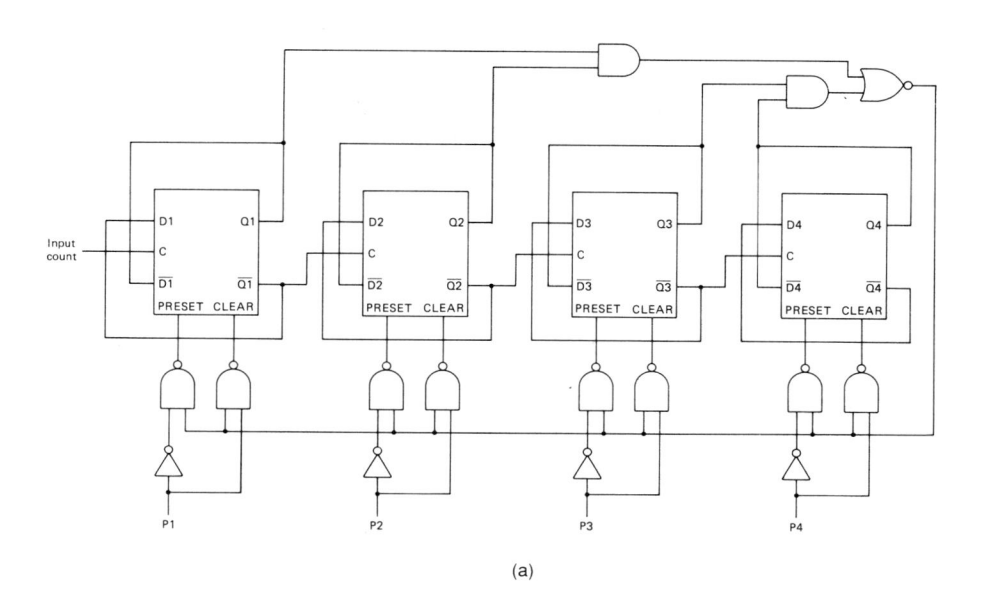

Preset inputs				Count	Modules
P4	P3	P2	P1		
0	0	0	0	Always 0	1
0	0	0	1	1 ⇒ 0	2
0	0	1	0	2 ⇒ 0	3
0	0	1	1	3 ⇒ 0	4
0	1	0	0	4 ⇒ 0	5
0	1	0	1	5 ⇒ 0	6
0	1	1	0	6 ⇒ 0	7
0	1	1	1	7 ⇒ 0	8
1	0	0	0	8 ⇒ 0	9
1	0	0	1	9 ⇒ 0	10
1	0	1	0	10 ⇒ 0	11
1	0	1	1	11 ⇒ 0	12
1	1	0	0	12 ⇒ 0	13
1	1	0	1	13 ⇒ 0	14
1	1	1	0	14 ⇒ 0	15
1	1	1	1	Always 0	1

(b)

Figure 3.32 Four-bit binary down-counter with variable modulus: (a) logic diagram and (b) counting sequence and modulus (assumed based on presettable states).

3.3.6 Lockout

In a nonbinary counter, there are states that are not used. For example, in the counter shown in Figure 3.31 the states $Q3Q2Q1 = 101$, and $Q3Q2Q1 = 010$ are not used. If the counter design does not provide for a preset function, then upon power-on, the counter can find itself in any one of the unused states. Such an event also can be the result of a fault in the circuit or noise affecting the operation of the flip-flops. Based on the counting sequence tables in Figure 3.31, we know the state that the counter advances to, from a known state. However, we do not know the next state for an unused state. Therefore, it is possible that when the flip-flops in a nonbinary counter assume an unused state, the counter might cycle from unused state to unused state, never reaching a used state. In this case, the counter is permanently faulty as far as its intended operation is concerned. If a counter's unused states can result in this mode of operation, then the counter is said to be susceptible to lockout. Therefore, it becomes important to provide the PRESET function in the design of a nonbinary counter so that it can be initialized in a known used state, thus ensuring proper operation.

If the counter is initialized to an unused state but eventually arrives at a used state either directly or after advancing through other unused states, then its operation may be acceptable for certain applications. During the time the counter shifts through unused states, its output is ignored as noise. In general, however, a sound design approach should avoid the possibility of lockout by providing the PRESET function (Figure 3.9) to the counter flip-flops. If this is unacceptable because of logic and interconnection overhead, the counter design must ensure that an unused state is always followed by a used state. This can be achieved by proper interconnection of the feedback signals that are used to define the unused states.

3.4 SEQUENCE GENERATORS

Sequence generators are synchronous circuits which generate a prescribed sequence of bits starting from some known initial state. For example, a synchronous counter is a sequence generator. In this section, we will discuss how shift registers can be designed to operate as sequence generators.

Shift register based sequence generators are widely used in two fields of special interest; namely, built-in self-test (BIST)[6] and cryptography.[7] These two fields will greatly benefit from the utilization of high performance GaAs sequence generators.

The basic design of a sequence generator is shown in Figure 3.33. There are n flip-flops connected in a shift register chain and some feedback function. The state of the shift register is a binary n-tuple, which can be regarded as the binary expression of an integer. Clearly, the shift register has 2^n possible states. As the clock advances, the contents of FFi are transferred into $FF(i + 1)$ for all i, such as $1 \le i \le n - 1$. However, to obtain the new value for $FF1$, we must compute the value of the feedback function of all the present terms and transfer

Sec. 3.4   Sequence Generators   119

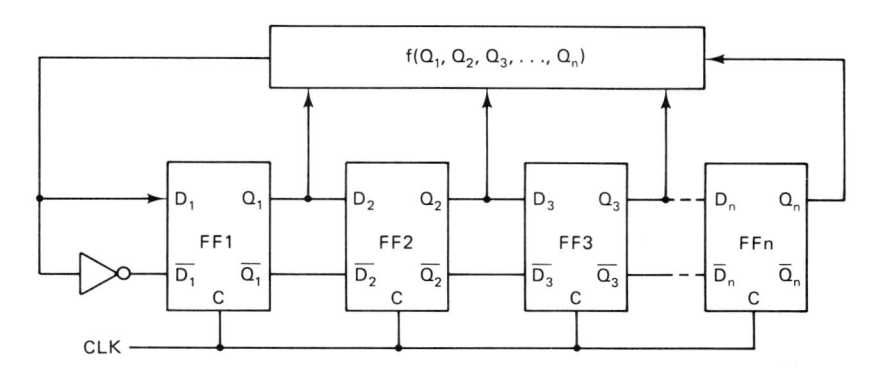

Figure 3.33 General configuration of a sequence generator.

this into *FF 1*. The output of the register can be read at any time at the Q terminals of the flip-flops. As drawn, the register in Figure 3.33 has no input. This design can be modified to allow for some input sequence, as shown in Figure 3.34. The flip-flops in both designs are based on the flip-flop design shown in Figure 3.2.

It is apparent from the way the shift register is drawn in Figure 3.34 that it operates as a finite state machine. The set of possible states, N, is the set of binary expansions for the integers from 0 to $2^n - 1$, inclusive. The number of successive states generated in the sequence before the pattern of states repeats is called the length of a sequence. Therefore, in order to implement a sequence generator capable of generating a sequence of length S, it is necessary to use, as a minimum, a number n of flip-flops where n is such that

$$S \leq 2^n - 1 \tag{3.4}$$

The length of the sequence is also called the sequence period. In most applications (especially in cryptography), the length of the sequence is more important than the exact pattern of repeated states. If the order of outputs in the

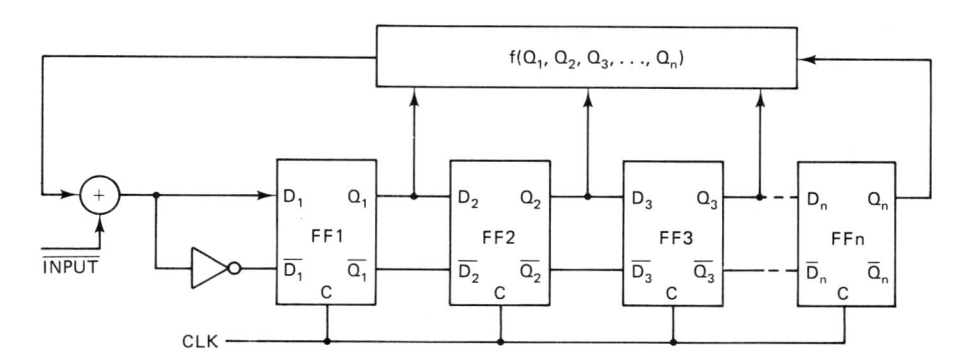

Figure 3.34 General configuration of a sequence generator with serial input.

sequence needs to follow a prescribed pattern, in general it will not be possible to generate a sequence of length S with a sequence generator implemented with the minimum number of flip-flops. However, a sequence generator implemented with n flip-flops can generate at least one sequence of maximum length $S = 2^n - 1$, for any n.

If the feedback function of the shift register in Figure 3.33 or 3.34 can be written in the form $f(Q_1, Q_2, Q_3..., Q_n) = (C_1 Q_1 \oplus C_2 Q_2 \oplus C_3 Q_3 \oplus ... \oplus C_n Q_n)$, where C_i is 0 or 1 and all addition is over $GF(2)$ (i.e., Galois Field [2] or mod-2 addition), then the shift register is called linear. The constants $C_1, C_2, C_3...C_n$ are called the feedback coefficients. The general implementation of a linear feedback shift register is shown in Figure 3.35. An open switch in this figure implies that the corresponding feedback coefficient is 0, while it is 1 when the corresponding switch is closed. In actual implementation, the $\oplus$ function is performed with an XOR gate, while a connection from a flip-flop output is used whenever $C_i = 1$, while a connection is omitted when $C_i = 0$.

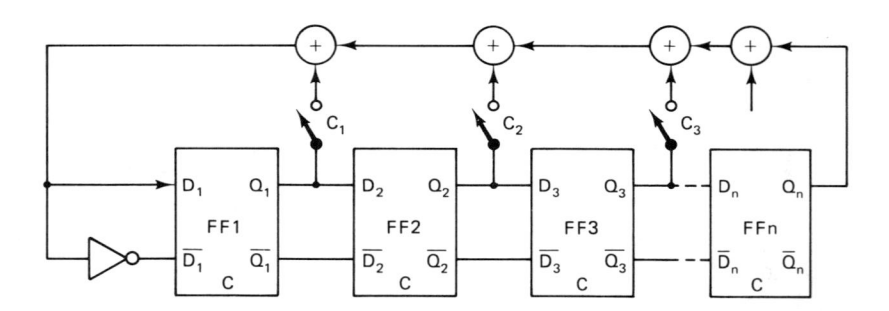

Figure 3.35 General configuration of a linear feedback shift register.

An example of a 4-bit sequence generator based on a linear feedback shift register configuration is shown in Figure 3.36. It is assumed that this circuit can be initialized in a state other than the all zero (null) state. In fact, if this circuit is initialized to the null state, the sequence generator will be unable to produce any other state. Initialization to the all-one state can be secured by providing the preset function to the flip-flops. In this case, the sequence provided by the sequence generator will be the one shown in Figure 3.36b.

Our discussion so far indicates that an n-stage linear feedback shift register is completely determined by its initial state, and its feedback coefficients $C_1, C_2,...C_n$. If S_t is the output sequence generated from an initial state of $Q_1, Q_2, Q_3...Q_n$, then the following recurrence relation of order n is satisfied:[6]

$$S_{t+n} = \sum_{i=1}^{n} C_i S_{t+i} \tag{3.5}$$

for $t = 1, 2, ...$. Associated with any such recurrence relation is a characteristic polynomial $f(x)$ defined by

Sec. 3.4 Sequence Generators 121

$$f(x) = 1 + C_1 x + \ldots + C_{n-1} x^{n-1} + x^n \qquad (3.6)$$

For a given characteristic polynomial, there are 2^n different possible initial settings of the register; consequently, the polynomial can be used to "generate" 2^n different sequences (one of which is the null sequence). As indicated above, the length of the generated sequence is of major interest and therefore, it is essential to know what coefficients should be used in order to secure the generation of sequences having maximum length (i.e., $2^n - 1$). It can be shown that if $f(x)$ is a

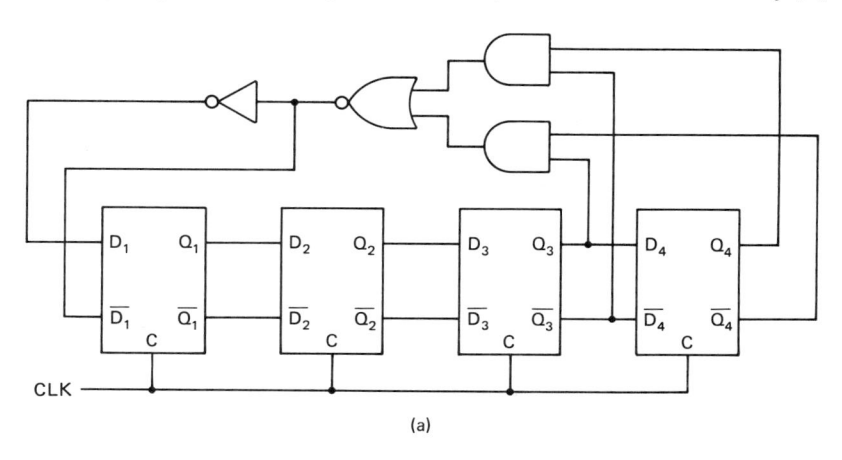

(a)

Clock	Q_1	Q_2	Q_3	Q_4	$D_1 = f(Q_1, \ldots, Q_4) = Q_3 \oplus Q_4$
1	1	1	1	1	0
2	0	1	1	1	0
3	0	0	1	1	0
4	0	0	0	1	1
5	1	0	0	0	0
6	0	1	0	0	0
7	0	0	1	0	1
8	1	0	0	1	1
9	1	1	0	0	0
10	0	1	1	0	1
11	1	0	1	1	0
12	0	1	0	1	1
13	1	0	1	0	1
14	1	1	0	1	1
15	1	1	1	0	1
1	1	1	1	1	0
2	0	1	1	1	0
.	.	.	.	.	.
.	.	.	.	.	.
.	.	.	.	.	.

(b)

Figure 3.36 Four-bit sequence generator: (a) logic diagram and (b) output sequence.

primitive polynomial, then every one of the non-null sequences generated by the linear feedback shift register has a period of $2^n - 1$, or a maximum length often called m-sequence. A polynominal $f(x)$ of degree n is called primitive when it gives a complete table with 2^n distinct symbols containing 0 and 1. Primitive polynomials of linear feedback shift registers of lengths up to about $n = 30$ are listed in Peterson and Weldon's work.[8]

From the implementation point of view, primitive polynomials which allow for the minimum number of feedback connections should be used. Based on the implementation of the XOR gate shown in the design of the sequence generator in Figure 3.36, D_1 will be stable three gate delays after Q_4 is stable, assuming the 2-AND-NOR function is implemented as a complex gate. Therefore, the feedback limits the toggle rate of the sequence generator and its effect should be kept minimal when high performance is required. Table 3.1 shows the feedback connections for maximum length sequences that can be generated by linear feedback shift registers with n up to 16:

TABLE 3.1 Feedback Connections for Maximum Length Sequences

n	Feedback Connections for $S = 2^n - 1$
2	$Q_1 \oplus Q_2$
3	$Q_2 \oplus Q_3$
4	$Q_3 \oplus Q_4$
5	$Q_3 \oplus Q_5$
6	$Q_5 \oplus Q_6$
7	$Q_4 \oplus Q_7$
8	$Q_2 \oplus Q_3 \oplus Q_4 \oplus Q_8$
9	$Q_5 \oplus Q_9$
10	$Q_7 \oplus Q_{10}$
11	$Q_9 \oplus Q_{11}$
12	$Q_2 \oplus Q_{10} \oplus Q_{11} \oplus Q_{12}$
13	$Q_1 \oplus Q_{11} \oplus Q_{12} \oplus Q_{13}$
14	$Q_2 \oplus Q_{12} \oplus Q_{13} \oplus Q_{14}$
15	$Q_{14} \oplus Q_{15}$
16	$Q_7 \oplus Q_{12} \oplus Q_{16}$

Maximum length sequences have two interesting properties. For example, examining the output sequence in Figure 3.36b, we can observe that 0s and 1s appear in the sequence with equal frequency. Also, before the sequence repeats itself, examining previous states is of no help in predicting the next state. These features are similar to a random event resulting from tossing a fair coin. In the case of a sequence, however, after it repeats itself we would be able to predict future states by comparison with corresponding states in the previous period. Because of the underlying periodicity and deterministic generation, the maximum length sequences are known as pseudorandom sequences. Of course, the longer the period, the more "random" the sequence appears to be. In practical designs, the period of a linear sequence generator can be lengthened by increasing the number of flip-flops in the shift register.

In applications where real-time error detection and correction is required, as well as in data encryption and decryption applications, the performance of the sequence generator is also critical in addition to its period. GaAs sequence generators can make major contributions in these areas. For example, a 10-bit sequence generator circuit reported in Reference 2 is used for error detection in a fiber-optic transmission link. This circuit is designed similarly to the sequence generator shown in Figure 3.36. To minimize the delay in the feedback path, however, the XOR function is implemented within the first flip-flop. This sequence generator operates with a 2 GHz clock generating a maximum length sequence $2^{11} - 1 = 1023$.

3.5 ARITHMETIC CIRCUITS

This section presents the design of circuits used to perform arithmetic operations. These circuits can be used in the design of general-purpose computer systems and high-speed signal processing systems. The selection of gates for designing these circuits is dictated by the implementation efficiency of these gates in a given GaAs logic family.

3.5.1 Adders

Adders are the basis for all arithmetic circuits; therefore, their design and implementation are of major importance. The binary addition of an addend and an augend digit results in a sum and a carry digit given by the following logic relationships:

$$\text{Sum} = A \oplus B = A \bar{B} + \bar{A} B \tag{3.7}$$

$$\text{Carry} = A \cdot B \tag{3.8}$$

where A and B are the addend and the augend digits respectively, and the above operations imply mod-2 arithmetic. In general, however, addition is performed with mod-2 numbers of N digits. In this case, if $A = A_N A_{N-1} \ldots A_1$ and $B = B_N B_{N-1} \ldots B_1$, then

$$\text{Sum}_i = C_{i-1} \oplus A_i \oplus B_i \tag{3.9}$$

$$\text{Carry}_i = A_i B_i + (A_i \oplus B_i) \cdot C_{i-1} \tag{3.10}$$

for $i = 1, \ldots N$.

The circuit that performs the addition implied by the above expressions is called a full adder. The logic design of a 2-bit full adder is shown in Figure 3.37.

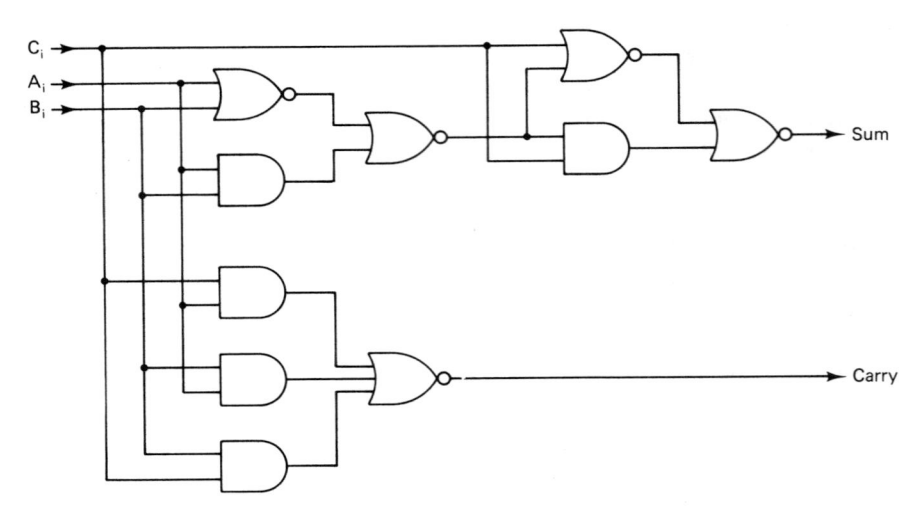

Figure 3.37 Logic design of a 2-bit full adder.

This circuit is suitable for implementation in BFL where all AND-NOR functions can be implemented as complex gates. As shown in Figure 3.37, the carry path has a shorter propagation delay than the sum path. However, in the case of an n-bit adder, the carry has to ripple through n stages before the output is valid; that is a speed limitation of the so-called ripple carry adder. This limitation is similar to the comparable limitation in ripple counters. The design of a 4-bit ripple carry adder is shown in Figure 3.38. The output buffers added to this circuit are not required for its logic function, but they are used for interfacing and improved load driving capability. The longest delay path, often called the critical path, is from C_{in} to the most significant bit of the sum S_4. On this path, the average fan-in to fan-out ratio is 5/3.

A circuit implementation of this adder design was reported in Reference 9. The technology used to implement this circuit is an extension of BFL called low-power BFL (lp-BFL).[10] Lp-BFL uses MESFETs with V_p between −0.9 V and −0.6 V, and two level shifting diodes in the source follower section of a gate. As an example, Figure 3.39 illustrates the circuit design of the AND-NOR gate that produces the carry in the full adder.

Test measurements on prototypes reported in Reference 9 indicate average add time of 1.9 nsec, at a total power dissipation of 45 mW. This corresponds to an average per gate propagation delay of 380 psec (with fan-in/fan-out = 5/3), an average per gate power dissipation of 1.5 mW, and thus a power-delay product of 0.6 pJ.

A design of the 4-bit ripple carry adder based only on NOR gates is shown in Figure 3.40a, while Figure 3.40b illustrates the logic design of an adder stage. This design is best suited for DCFL implementation, where the efficient imple-

Sec. 3.5   Arithmetic Circuits

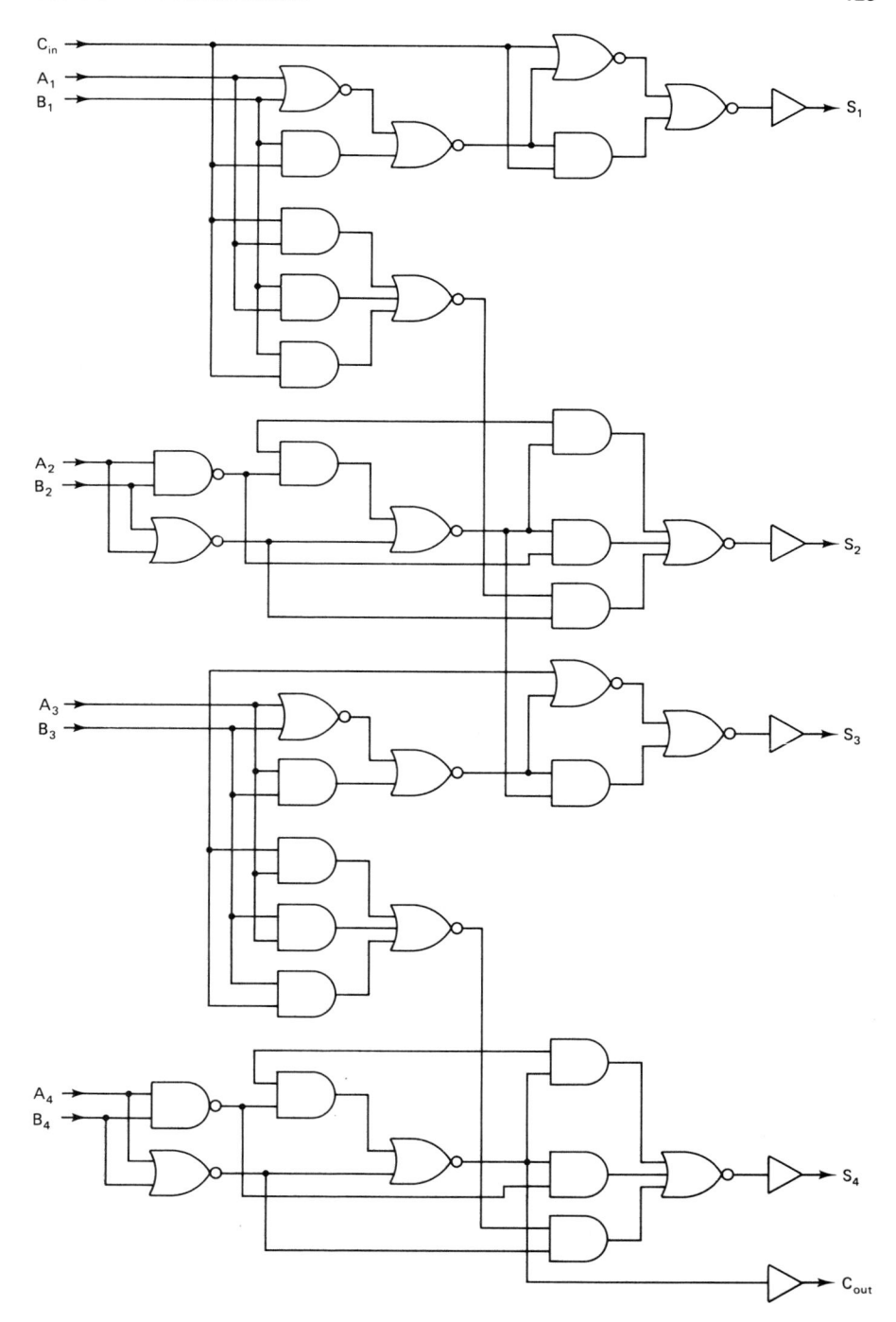

Figure 3.38   Four-bit ripple carry adder.

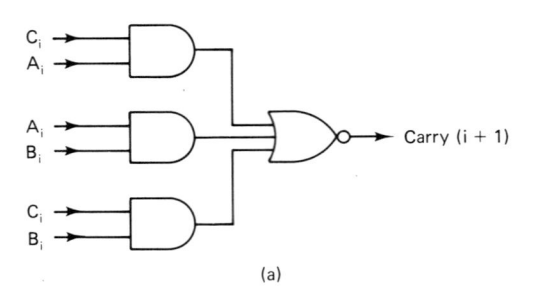

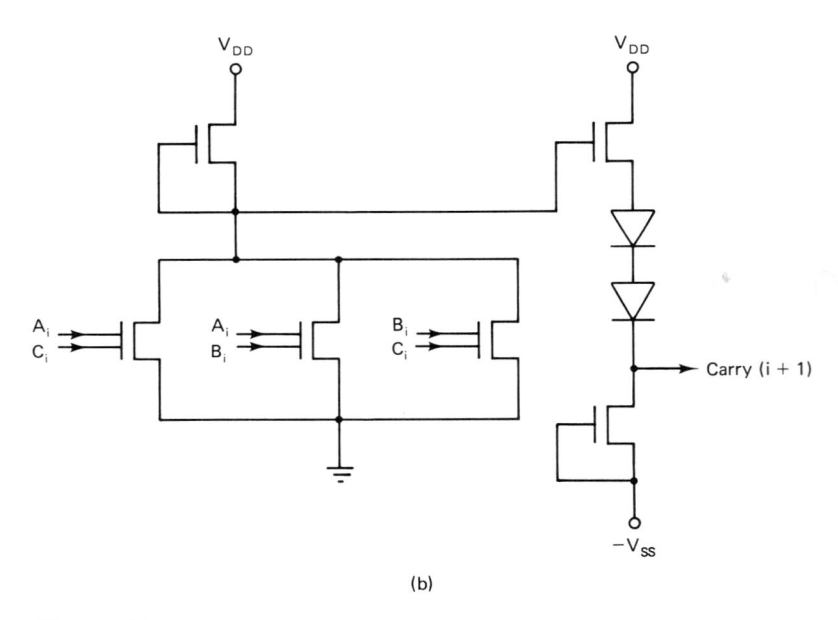

Figure 3.39 Logic design of: (a) the carry circuit and (b) its circuit implementation lp-BFL.

mentation of NOR gates without requirements for level shifting results in fast and yet low-power circuits. A DCFL implementation of this adder circuit is reported in Reference 11. As shown in Figure 3.40b, the carry path of the adder cell has two gate delays, while the sum path has either two or three gate delays depending on the logic levels at the inputs. For example, for $BC = 00$ or 11, the path from A to sum has two gate delays, while $BC = 01$ or 10 results in three gate delays for the same path. The propagation delay of the adder cell implemented in Reference 11 was measured to be between 210 psec and 260 psec depending on the input code. This results in 1.3 nsec add time for the 4-bit adder (Figure 3.40a) in the worst case. This performance is achieved with the circuit biased at 1.5 V, where power dissipation was measured as 0.36 mW/gate or 17.28 mW for the 4-bit

Sec. 3.5   Arithmetic Circuits   127

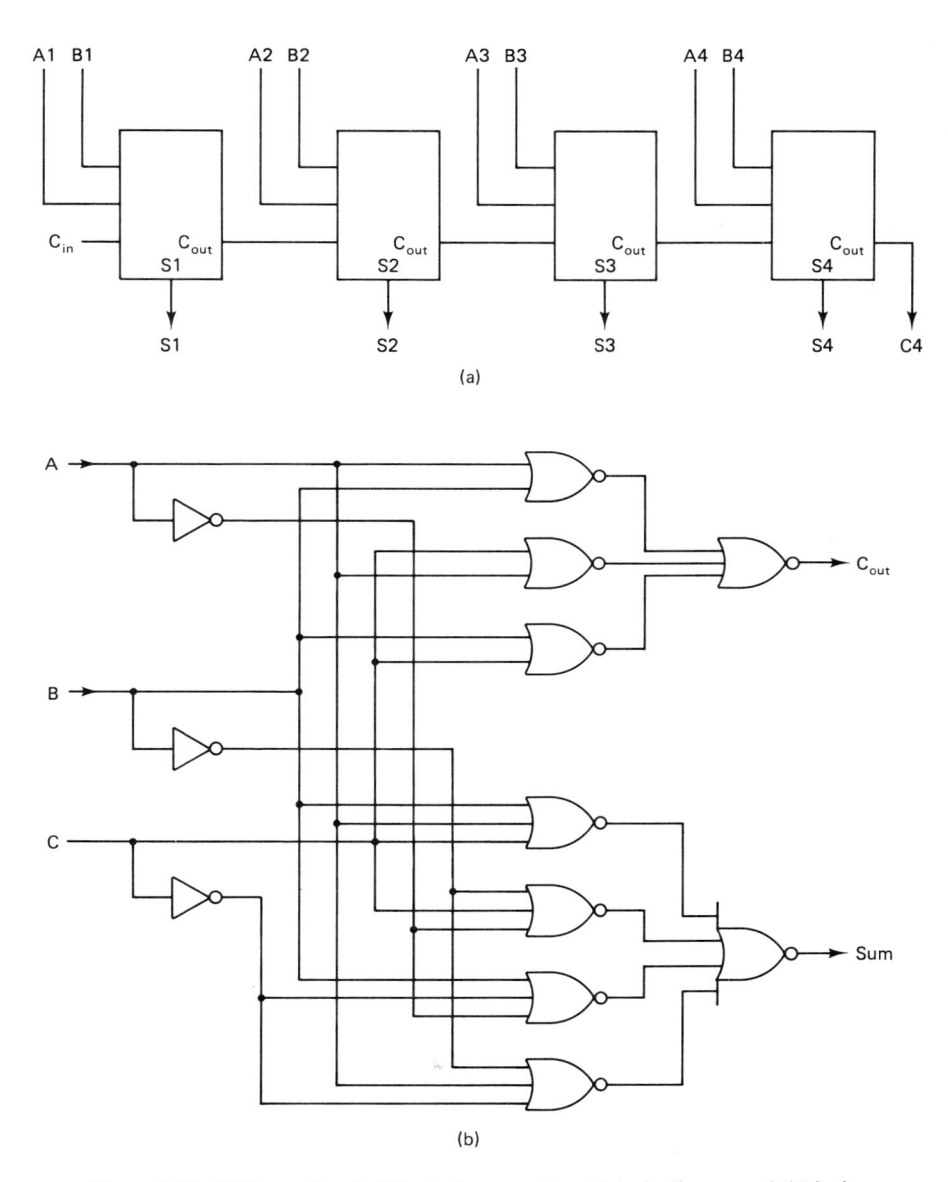

Figure 3.40 NOR gate-based, 4-bit ripple carry adder: (a) logic diagram and (b) logic gate design of an adder stage.

adder. Although these results show an improved power dissipation performance compared to the lp-BFL adder discussed above, selection of either circuit should include additional performance metrics such as noise margins and load driving capability.

Implementation of the 4-bit ripple carry adder in SDFL requires modification of the design of the basic adder stage so that a structure best suited for this technology can be achieved. The design of the basic adder stage best suited for implementation in SDFL is shown in Figure 3.41. The interconnection of adder stages is the same as was illustrated in Figure 3.40a. Note that the carry output of an adder stage (Figure 3.41) is actually the complement of the true signal. However, when an even number of stages is used, the final carry output is the true signal. In adders with an odd number of stages, the final carry output should be inverted. By using 2-level and 3-level complex SDFL gates, the full adder in Figure 3.41 produces the carry output and the sum output after one and two gate delays, respectively. Implementation of such an adder design is reported in Reference 12. With gate propagation delays of 172 psec, the add time for the 4-bit ripple carry adder is 1.72 nsec.

The speed of the carry ripple adders discussed above is limited by the carry propagation time. The carry that is generated at the least significant bit position has to propagate through all adder stages before the final result of the addition is valid. One way to avoid the propagation of carry through all adder stages is to use a carry bypass, or a carry lookahead scheme. A carry lookahead scheme allows the carry generated by a group of bits to be propagated to a higher order

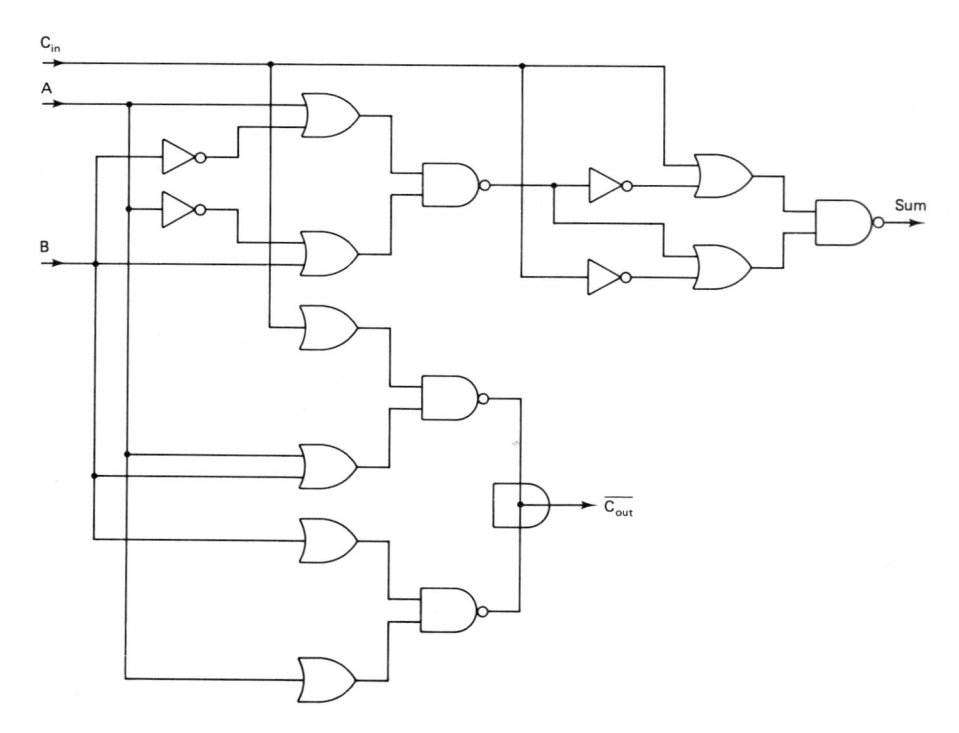

Figure 3.41 Basic adder stage for SDFL implementation.

group of bits by passing through only one level of logic instead of propagating through the individual stages of the group. An individual binary position stops a propagating carry when both input operands are 0, while it generates its own carry when both operands are 1. A carry is propagated through one position only when one operand is 1 and the other 0. The size of the group of bits depends on design requirements and implementation trade-offs. Small groups allow short time between the generation of a carry within the group and its appearance as valid output of that group. Larger groups result in reduced overall propagation time due to the reduced number of total groups. A major consideration in every case is the additional gates needed to implement the carry lookahead function. Additional gates imply greater area and power dissipation requirements, which must be considered against the performance improvements before a decision is made about selecting a group size.

A 2-bit adder with a carry lookahead circuit is shown in Figure 3.42. The carry lookahead circuit generates or propagates a carry signal to upper bit positions in anticipation of a carry signal from lower bit positions. This adder circuit is suitable for BFL and DCFL implementation. Utilization of a 2-bit adder, as a building block, to realize a 32-bit adder, is reported in Reference 13. The carry lookahead operation is carried out over three levels in this circuit, namely every 2-bit adder in the first level, every 4-bit adder made of 2-bit adders in the second level, and every 8-bit adder made of 4-bit adders in the third level. Device widths vary between 10 µm and 20 µm and FET pinch-off voltage is −0.5 V, so a BFL implementation of this 32-bit adder yields a performance of maximum add time of 2.9 nsec. At this performance level and with bias of $V_{DD} = 2.56$ V and $V_{SS} = 1.32$ V, the power dissipation is 1.1 W.[13]

One adder design of particular interest is the bit-serial adder. This adder utilizes a single adder stage (i.e., 2-bit adder) and performs addition of bit streams, two bits at a time, least significant bits (LSB) first. The interest in this design lies in its small area requirements and low power dissipation at the expense of computational delay. Figure 3.43 illustrates the design of a bit-serial full adder. The full-adder circuit in this figure is the one shown in Figure 3.37, while the flip-flop is the circuit shown in Figure 3.2. The two words to be added are applied to the inputs A and B in bit synchronism, least significant bit first. The feedback logic on the carry terminal takes the carry from one bit position and applies it to the next more significant position as required. To prevent interference between words, it is necessary to inhibit the carry propagation during the arrival of the least significant bit of each new word. This is accomplished by the LSB signal, which is combined with the output carry by use of a NOR function. The LSB signal is valid "high" and coincides with the least significant bits of A and B, thus forcing C_{in} to be 0. LSB remains valid for one clock period and becomes valid again during the addition of the least significant bits of the next words to be added. The sum is computed every clock period, one bit at a time, least significant bit first. The time to add two words with this circuit is $n \cdot T$, where n is the number of bits in the words to be added and T is the clock period.

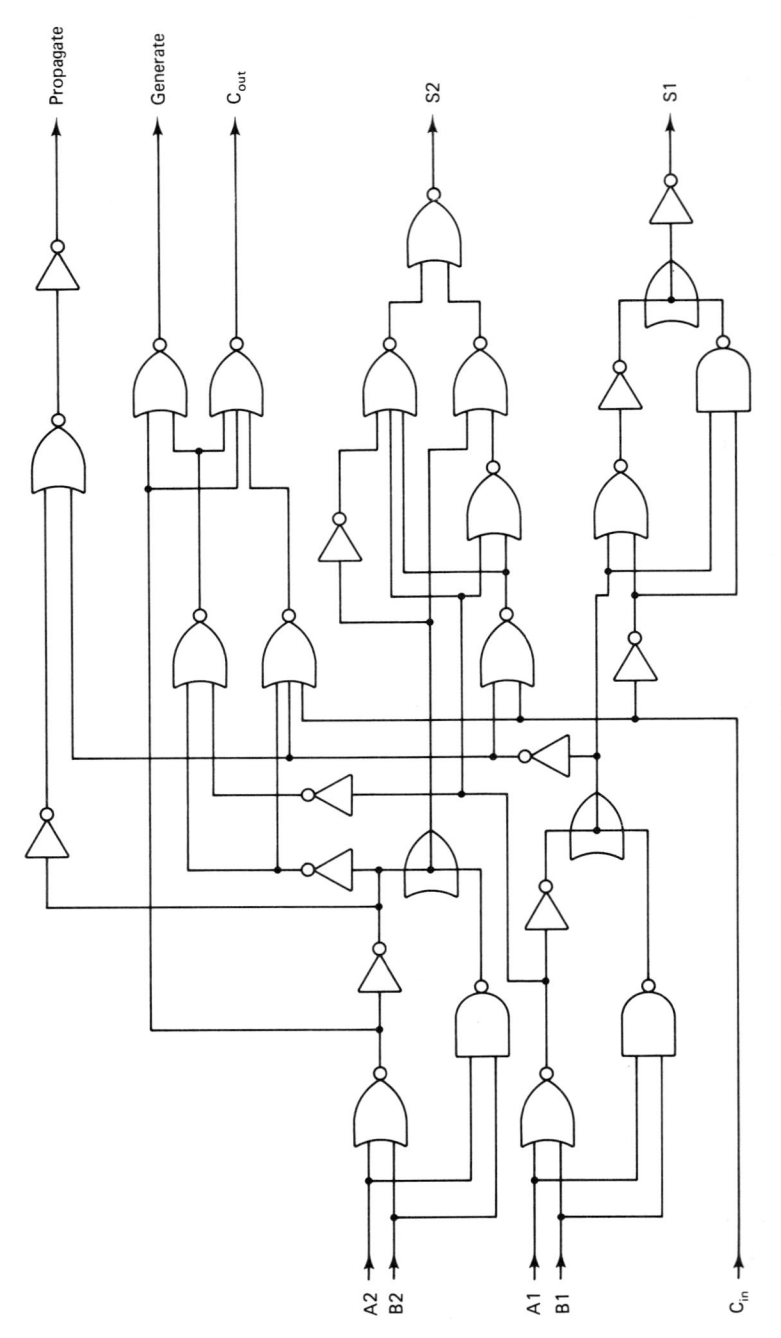

Figure 3.42 Two-bit full adder with carry lookahead.

Sec. 3.5   Arithmetic Circuits   131

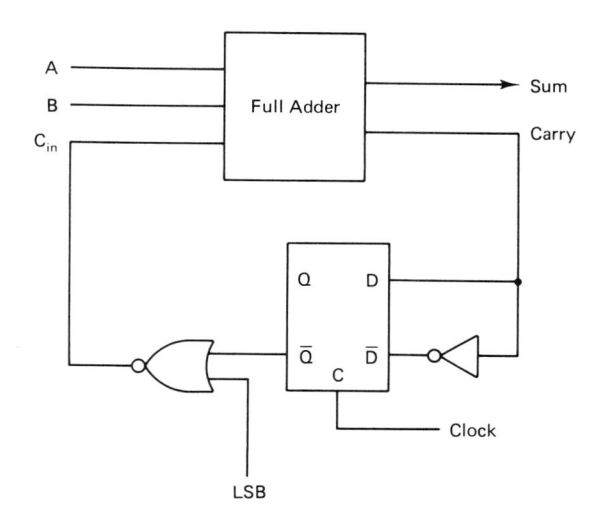

Figure 3.43   Bit-serial full adder.

3.5.2 Multipliers

There are many different techniques for performing multiplication, and their implementation depends on the number representation used for the operands and implementation trade-offs dictated by performance requirements and system specifications.

The conventional multiplication algorithm for two positive integers x and Y, resulting in a product P, can be represented as:

$$P = \sum_{i=0}^{n-1} x_i Y \cdot 2^i \qquad (3.11)$$

where Y is the multiplicand, x is the multiplier, and n is the number of bits in the multiplier. Equation 3.11 implies:

1. Y is the multiplicand.
2. $Y \cdot 2^i$ is the multiplicand shifted i bits.
3. x_i is the ith multiplier bit.
4. If x_i is 1, then $Y \cdot 2^i$ is added to the partial sum, and if x_i is 0, $Y \cdot 2^i$ is not added.
5. The n partial products are combined by carrying out the summation from $i = 0$ to $i = n - 1$, to yield the final product P.

The operation described above indicates how multiplication is done in paper-and-pencil calculations. Equation 3.11 can be simplified to show that $P = X \cdot Y$, in the following manner:

$$P = \sum_{i=0}^{n-1} x_i Y \cdot 2^i = Y \sum_{i=0}^{n-1} x_i 2^i \tag{3.12}$$

where $\sum_{i=0}^{n-1} x_i \cdot 2^i = x_0 2^0 + x_1 2^1 + \ldots + x_{n-1} 2^{n-1}$ which by definition is equal to x; therefore Equation 3.12 yields $P = X \cdot Y$. A 4-bit multiplication example based on Equation 3.11 is shown in Figure 3.44a, while Figure 3.44b illustrates a numerical implementation of this example.

A similar procedure as above can be devised for 2's complement numbers. In this case the product P of two non-zero numbers can be represented as:

$$P = X \cdot (-Y) = \sum_{i=0}^{n-1} x_i (2^{2n} - Y \cdot 2^i) \tag{3.13}$$

where X is the multiplier, Y is the multiplicand, and n is the number of bits in the X and Y operands. A different way of writing Equation 3.13 is:

$$P = X \cdot (-Y) = 2^{2n} \sum_{i=0}^{n-1} x_i - Y \sum_{i=0}^{n-1} x_i 2^i \tag{3.14}$$

and since by definition $X = \sum_{i=0}^{n-1} x_i 2^i$, Equation 3.14 becomes

$$P = X \cdot (-Y) = 2^{2n} \sum_{i=0}^{n-1} x_i - Y \cdot X \tag{3.15}$$

Based on our initial assumption that both X and Y are non-zero, two cases can be considered for the term $2^{2n} \sum_{i=0}^{n-1} x_i$ of Equation 3.15.

				y_3	y_2	y_1	y_0	
				x_3	x_2	x_1	x_0	
				P_{03}	P_{02}	P_{01}	P_{00}	$x_3 Y$
			P_{13}	P_{12}	P_{11}	P_{10}		$x_2 Y2$
		P_{23}	P_{22}	P_{21}	P_{20}			$x_1 Y2^2$
	P_{33}	P_{32}	P_{31}	P_{30}				$x_0 Y2^3$
P_7	P_6	P_5	P_4	P_3	P_2	P_1	P_0	$\sum_{i=0}^{3} x_i Y2^i$

(a)

				1	0	1	0	Y
				1	1	0	1	X
				1	0	1	0	
			0	0	0	0		
		1	0	1	0			
	1	0	1	0				
1	0	0	0	0	0	0	0	

(b)

Figure 3.44 Four-bit multiplication example: (a) general computation and (b) numerical implementation.

Sec. 3.5 Arithmetic Circuits

Case 1: There is only one i such that $x_i = 1$. In this case, $2^{2n} \sum_{i=0}^{n-1} x_i = 2^{2n}$.

Case 2: There are m 2s ($0 < m \leq n-1$) for which $x_i = 1$. In this case the addition of partial products will result in $m-1$ carry-outs. However, these carry-outs are neglected in 2's complement arithmetic and thus $2^{2n} \sum_{i=0}^{n-1} x_i = 2^{2n}$.

Using the above analysis, Equation 3.15 yields:

$$P = X \cdot (-Y) = 2^{2n} - X \cdot Y \tag{3.16}$$

which is the 2's complement of the product, as expected. The implication of Equation 3.16 is that direct multiplication of 2's complement numbers can be performed in a way similar to the multiplication of two positive numbers. In fact, the operations implied by Equation 3.13 are identical to the operations implied by Equation 3.11, but in the 2's complement number system.

Multiplier designs based on the shift-and-add operation described by Equations 3.11 and 3.13 are called array multipliers and are implemented with combinational logic. Each partial product, P_{ij}, can be formed from the Y_i multiplicand bits and x_j multiplier bits using an AND function. In addition, an array of full adders is required to sum the partial products. The array of full adders can be arranged in a systematic way, as the example of Figure 3.44 indicates. The first row of adders for this example will combine the partial products of the top two rows. The sum outputs of one row of adders provide input to the adder one row below, while the carry outputs are shifted one bit to the left. The last row of adders works in a ripple-carry configuration in forming the last P_{nn} partial product. This carry-save scheme allows for a systematic structure of a parallel array multiplier with regular and repetitive layout. The critical path in the array multiplier is the path from the partial product $x_n Y_0$ to the final product bit P_{nn}. This critical path also defines the maximum multiplication time by the array multiplier. From the circuit design point of view, however, note that each row of adders provides signals only to the subsequent row of adders—thus minimizing interconnection propagation delays from stage to stage.

Figure 3.45 illustrates the design of a 4 × 4 bit array multiplier. NOR gates are used to obtain the partial products to be added; this creates the need for the input inverters in the circuit. NOR gates are preferred over AND gates because of their simplicity in terms of both design and process, their high homogeneity, and their run-to-run reproducibility. The half-adders (HA) are like the full-adder circuits discussed in Section 3.5.1, except they do not have a carry-in input. The design of the adders is based on all NOR gates (Figure 3.40) or NOR and AND-NOR gates (Figure 3.37). Since high-performance multipliers are critical for implementing high-throughput signal processing systems, several efforts have been reported for designing and fabricating GaAs array multipliers. A 4 × 4 bit array multiplier designed by Toshiba using DCFL technology has a typical multiplication time of 6.5 nsec while dissipating 39 mW. The FETs used in the design

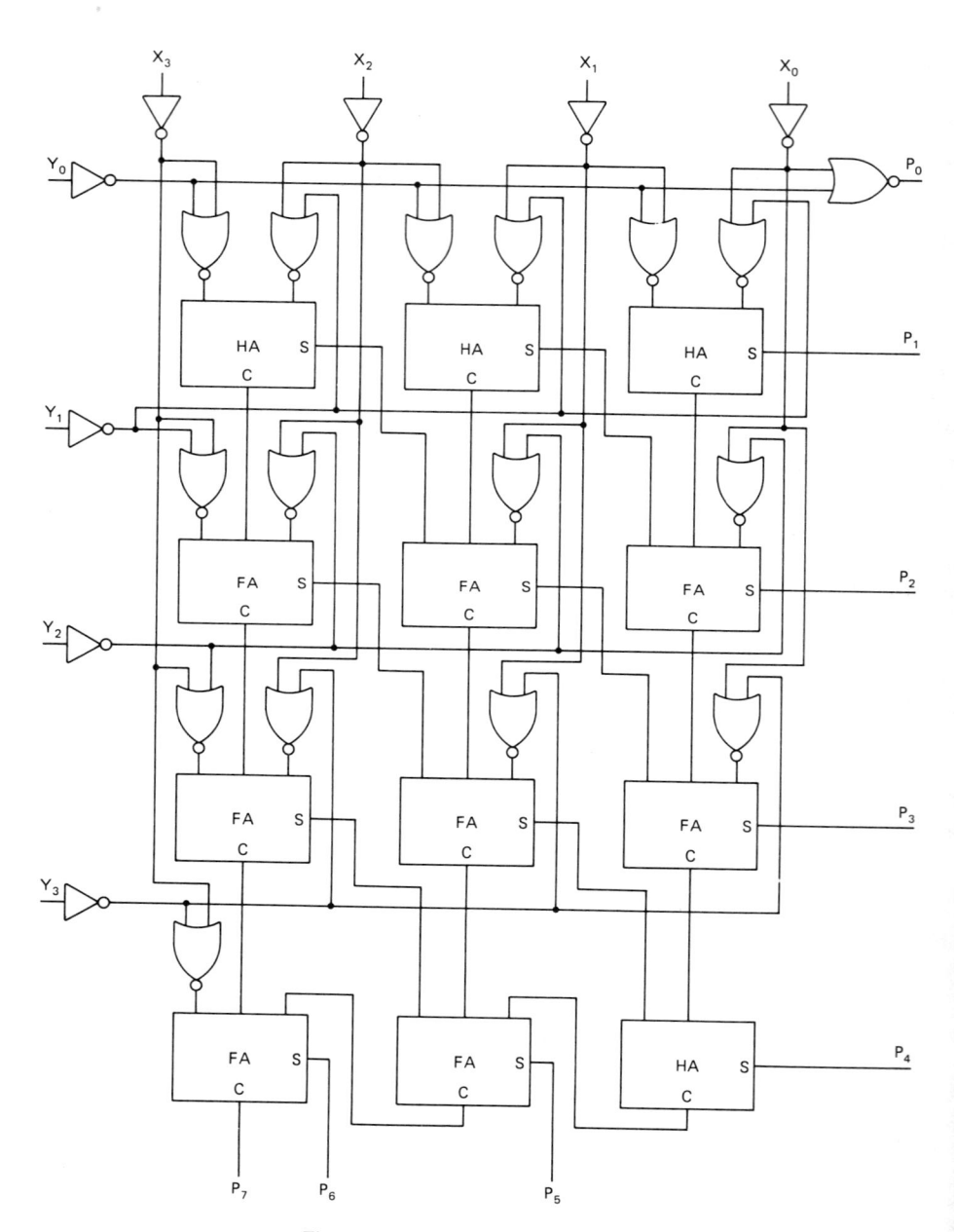

Figure 3.45 4 × 4 bit array multiplier.

of this circuit have 1.4 μm gates and the logic swing is on the order of 0.6 V.[14] A 4 × 4 bit array multiplier with 3.5 nsec multiplication time has been designed by Kato for Sony.[15] This is also a DCFL circuit, which at the indicated multi-

Sec. 3.5 Arithmetic Circuits   135

plication speed dissipates 70 mW of power, having an output logic swing of 0.8 V. A higher performance 4×4 bit array multiplier has been designed by AT&T Laboratories, using selectively doped heterostructure transistors (GaAs/AlGaAs).[16] This multiplier was also implemented using DCFL and resulted in multiplication time of 1.6 nsec while dissipating 55 mW of power with an output logic swing of 1 V. The most complex array multiplier design, however, has been performed by Fujitsu, also using DCFL.[17] This is a 16 × 16 bit multiplier capable of 10.5 nsec multiplication time with a power dissipation of 952 mW at a supply voltage of 1.6 V. The FETs used in the design have 2 μm gate lengths.

As noted in the discussion above, all multipliers reported are DCFL circuits with low-power dissipation and low-bias voltage. A multiplier circuit implemented with depletion-only devices has been designed by Rockwell. This circuit is an 8 × 8 bit array multiplier designed in SDFL using the NOR gate based adder configuration.[18] This multiplier performs an 8 × 8 bit multiplication in 5.25 nsec and dissipates 2.2 W when it is biased at V_{DD} = 2.7 V and V_{SS} = −2.0 V.

A very efficient multiplication algorithm for all 2's complement numbers was developed by Booth.[19] This algorithm treats positive and negative numbers uniformly and also takes advantage of strings of 0s and 1s in the multiplier, which it skips over without addition or subtraction, thus increasing the speed of multiplication. The algorithm is based on the fact that a multiplier consisting of K consecutive 1s can be implemented by performing a sequence of K additions and shifts. However, K 1s equals $2^k - 1$ and, therefore, a subtraction first, then K shifts, and an addition can be performed to yield the desired result. For example, consider the decimal number 127 = 001111111 (binary 2's complement). Straightforward multiplication would require seven additions, and six shifts of the multiplicand. Based on the above discussion, however, we can write:

$$127 = 128 - 1 \text{ (decimal notation)}$$
$$001111111 = 2^7 - 2^0 \text{ (binary notation)}$$
$$= 010000000 - 000000001$$
$$\overset{+\ \ \ \ \ -}{= 01000000\overline{1}}$$

and thus the multiplication process reduces to two addition/subtractions and one shift of seven bits. It can be seen that this technique offers a significant reduction in the number of operations compared to straightforward multiplication. The worst case occurs when the multiplier consists of a string of alternating 1s and 0s.
For example, $01010101 = 0\overset{+}{1}0\overset{+}{1}0\overset{+}{1}0\overset{+}{1}$. In this case, the Booth algorithm has the same performance as straightforward multiplication. In general, however, a multiplier word will not consist of alternating strings of 1s and 0s.

A direct implementation of Booth's algorithm into hardware would be extremely complex because in general it will require variable shifts that are not easy to handle. However, if the number of bits to be shifted is made constant, the logic that controls the shifting can be easily implemented in hardware. In most implementations a constant shift of two bits is considered between examinations of multiplier bit sets. Each examination checks the present two multiplier bits X_i and X_{i+1}, and the previous bit X_{i-1}. Using Booth's algorithm one can derive the decoding logic required. If, say, $X = X_{i+1} X_i X_{i-1} = 000$ or 111, then no action is taken. In this case, it is assumed this set is part of a string of 0s or a string of 1s. The other cases are:

$x = 001 = 0\overset{+}{1}0$: at MSB end of a string; add multiplicand

$x = 010 = 0\overset{+}{1}0$: single digit string; add multiplicand

$x = 011 = \overset{+}{1}00$: at MSB end of a string; subtract 2 × multiplicand

$x = 100 = \overline{1}00$: at LSB end of a string; subtract 2 × multiplicand

$x = 110 = 0\overline{1}0$: at LSB end of a string; subtract multiplicand

$x = 101 = \overset{-+}{1}10 = 0\overline{1}0$: at LSB end of a string and at MSB of next string; subtract multiplicand

The 1× and 2× the multiplicand are performed by shifting or not shifting the multiplicand.

This technique also can be applied to constant shifts of three bits or more. In the case of a three-bit shift, three bits of the multiplier plus the previous one are examined. The immediate implication in this case, however, is more complex hardware implementation in terms of decoding logic and interconnection lines, because its adder in the array will require, in addition to the multiplicand and partial product, an input representing 3x the multiplicand.

A formal proof for Booth's algorithm is presented in Reference 20.

Although Booth's algorithm can be used to increase the performance of multiplier circuits, its utilization in GaAs array multipliers should be considered in light of the hardware overhead required for its implementation. However, Booth's algorithm is very well suited and should be seriously considered for the design and implementation of bit-serial multipliers.

Figure 3.46 shows a block diagram of a pipelined bit-serial multiplier which accepts 2's complement operands (X and Y) at the input, and performs the operations that are required to yield the serial product (P) in 2's complement form at the output. The various operations are controlled by the LSB signal, which is a pulse coincident with the least significant bits of the two input operands. The basic steps that are carried out to obtain the 2's complement product are the following:

1. Each two-bit slice of the Y operand, designated by $Y_{i+1}Y_i$, is recoded and the appropriate action (addition/subtraction of 0, or X, or $2X$, to/from the

Sec. 3.5　Arithmetic Circuits

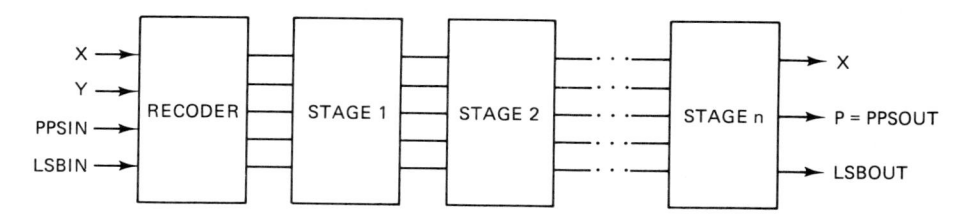

Figure 3.46　Block diagram of a pipelined bit-serial multiplier.

partial product sum *PPSIN*) is taken. This recoding process is based on Booth's algorithm discussed above. The logic of the recoder, shown in Figure 3.47, essentially performs this recoding process in two successive clock cycles. The operation in the two clock cycles is explained in Table 3.2.

2. As the recoded bits of step 1 are formed, they are stored in the various multiplier stages. The least significant two bits of *Y* are recoded and stored in multiplier stage 1, the next two bits are recoded and stored in stage 2, and so on. The logic of a multiplier stage is shown in Figure 3.48. It is the timing of the LSB pulse that ensures the proper storage of the appropriate recoded bits in each multiplier stage. The positive LSB pulse arrives at the input of each multiplier stage at a time when the appropriate recoded bits are available at the 1-inputs of the three multiplexers, *M*1, *M*2 and *M*3 (i.e., at the inputs that are passed through to the output when LSBIN is "high"). At all the other times, the LSBIN node is "low," and hence the appropriate recoded bits are retained in each multiplier stage.

3. As the *X* operand passes through each stage, a partial product is formed and combined with the partial product sum from the preceding stage. A controlled complementer (exclusive-or gate D1) on the PPSIN input is used to implement subtraction serially. Each multiplier stage truncates precisely

TABLE 3.2　Operation of the Recoder

	Y-RECODE 1 OUT	Y-RECODE 2 OUT
First Clock Cycle (Y_{i+1}, the most significant bit of the two bit slice $Y_{i+1}Y_i$, enters the recoder)	1 → Subtraction of zero or *X* or 2*X* from the partial product sum is to be performed; 0 → addition of zero or *X* or 2*X* to the partial product sum is to be performed	1 → addition or subtraction of *X* is to be performed. 0 → no addition or subtraction of *X* is needed
Second Clock Cycle (Y_{i+2}, the least significant bit of the next two bit slice $Y_{i+3}Y_{i+2}$, enters the recoder)	don't care	1 → addition or subtraction of 2*X* is to be performed. 0 → no addition or subtraction of 2*X* is needed

138 GaAs Logic Circuits Chap. 3

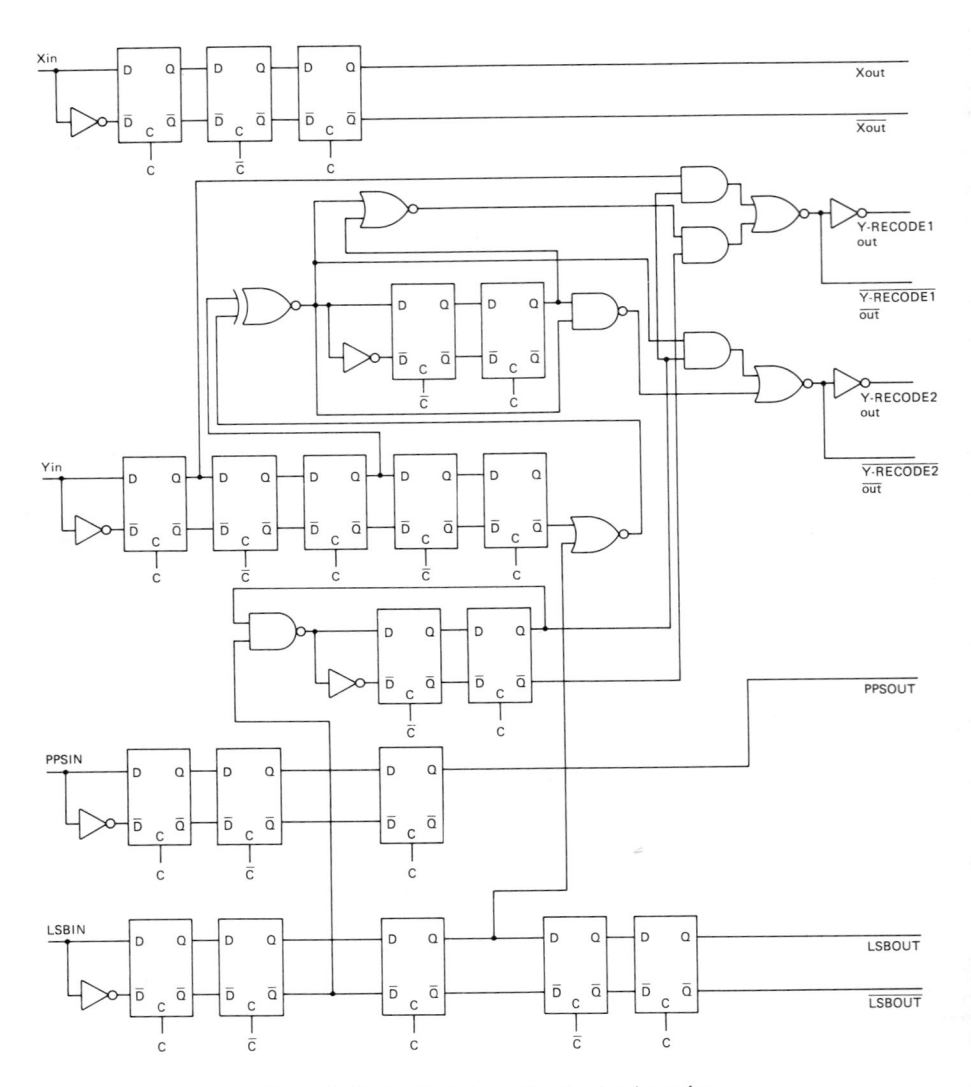

Figure 3.47 Logic design of the five-level recoder.

two least significant product bits of the partial product sum before the next partial product is added at the next multiplier stage.

This truncation is necessary to achieve the pipelined operation in which the multiplier can accept a continuous stream of X and Y operands. In some cases, in order to guarantee correct timing, one extra bit-time separation of data words may

Figure 3.48 Logic design of the multiplier stage.

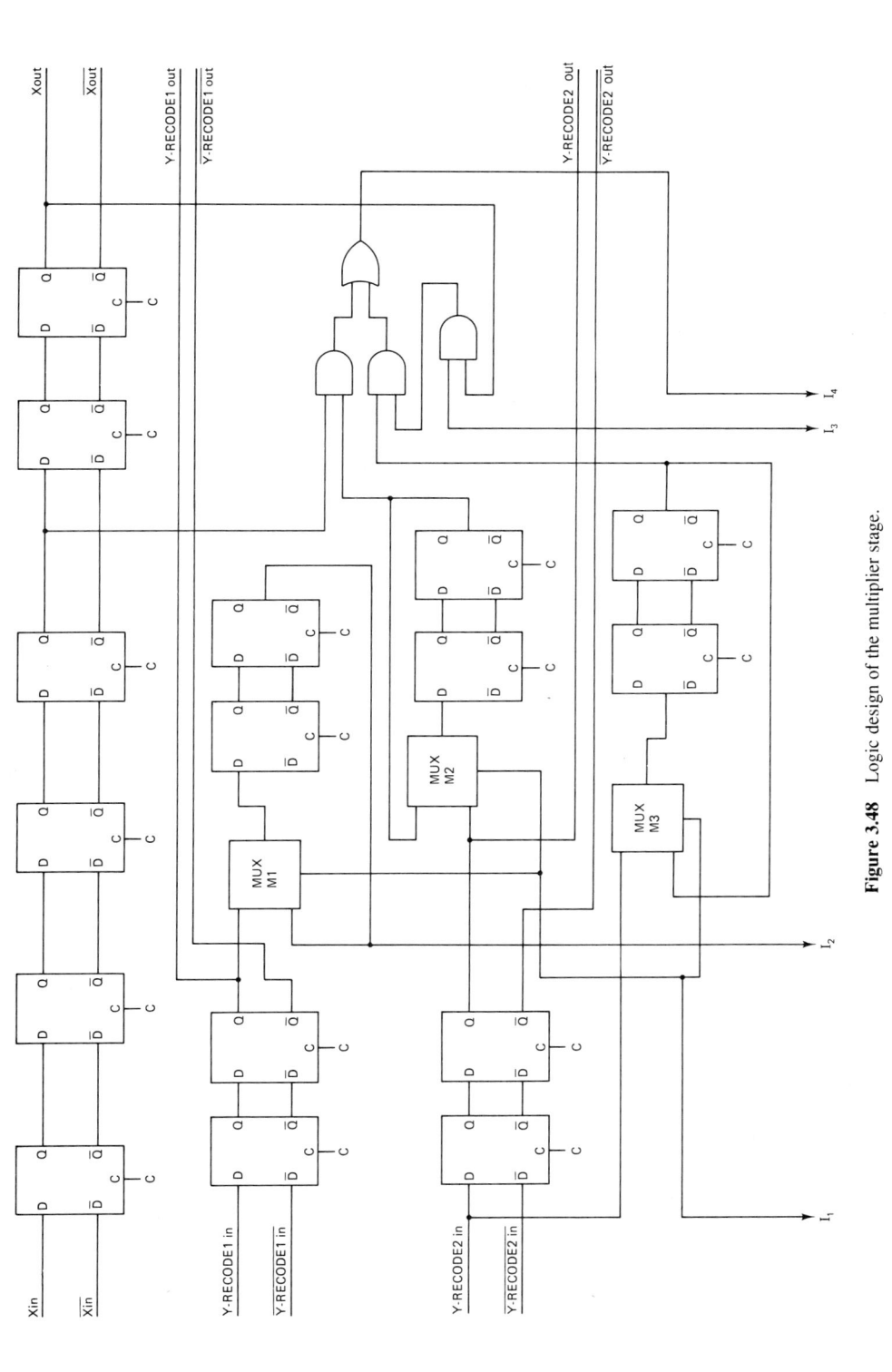

Figure 3.48 Second Part of Figure 3.48

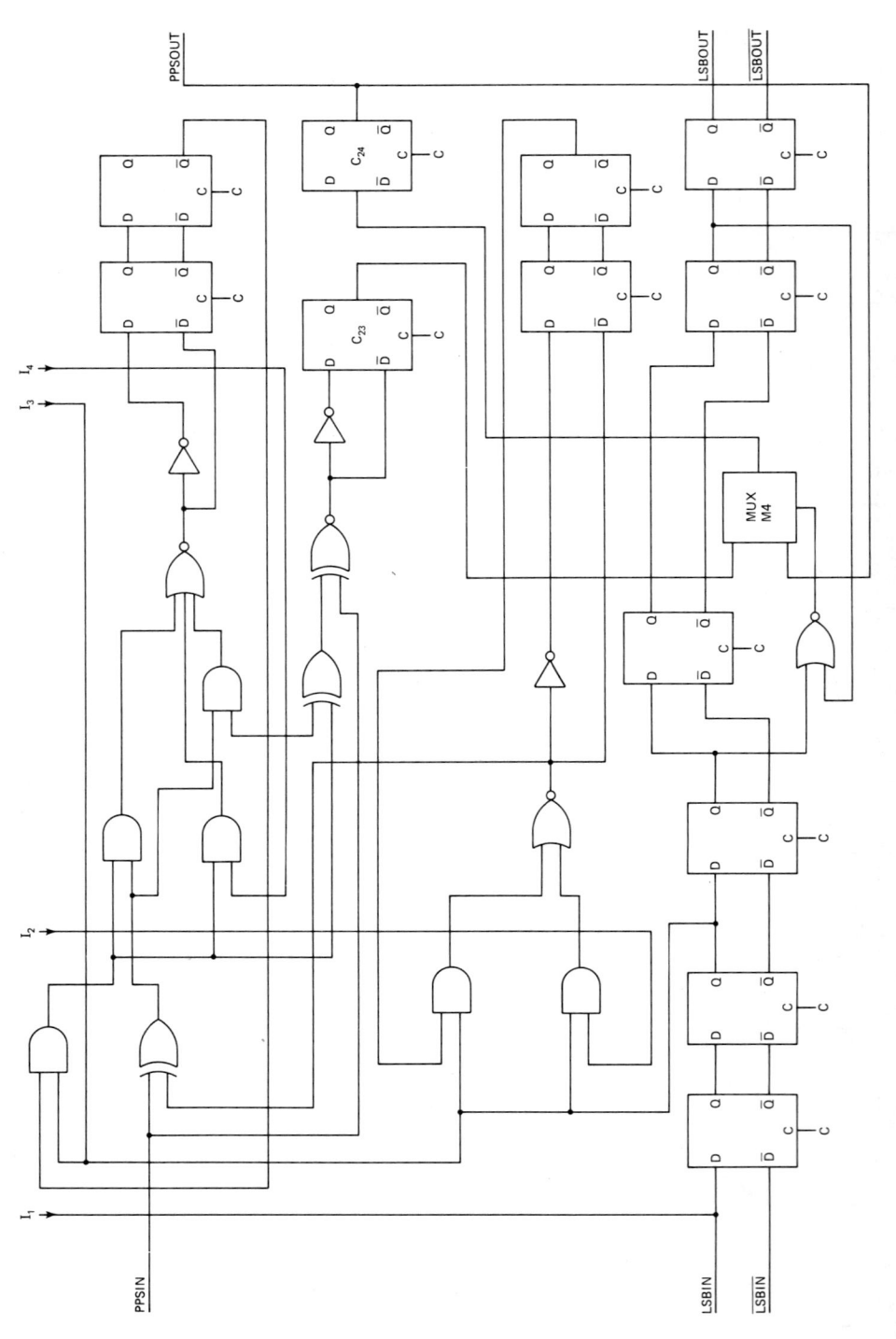

be required. Flip-flops C_{23}, C_{24} and multiplexer *M4* are used to implement this truncation/sign extension logic in every multiplier stage.

A $2n$-bit serial multiplier employing five-level multiplier recoding uses n multiplier stages. This results in a circuit occupying relatively small area. The area is also minimized by the fact that interconnections are minimal in this circuit due to its bit-serial nature. In addition to the product output, this multiplier also outputs the *X* operand without modification. This feature can be very useful especially for functional verification testing purposes. The flip-flops shown in Figures 3.47 and 3.48 are only the master portion of the flip-flop illustrated in Figure 3.1.

3.6 MULTIPLEXERS/DEMULTIPLEXERS

Multiplexers are data selection circuits that route data from two or more inputs to a single output. The data selection can take place in any sequence, one input at a time. A general schematic diagram of a multiplexer is given in Figure 3.49. This circuit configuration has 2^n inputs which are selected using $(n-1)$-bit address, and the output is available in both true and complementary forms. The logic implementation of a multiplexer circuit is rather simple when the data selection address is not part of the circuit and the number of data inputs is relatively small. Figure 3.50 illustrates the logic design of a 4-to-1 multiplexer. This circuit can be implemented as a complex AND-NOR gate, a very efficient approach when using BFL or DCFL technologies. However, the designer should pay attention to the three inputs of the AND gate. These inputs require a dual-gate FET device and a single-gate FET device to be connected in series; this becomes a design issue due to threshold drop, as was discussed in Chapter 2. Therefore, in case of larger input requirements, a different scheme for the data selection control should be used.

In addition to data selection, a multiplexer also can be used as a function generator, which produces an output in specified sum-of-products terms. For ex-

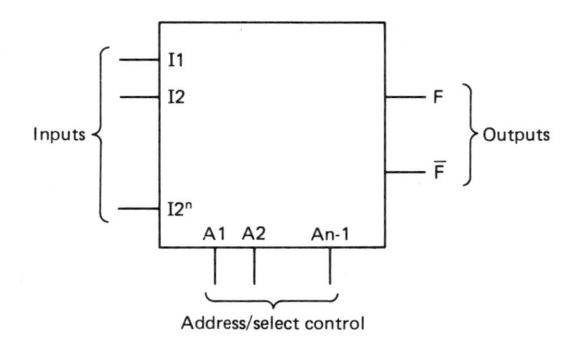

Figure 3.49 General functional configuration of a multiplexer.

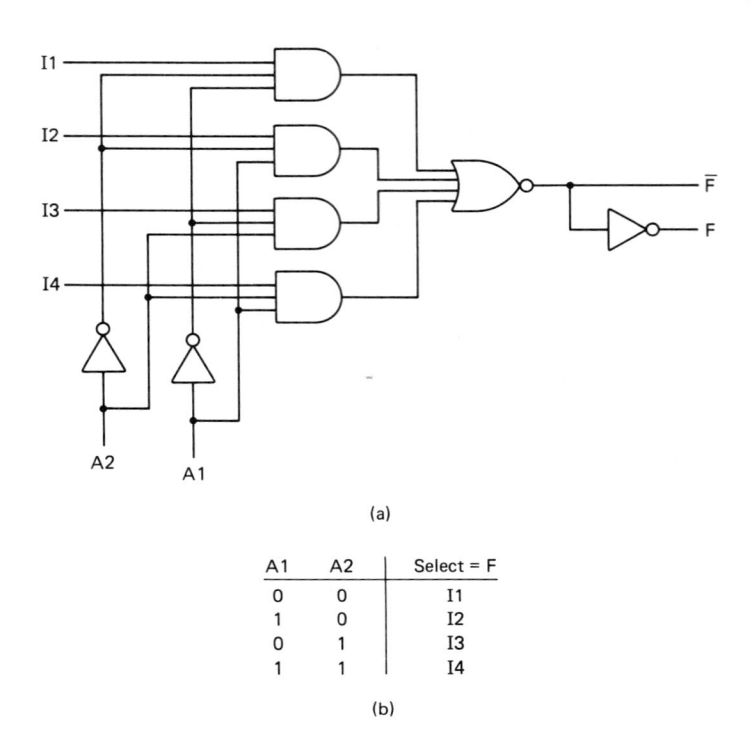

Figure 3.50 Four-to-1 multiplexer: (a) logic design and (b) truth table of data selection.

ample, consider the Boolean expression $F = A\overline{BC} + \overline{AB}C$. This function can be implemented in a straightforward manner by means of two 3-input NAND gates, assuming the A, B, C inputs are available in both true and complementary forms. However, the same function easily can be implemented using the 4-to-1 multiplexer in Figure 3.50 with appropriately connected inputs, as illustrated in Figure 3.51. Although this example illustrates the implementation of a particular function, the fact is that any logic function can be implemented with a multiplexer circuit having one less address input than variables in the logic expression. This, of course, is not to say that every logic function should be implemented with a multiplexer, but this multiplexer technique becomes advantageous and should be considered whenever there are a relatively large number of input variables and a relatively small number of outputs.

While the task of a multiplexer circuit is to select any one of a number of inputs at any given time, a demultiplexer enables one input to appear at any one of a number of outputs at any given time. The logic design of a 1-to-4 demultiplexer is shown in Figure 3.52. The NOR implementation of this circuit is preferred when BFL or DCFL technology is used. If the data input I is not in-

Sec. 3.6   Multiplexers/Demultiplexers   143

$F = A\overline{BC} + \overline{A}B\overline{C} + ABC$

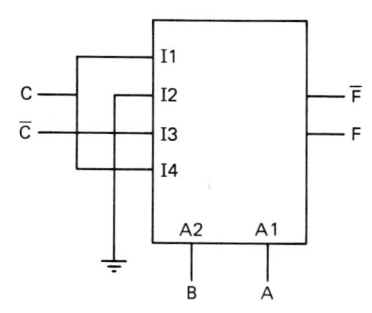

Figure 3.51 Implementation of a logic function using the 4-to-1 multiplexer.

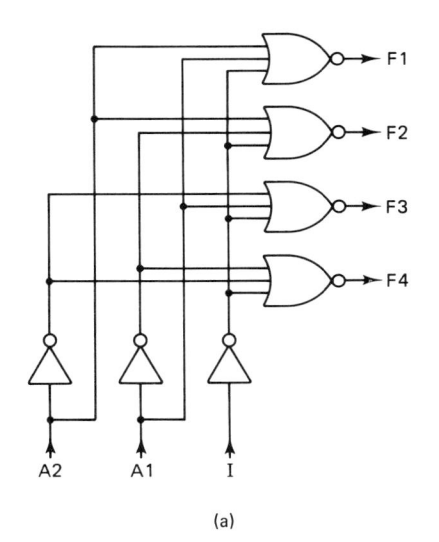

Address		Outputs			
A1	A2	F1	F2	F3	F4
0	0	1	0	0	0
0	1	0	1	0	0
1	0	0	0	1	0
1	1	0	0	0	1

(b)

(a)

Figure 3.52 One-to-4 demultiplexer: (a) logic design and (b) truth table of logic function.

verted before it is applied to the NOR network, then an inverted version of I will appear at the output enabled.

Manufactured GaAs multiplexer and demultiplexer circuits use frequency dividers to produce the address that controls the function of these circuits. The frequency dividers change state with the clock and enable a different path every clock cycle. This time-division multiplex is particularly useful for multiplexing and demultiplexing a number of channels for data transmission over a single path. Various technologies have been used for the implementation of these circuits. In Reference 21, an 8-bit multiplexer implemented in BFL was functionally demonstrated at data rates of 1, 2, and 3 Gbits/sec. However, this circuit does not correctly generate certain bit patterns due to a timing problem which was not entirely understood from the test data. A circuit that combines both the multiplexer and demultiplexer functions is reported in Reference 22. This is also an 8-bit circuit implemented in BFL, successfully tested with data rates of up to 2 Gbits/sec. A 12-bit multiplexer/demultiplexer circuit demonstrated at 1.3 Gbits/sec is presented in Reference 23. This circuit is implemented in DCFL and has very low power consumption, 100 mW.

3.7 MEMORY CIRCUITS

Memory circuits are an integral part of any computer system. Apart from computers, memory circuits are used in all manner of digital logic applications. Memory integrated circuits account for most of the sales of integrated circuits in the international semiconductor markets. The high utilization of memory chips and therefore the high demand in the semiconductor market are the reasons for employing the state of the art design and fabrication techniques to memory circuits first. In fact, memory designs reflect the advancements in semiconductor technology. Since Intel introduced a 1 K Random-Access Memory (RAM) chip in 1971, the memory development has made rapid advancements yielding 1 M bit chips in the late 1980s. Although these very dense memory circuits are implemented with Si metal oxide semiconductor technologies, the development of GaAs circuits parallels this trend. GaAs memories have the highest level of integration achieved in this technology, and memory designs are part of the inventory or the plans of any manufacturer with interest in GaAs technology.

This section will concentrate on two types of memory: the random access memory (RAM), which is widely used in general-purpose computer applications, and the first-in-first-out (FIFO) memory, which is more common in signal processing applications where processing of data is performed in a predetermined sequence. Some functional descriptions are given to illustrate the general principle of operation, but the main concentration of this section is on memory architecture, structure, circuit design, technology, and performance.

3.7.1 Random-Access Memory

There are three basic characteristics of a memory circuit—size, organization, and control. *Size* refers to the total number of bits the memory can store when it is full. *Organization* refers to the way bits are stored in the memory. If a memory has an n-bit address, and the full address space is used, then there are 2^n memory locations, each of which can store a k-bit word, leading to a total memory size of $2^n \cdot k$ bits. Frequently, memory circuits are referred to by a rounded quotation indicating their capacity. Therefore, a 1 K RAM notation implies a capacity of 1024 bits, but it does not give any information on how these

Sec. 3.7 Memory Circuits   145

bits are stored. When, however, a memory is referred to as $A \times B$ bits, then A indicates the number of memory locations while B indicates the number of bits that can be stored in each location. *Control* refers to the way data is read or written into the memory. If any memory location can be read or written at any time, then the memory is classified as random-access. A memory is classified as m-port when data can be read and/or written through m different ports. Although an m-port memory allows concurrent "read" and "write" operations, the RAM function is subject to first-order dependencies; that is, concurrent "read" and "write" operations cannot be executed for the same memory location.

Figure 3.53 shows a typical RAM architecture and organization. The memory cells are arranged in an array of columns and rows according to the memory organization required. The address is divided into row and column segments accordingly. The schematic in Figure 3.53 applies to a single port memory where data input and output use the same data bus. There are two control signals in this case, $R-\overline{W}$ and $\overline{CS}$. The $R-\overline{W}$ control is a single signal enabling both the "read" and "write" operations. A "read" operation is enabled when $R-\overline{W}$ is "high," while the "write" operation is enabled when $R-\overline{W}$ is "low." The chip-select ($\overline{CS}$) control is used to enable the operation of the overall circuit. When $\overline{CS}$ is "low" the memory operations can take place; while $\overline{CS}$ "high" disables the memory by putting the tristate data latches into a high impedance state. The address latch stores the address, indicating the memory location to be read or written. This address is stored for the entire duration of a "read" or "write" operation. The data latches store the data temporarily before it is transferred to the storage cell ("write") or to the outside world ("read").

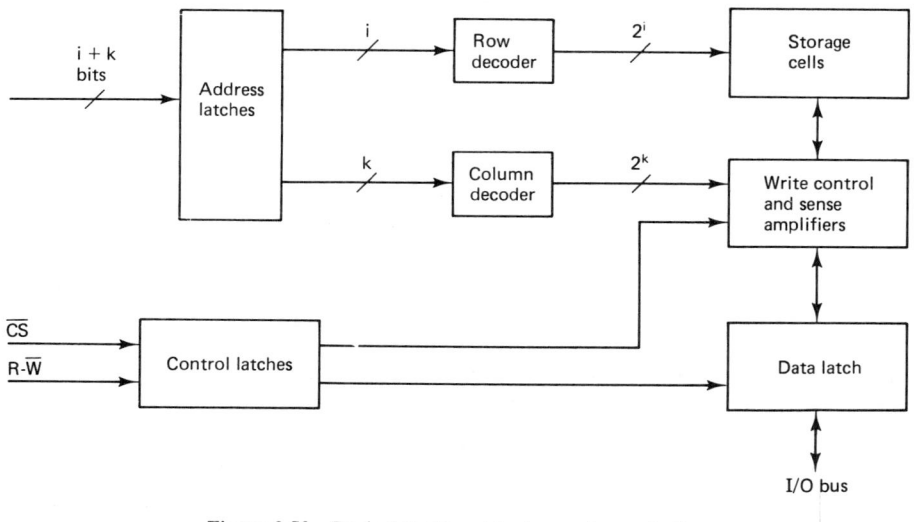

Figure 3.53 Typical RAM architecture and organization.

The row and column decoders enable the memory location, indicated by the address, to be accessed. Depending on the memory organization one decoder circuit may suffice. The sense amplifiers are probably the most critical part of the memory design. These circuits are differential amplifiers that sense the charge in the storage cells and amplify the signal to be stored in the data latch during a memory-read cycle. Given the low bias of GaAs circuits and leakage currents from the storage cells, it becomes critical for the sense amplifiers to be able to sense voltage differences less than 100 mV and subsequently to produce voltage levels required for driving the output latch.

The RAM configuration shown in Figure 3.53 is typical for memory circuits implemented in a single chip. In this case, the $\overline{CS}$ signal is used to enable different sections (i.e., pages) of a memory subsystem implemented using many of these memory chips. If the memory, however, is part of a larger design in a single chip, then the $\overline{CS}$ may not be required.

Unlike Si RAMs, all GaAs RAM designs reported up to this point are static. This is due to the refresh requirements for dynamic memories and their inherent low noise margins. As was discussed in Section 3.2, dynamic operation in GaAs circuits has a minimum clock frequency requirement. Therefore, a GaAs dynamic RAM will require high-frequency refresh cycles, which will greatly increase power requirements. Most GaAs static RAMS are designed using DCFL because this technology can provide the needed complexity with a relatively simple circuit structure. The difficulty with the threshold voltage control, however, can limit the yield of such designs. Static RAMS based on depletion-mode MESFET technology can have higher yield at the expense of a more complicated structure that limits the memory size and can be integrated on a single chip. Utilization of this memory cell results in power dissipation, because the cell operates on a higher current level during "read" than in the static mode.[24] As shown in Figure 3.54, the bit sense line, instead of being connected directly to the storage cell, is connected to it through a diode and the drain of a sense transistor whose gate and source are connected in parallel with FET in the memory cell. The sense FET will turn "on" only when a 0 is stored in the memory cell. In the static mode, the sense FET does not conduct current because either the column is not being addressed (and therefore there is no current supplied to the sense line) or the column is being addressed but not the particular row (i.e., word line). As a result, the "high" voltage on the word line prevents the diode connected to the bit sense line from conducting current. When both row and column are enabled, the word line voltage goes "low" and the state of the sense FET, "on" or "off," will determine the voltage on the bit sense line which is the memory output. The row and column decoding is provided by a NOR array and the output is buffered. A 256-bit static RAM based on this design was demonstrated by Rockwell.[24] The access time of this memory was measured at 5.5 nsec (typical) and 1 nsec (minimum) at 267 mW power consumption. One parameter that is critical for low-power static RAMS is subthreshold leakage current. This is the current that still flows be-

Sec. 3.7 Memory Circuits   147

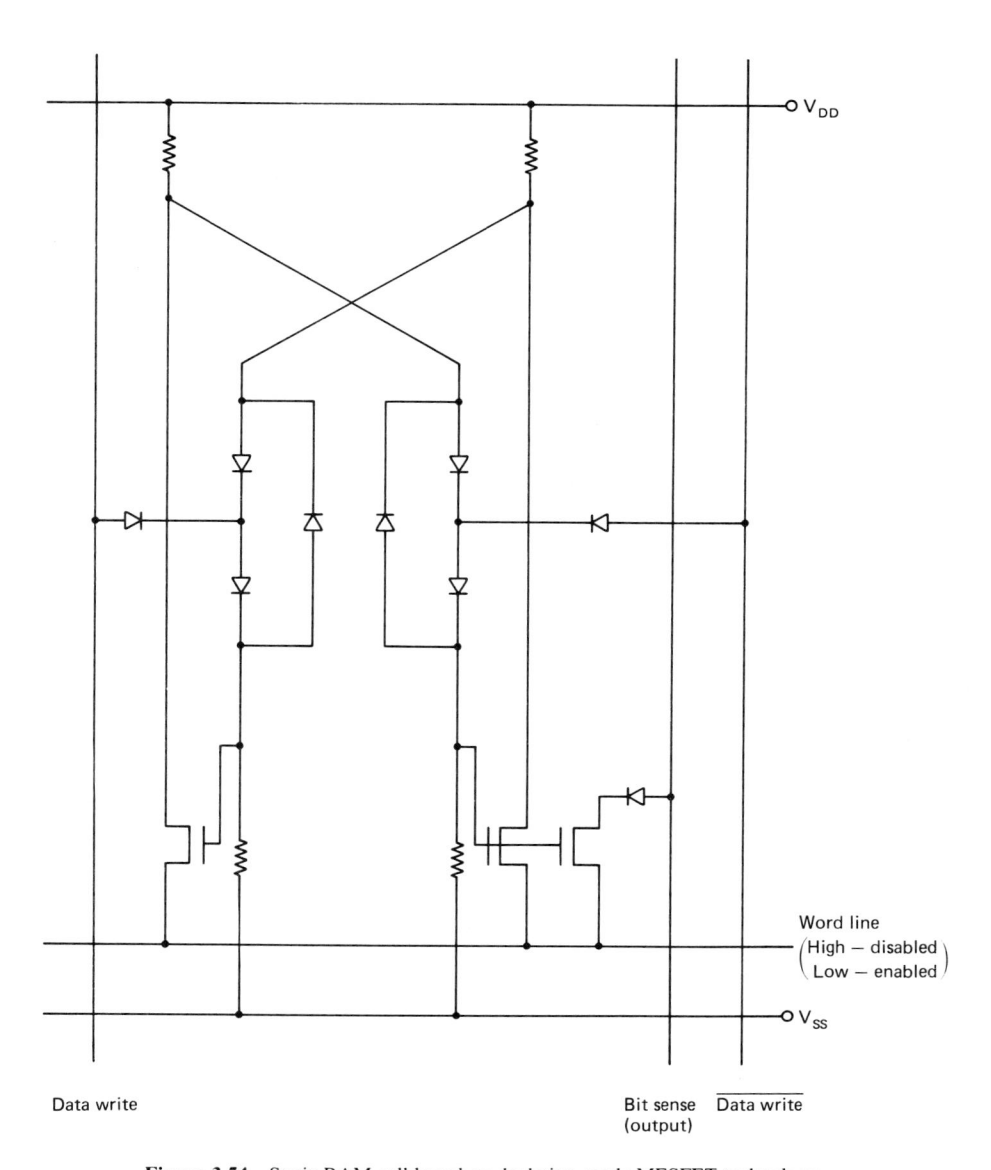

Figure 3.54 Static RAM cell based on depletion-mode MESFET technology.

tween drain and source when the FET should be turned "off." Leakage currents can cause excess current flowing through the pull-up resistors that will result in latching failure of the memory cell. Higher power dissipation will result if the device threshold is increased and the resistor values are decreased. This in turn will result in shorter access times.[24] The device threshold, however, can be increased only up to a point because arbitrary increase of the threshold voltage can cause failure of the cell to latch. Threshold uniformity is also an important parameter for proper circuit operation. Large threshold variations across the circuit will prevent logic gates from properly driving other logic gates.

A memory design of the same size (256-bits), also utilizing pull-up resistors, but n-type JFETS instead of depletion mode transistors, was developed by McDonnell Douglas.[25] The goal for this design, too, is low power. With a best-case access time of 6.5 nsec (10 nsec typical), this memory dissipates only 30 mW and 1.5 mW when it operates in standby mode. Another interesting characteristic of this design is its radiation resistance, which is a very desirable feature for space-based applications. No degradation of the circuit performance was observed at a total ionizing dose of 5×10^7 rads, while the first memory cell errors were observed at a dose rate of 6×10^9 rad/sec of pulsed ionizing radiation.[25] At 10^{10} rad/sec, half the memory cells exhibit errors.

Static RAMS designed using DCFL technology have essential features such as high speed, low power dissipations, and high storage capacity. Although memory organizations vary, the memory architecture and the circuit design of the memory cells and the control circuitry are very similar among different RAM designs reported by semiconductor manufacturers. Figure 3.55 illustrates the basic circuits of a DCFL static RAM design. The memory cell circuit is the same as the well known 6-transistor RAM cell used in static MOS memory designs. The word line address drives the row decoder through a buffer. The decoded signal, in turn, drives the word line through a source follower, which is used because it can drive higher capacitive loads than a buffer with normally-off FETS and, therefore, can provide faster response times. If a source follower is not used and the word line is using E-mode FET pass devices, controlled through a DCFL buffer, the output of the buffer should also be connected to ground through a clamp diode. This approach is taken to prevent the word line drivers from over-driving the pass device E-mode FET in the memory cell. If this pass device becomes forward biased, then forward current is injected into the memory cell resulting in an increase of the "low" level of the cell. As a consequence, the memory cell will malfunction and this malfunction can be manifested as data loss and/or inability to perform the "read" operation. In Figure 3.55, the pass devices in the memory cell are D-mode FETs, and this is the reason for the level shifting diode in the source follower driving the word line.

The sense amplifier design is critical for the operation of the memory. This circuit detects a small bit line voltage difference which constitutes the logic level stored in the memory cell, and subsequently amplifies this small voltage difference to appropriate levels required for driving the output buffers. There are different sense amplifier designs that can be used to satisfy design requirements in terms of performance, power dissipation, area, and noise margins. Figure 3.56 illustrates four sense amplifier circuit configurations.[26] In configuration (a), the (W/L) ratio of the FETs is such that the load current is smaller than the current

Sec. 3.7 Memory Circuits 149

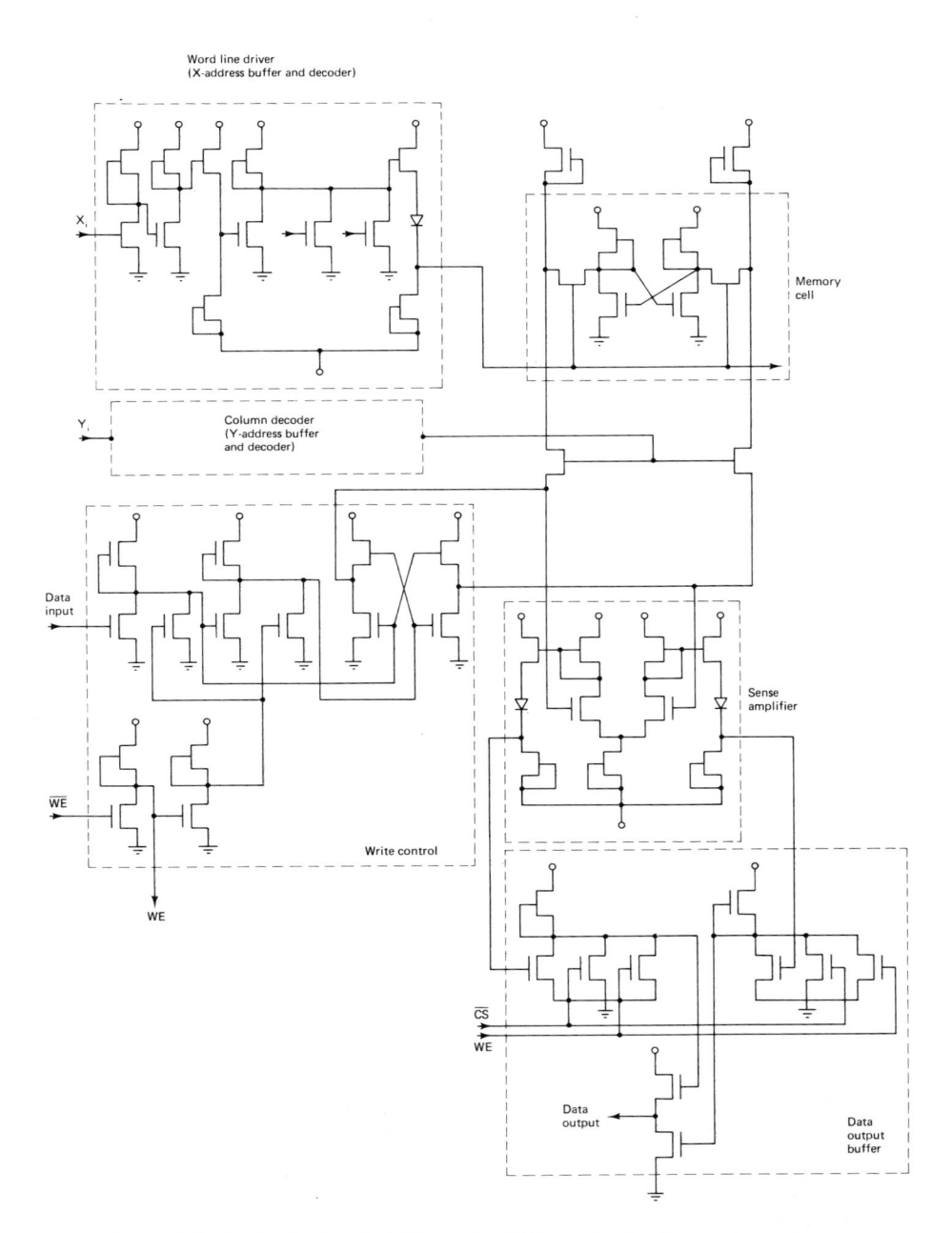

Figure 3.55 Basic circuits of DCFL static RAM (reprinted with permission from IEEE).

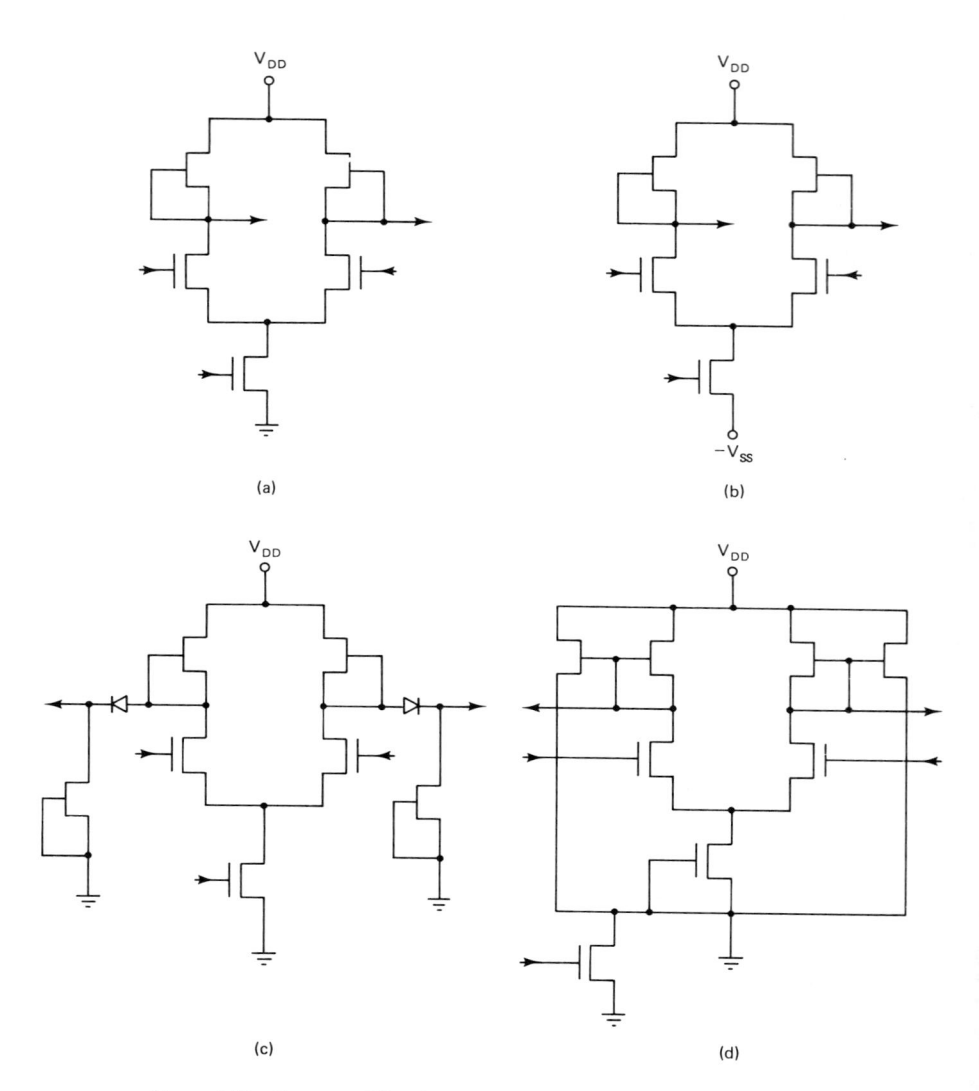

Figure 3.56 Sense amplifier circuits (reprinted with permission from IEEE).

driving capability of the switching FETs. The common source voltage in this case is almost zero independent of the input voltage and, therefore, the circuit in this configuration operates as a source-coupled inverter pair. Configurations (b) and (c) have load and switching FETs with equal (W/L) ratios and thus operate as source-coupled differential amplifiers. In these configurations, the output of the sense amplifier is pulled down, to appropriate levels for driving the output buffer, by biasing the current sink FET to $-V_{SS}$ (b) and by adding output voltage shift diodes (c). In configuration (d), the output voltage is fed back to obtain an optimum sensing condition. This last configuration is well known from its utilization in Si NMOS SRAMS. The voltage levels required for driving the output buffer determine the minimum voltage difference that the sense amplifier must be able to sense for correct circuit operation.

The output buffer is designed as a push-pull stage, using E-mode FETs with large (W/L) ratios so they are capable of driving the maximum load specified by the design and a standard 50 Ω resistance which usually terminates the output.

One of the most effective ways of achieving high-speed memory operation is to reduce the voltage swing at the bit line. This can be achieved by having a large parasitic capacitance at this line. However, a small logic swing means a very small voltage difference at the input of the sense amplifier and thus more stringent design requirements for this circuit. In many applications, a pre-sense amplifer is used at the column to sense the small voltage difference and subsequently to drive the main sense amplifier, which can have multiple stages. Utilization of D-mode FET pass devices in the memory cell speeds up discharging of the bit line capacitance (Figure 3.55) resulting in faster read-out operation. In this case, however, a level, shifting diode is required at the decoder buffer driving the pass D-mode FET, which results in greater power dissipation since power supply levels must be increased to provide the necessary bias.

The memory cell layout and the memory array size affect the loading capacitance on the word line. This capacitance, attributed to fan-out, interconnection line, and crossover capacitances, affects the memory response time and determines the design of the decoder buffers driving the capacitive load. Every effort should be made to reduce the capacitance on the word line to achieve high-speed response from the memory circuit.

The organization of most static RAMS implemented is 1×1 bit; however, there are few circuits with more than one bit per addressable location. Reference 26 states that a 1 Kb static RAM was reported to have an access time of 6 nsec while dissipating 38 mW. This circuit was implemented using a Self-Aligned Implantation for N^+-Layer Technology (SAINT) FET, which demonstrated lower series resistance than the conventional flat FET devices, thus resulting in better speed performance and wider allowable threshold voltage variation. A circuit having the same access time of 6 nsec but lower power dissipation, 30 mW, is reported in Reference 27. This performance is achieved by minimizing the voltage swing at the bit-line to 0.17V. To ensure correct operation with this small voltage swing, this circuit employs differential pre-sense amplifiers, which drive the three-stage sense amplifier. A faster static RAM of the same size and organization as the two circuits above (1024 $\times$ 1 bit) is reported in Reference 28. This circuit, although implemented with longer gate devices (2 µm versus 1 µm) than the previously discussed circuits, has an access time of 4 nsec but dissipates 68 mW of power. Even higher speed (3 nsec) but also higher power dissipation

(1.2W) is reported in Reference 29 for a similar circuit. A 1 Kb static RAM organized as 256 × 4 bits is reported in Reference 30. The performance of this circuit is 4 nsec access time at 160 mW power dissipation.

In most of these circuits the bulk of the power dissipation is attributed to decoder and buffer circuits and very little is consumed by the storage array. This is due to the fact that the control circuits utilize larger devices in order to drive the large capacitive loads on the word and bit lines. In addition to utilizing large devices, the control circuits are also biased at higher voltages than the memory storage array, so they can drive their capacitive load at required speeds.

Larger static RAM circuits implemented in DCFL are reported in References 31, 32, and 33. All of these circuits can store 4 Kb and are organized as 4096 × 1 bit RAMs. Their performances are also very similar, ranging from 3 nsec access time at 700 mW power dissipation[21] to 2.6 nsec access time at 1.8 W power dissipation.[33]

A memory architecture aimed at providing sustained worst-case performance in the presence of different board layouts and clock skews was introduced by GigaBit Logic in Reference 34. This static RAM is organized as a 256 × 4 storage array and uses input and output registers to sample the data before it is stored or retrieved from the array. In addition, a clock signal generated on the chip synchronizes the input and output registers and enables the "read" and "write" signals. The addition of the input registers reduces the sum of the input set-up and hold time from that of the traditional RAM to that of the input register, which is significantly smaller. The addition of the input register and its clocking with the same clock that enables the memory functions allows for a fully-pipelined RAM operation. As proposed, this memory can also be used as a register file. This RAM is implemented using CDFL technology, has access time of 2 nsec, and dissipates 3.4 W of power with an output signal voltage swing of 1.5 V.

The highest performance of RAM circuits in the late 1980s is exhibited by circuits designed and implemented using High Electron Mobility Transistor (HEMT) technology. Although fabrication of these devices, process control, and operating conditions are still in the experimental stage with a multitude of issues and problems have to be solved, the circuits produced demonstrate superior performance. A 1 K × 1 bit static RAM using HEMT devices in DCFL configuration exhibited an access time of 900 psec with operating power of 360 mW.[35] The gate length of the devices in this circuit is 1.6 μm, in constrast to 1 μm devices used in most of the RAM circuits discussed above, and this performance is achieved when the circuit operates at a liquid nitrogen temperature of 77° K. At room temperature, the RAM access time was measured to be 3.4 nsec. A still better performance of a similar architecture and design circuit is reported by Sheng et al. in Reference 36. This circuit has an access time of 800 psec with operating power of 450 mW. The HEMT gate length in this case is 1 μm and this circuit can exhibit such performance at room temperature. However, the produced prototypes yielded only 99.7 percent functionality since three bits failed to function. The functional stability of the memory cell is obtained by keeping the width of the switching transistors constant and then optimizing the width ratio of the load device over the pass device (Figure 3.55).

3.7.2 First-In, First-Out Memory

A First-In, First-Out (FIFO) memory is a read/write circuit which automatically keeps track of the order in which data is entered into the memory and reads the data out in the same order. The memory functions like a parallel-in, parallel-out register whose length is always exactly equal to the number of words stored.

The most common application of a FIFO is as a buffer memory between two digital devices operating at different speeds. Even when two devices operate at the same data rate, it is not always possible for both to be operated synchronously. The FIFO provides the necessary data buffering to achieve synchronization—often a design requirement.

The FIFO design is very modular and is composed of two basic components: the memory cell and the control. These two circuits can be combined to design FIFOs of different sizes and with different organizations. However, as in RAM design, emphasis should be placed on keeping the load that the control has to drive as small as possible, because this load determines the performance of the circuit.

A circuit design using pass devices offers an area-efficient FIFO design at the expense of having minimum frequency requirements for the clock to ensure correct operation (as was discussed in Section 3.2.) Figure 3.57 illustrates the logic design of a FIFO. Data is stored in the shift registers connected so that the output of one feeds the input of the next. Although the number of shift register stages connected in series does not present a design problem other than active area utilization, the number of register stages in parallel demands significant driving capability from the control stage and it is recommended to keep this number smaller than eight.

The operation of the FIFO is performed by clocking each register independently so that data can be selectively shifted through the registers. Each register shifts independently based on the output of the cross-coupled NOR gates associated with each register, which determine whether or not that register contains valid data (see Figure 3.57).

Initially, the FIFO is reset; no data is stored in it. The FULL-j bits are all reset to 0. When the LOAD EN signal becomes 1, a data word can be entered into the first register and the FULL-0 bit is set at 1, indicating that valid data is present in the first register. The FULL-1 bit of the second register is 0, and this causes it continually to monitor the FULL-0 bit of the first register looking for a 1. When the data is stored in the first register, the control sees the 1 and generates a SHIFT-1 pulse which shifts the data from the first register into the second, sets the FULL-1 bit to 1 and resets the FULL-0 bit. The same process is

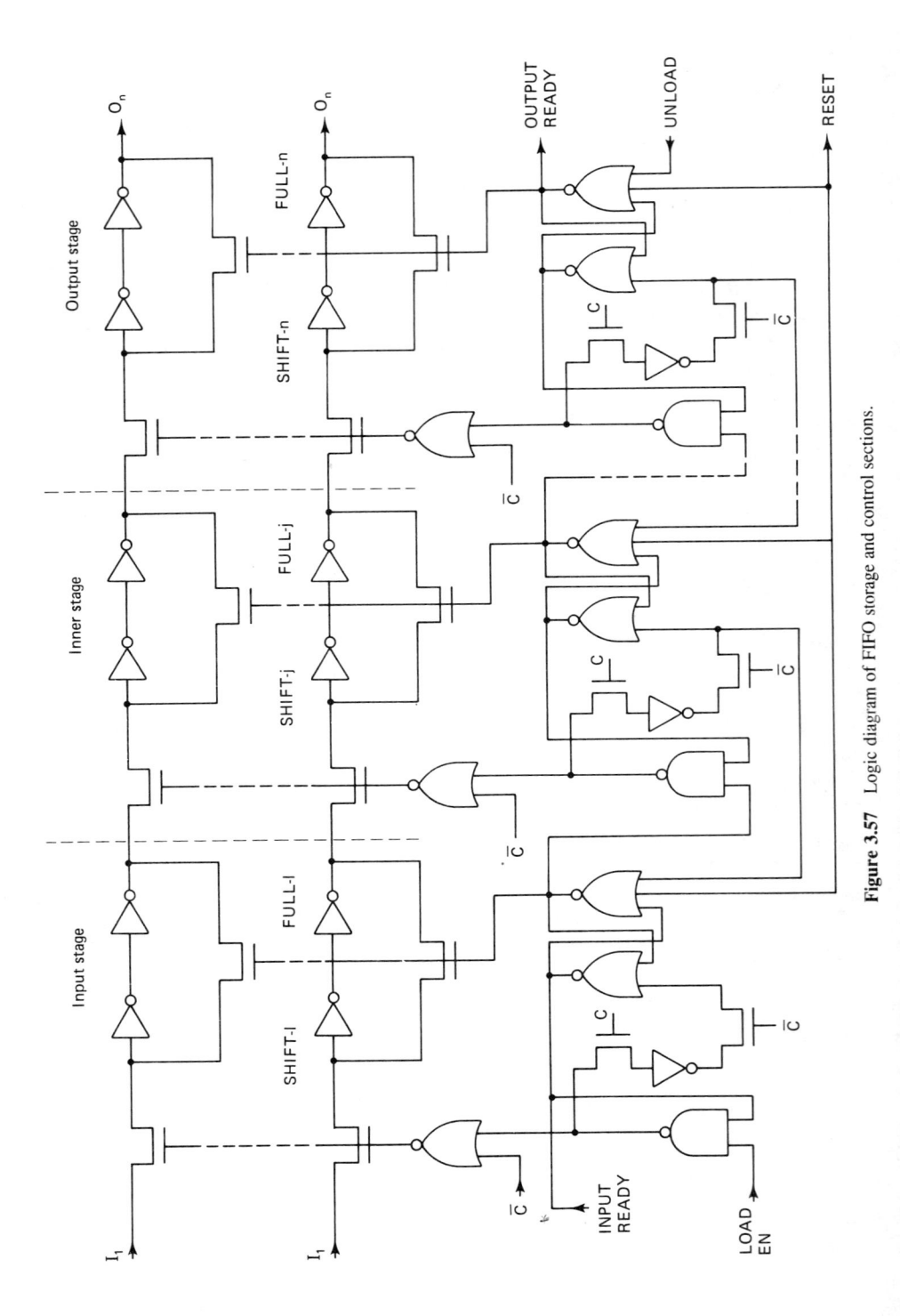

Figure 3.57 Logic diagram of FIFO storage and control sections.

Sec. 3.7   Memory Circuits   **155**

repeated until the data arrives at the last register. At this point, only FULL-n (OUTPUT READY) is set at 1, the others having all been reset as the data was shifted into the next register.

As soon as the data moves from the first register to the second, the FULL-0 bit is reset to 0. A new data word now can be shifted into the first register. The new data shifts through the registers as long as their FULL bits read 0. Eventually the data reaches the register immediately preceding the one containing data and is stored in that register since no further shifting is possible, and the process is repeated until all data words are entered.

When the UNLOAD line at the output goes "high," it causes the FULL-n bit to reset, indicating that the last register is empty. The next to the last word is shifted into the last register and the 0, on the FULL-n line (OUTPUT READY) moves back toward the FULL-0 as the data words move down one register. This process can continue until all data has been shifted out of the FIFO. When the last word has been read, the FULL-n (OUTPUT READY) bit remains 0, indicating that there is no data available at the output.

This scheme allows the reading and writing of data to occur completely independently. Data can be written into the FIFO as rapidly as one write per two clock cycles, after the LOAD EN line goes "high." A timing sequence example for a three word FIFO is shown in Figure 3.58.

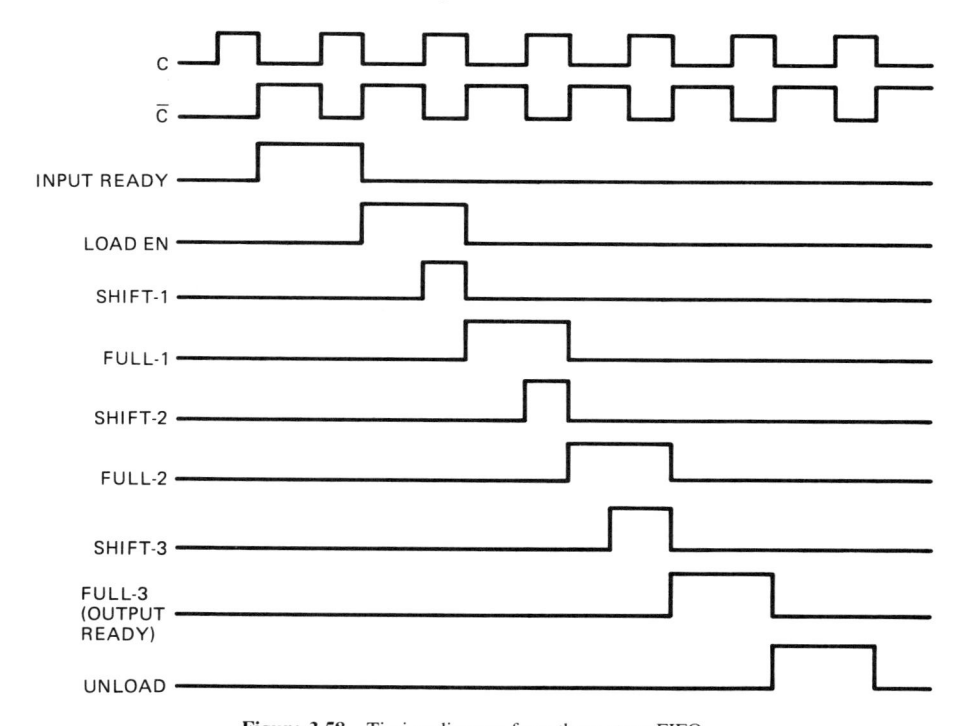

Figure 3.58 Timing diagram for a three-stage FIFO.

The amount of time required for the first data word to ripple through the registers has been defined as data latency. The data latency can be computed by multiplying the clock period by the shift register stages.

FIFO designs have been demonstrated in Si Technologies,[37] but no such designs have been implemented in GaAs as of this printing. However, the FIFO structure is well suited for GaAs implementation, because it is very regular with requirements for short interconnections and minimal fanout requirements as long as the number of parallel register stages is kept small.

3.8 PROGRAMMABLE LOGIC ARRAYS

A Programmable Logic Array (PLA) can be viewed as a nonvolatile memory in which the content is fixed, and the data pattern as well as the address pattern are specified. The output of a PLA is the sum of products of the applied inputs; the products formed according to a prespecified pattern. Therefore, a PLA is implemented as two-plane logic, namely, an AND-plane that forms the partial products and an OR-plane that sums them. The inputs to a PLA are uncomplemented variables. The circuit generates all the complemented variables of the applied inputs internally. Figure 3.59 illustrates the general design of a PLA. The dashes in this figure imply the presence of a FET whose gate is driven by the input signal. All FETs in a column of a plane are connected in parallel, while the output of each column of the AND-plane drives the FETs in the same column at the OR-plane. For simplification purposes, Figure 3.59 does not show line terminating resistors and power supply connections.

A characteristic feature of a PLA is the flexibility of using a product term repeatedly for any of the outputs. PLAs offer a structured design alternative to random logic for implementing multiple output combinatorial functions. In addition, if combined with some registers, PLAs can be used to implement finite state machines. The regularity of a PLA structure is a desirable feature, since it simplifies interconnection schemes and therefore reduces the cost of routing and design verification. However, the speed performance of a PLA is strongly affected by the crossover stray capacitance, so topological optimization is often required to meet performance requirements. The decision for implementing a function with a PLA should be based on performance-cost trade-offs. Cost in this case is the required PLA size that depends on the number of input variables, the number of outputs, and the number of product terms. The minimum number for these parameters can be obtained by using a Karnaugh map or other simplification techniques for a given function to be implemented with a PLA.

A low-power BFL PLA design is reported in Reference 38. This PLA has a NOR-wired-OR organization and takes full advantage of the multilevel logic capa-

Sec. 3.8   Programmable Logic Arrays   157

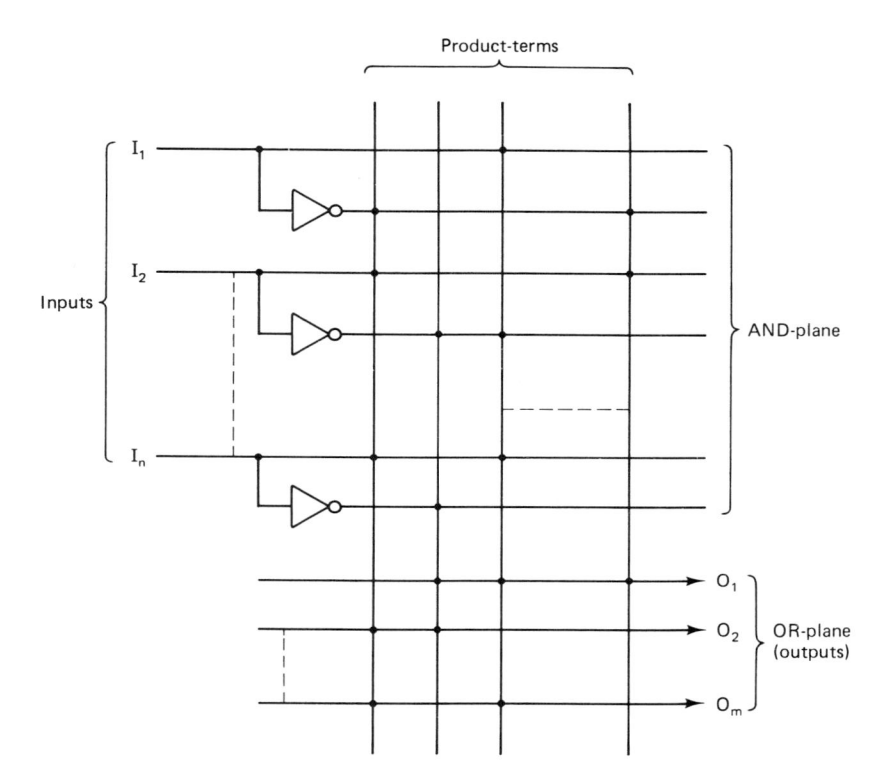

Figure 3.59 General configuration of a PLA.

bility offered by the BFL logic family. Figure 3.60 illustrates the logic and circuit configuration of this PLA design. Note that the output of the AND-plane (i.e., the output of the NOR gates) can be used to perform several wired-OR operations. Topological implementations are required to maintain high performance of this design, because the crossover stray capacitance (i.e., wire-wire, and the FETs loading the line) dominates the speed of this circuit. To demonstrate its performance in terms of area and speed, the authors in Reference 38 designed a divide-by-2 frequency divider using both the PLA approach and random logic. The circuit designed using random logic was tested at maximum clock frequency of 2.2 GHz, while the maximum clock frequency for the PLA design was 1 GHz. The reduced maximum frequency for the PLA circuit is attributed to greater crossover stray capacitance present in the PLA. Although minimization techniques can be used to reduce the number of crossovers, the PLA design will, due to its structure, have more crossovers than the random logic circuit. In terms of circuit area, however, the PLA design was found to be 30 percent smaller than the random logic circuit. This is attributed to the regular structure of the PLA, which requires

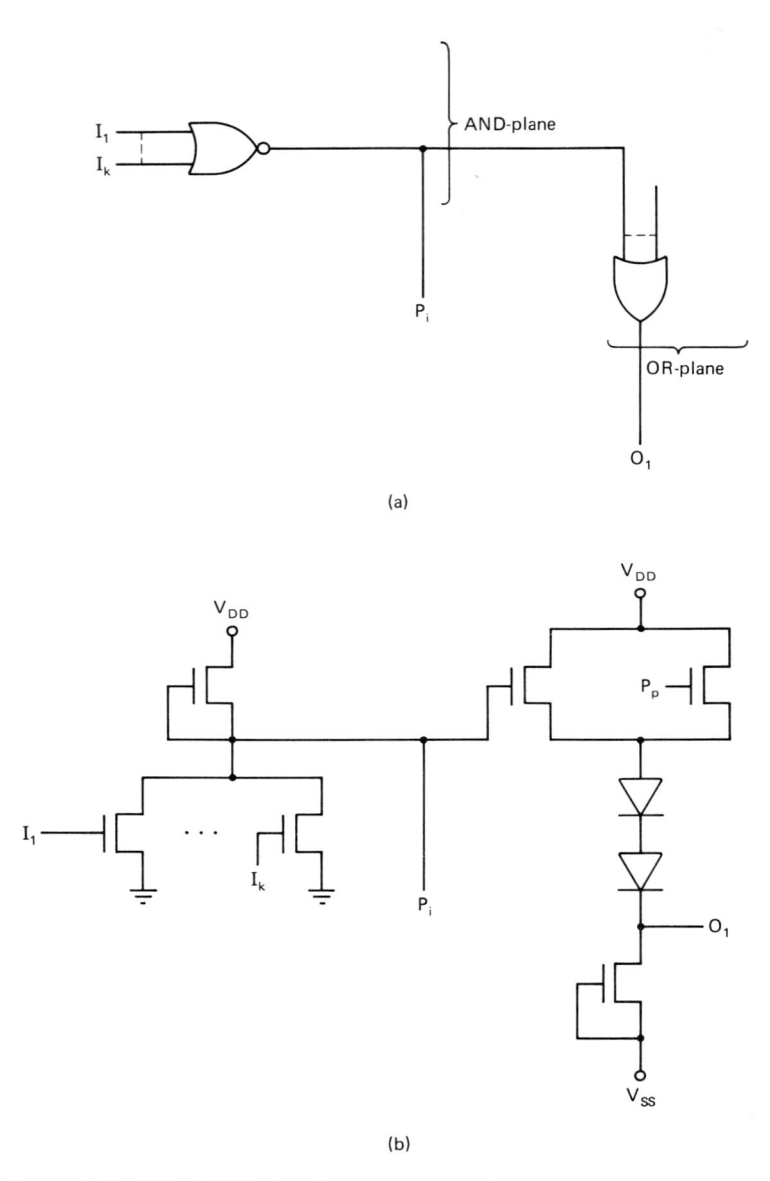

Figure 3.60 NOR-WIRED-OR PLA configuration for one product term: (a) logic design and (b) circuit design.

relatively simple interconnection routing. These comparisons, however, do not represent general results, because the compared quantities depend on circuit complexity and PLA optimization techniques. Further, a major design consideration

Sec. 3.9   Data Conversion Circuits

is the interface of either circuit (in terms of interconnection capacitance and load driving capability) to the rest of the logic that may be included in the same chip. Therefore, the designer should select an approach not only after comparing two circuit configurations but also after determining how each circuit configuration affects the overall design of the integrated circuit under consideration.

3.9 DATA CONVERSION CIRCUITS

Data converters are one of the largest sectors of linear integrated circuits. With information processing technology becoming digital to the greatest possible extent, Analog-to-Digital Converters (ADC) and Digital-to-Analog Converters (DAC) are the core of a data acquisition system, operating as a peripheral to a data processing computer.

Although much work has been performed in the design and implementation of Si data converters, and many commercially-available monolithic data conversion circuits exist, many system applications in the areas of instrumentation, signal processing, and telecommunications require performance levels even higher than what is available today in Si. GaAs data conversion circuits are aimed at addressing these applications with very high performance requirements.

This section is confined to linear converters in which the digital signal is directly proportional to the amplitude of the analog signal. Nonlinear converters are special circuits, offering high signal-to-noise ratio at low signal levels, and are used primarily in telecommunication applications. The design of nonlinear data converters is beyond the scope of this book.

The data conversion process can be described by the expression

$$V_a = V_{ref} (b_1 2^{-1} + b_2 2^{-2} + \ldots b_{n-1} 2^{n-1} + b_n 2^n) \qquad (3.17)$$

which can be applied for both analog-to-digital (A-D) and digital-to-analog (D-A) conversion. In the former case, an analog voltage V_a is converted into an n bit binary number. V_{ref} is, in this case, the "high" logic level of the system.

For D-A conversion, the input $b_1 b_2 \ldots b_n$ is converted into an analog voltage by the scaling factor V_{ref}, which represents a reference voltage. The accuracy of V_{ref} is crucial in obtaining the required resolution for the conversion. According to Equation 3.17, the conversion resolution is ± 1/2 LSB (least significant bit) resulting in a quantization error ± δV_a, where

$$\delta V_a = V_{ref} \cdot 2^{-(n+1)} \qquad (3.18)$$

The sampling of V_a in A-D conversion introduces another error due to the time involved in performing the sampling. This error $\delta_t V_a$ is given by

$$\delta_t V_a = \int_0^{t_a} \delta V_a / \delta_t \, dt \qquad (3.19)$$

where t_a is the sampling time of measurement.

The quantization error, often called quantization noise, computed by Equation (3.19), can be used to define the signal-to-quantization noise ratio (S/QN). The S/QN is given by:

$$S/QN = V_a \cdot (V_{ref})^{-1} \cdot 2^{n+1} \qquad (3.20)$$

As the above expression indicates, S/QN increases linearly with V_a, having its minimum value when $V_a = V_{ref}\ ^{2-n}$ or $S/QN = 2$.

The prime parameters characterizing the performance of data converters are resolution, accuracy, and dynamic response. It is convenient to define three parameters for DACs and subsequently relate these definitions to ADC parameters, which are very similar. *Integral linearity* or *relative accuracy* is defined as the straightness of the transfer function of the converter. It expresses the deviation of points of the converted signal from a reference straight line which is an empirical fit to result in the best linearity specification. The integral linearity expresses the maximum deviation in terms of an LSB fraction.

Differential linearity or *differential nonlinearity* expresses the deviation of the analog output from the ideal for a unit change in input. Ideally, this change is constant and equal to 1 LSB for a single bit change at the input. Differential nonlinearity is measured in LSBs and when it is ± 1 LSB or less, monotonic behavior of the DAC is guaranteed. That is, the signs of both input and output changes are the same. Generally, DACs must be monotonic, especially when used for control systems, because nonmonotonicity can lead to positive feedback and subsequently to loop instability. Zero output change, for some input change, is still considered monotonic.

Integral and differential linearity are related in the sense that the maximum differential is smaller than or equal to two times the integral linearity. If a DAC specification contains information for one of these parameters, then this rule can be used to assume the worst case specification for the other.

DAC settling time is defined as the total time interval between application of a new input code and the point where the analog output settles to within a specified error band around its final value. The most common specification for this value is ± 1/2 LSB. The maximum throughput rate is the maximum number of conversions per second that the DAC can perform.

One operational characteristic of DAC circuits is a transient spike, known as a glitch, at the output when the digital input is switched. This phenomenon occurs because the devices in the circuit do not switch on and off at the same speed. The glitch is particularly severe at half-scale, when the Most Significant Bit (MSB) makes the transition from 011....11 to 10....00. Fast devices and good matching of transistors within the DAC circuit can minimize this transient.

The definition of A-D conversion characteristic parameters is very similar to the definitions given above for D-A converters. Integral and differential linearities are the same as previously defined and most circuit techniques for ADCs guarantee monotonic operation. There is, however, a potential problem—the possibility of missing codes. This phenomenon manifests itself as a change at the output code of more than one bit for a unit increase (or decrease) at the input. An integral linearity of ± 1/2 LSB guarantees no missing codes for ADCs.

The dynamic performance of an ADC circuit is specified by the conversion time, which is defined as the time required to perform a single A-D conversion.

3.9.1 D-A Converters

A simplified block diagram of n-bit DAC is shown in Figure 3.61. The reference current generators are constant sources. Current switches are voltage controlled switches that steer the reference current away from or to the resistor network according to the digital inputs. The resistor network is often a resistor ladder structure that converts the steered current into voltage.

The two most important parameters of a DAC are differential linearity and settling time. The differential linearity depends on the construction and matching of the resistor network and the reference current generators. The resistor network is, in most cases, an R-2R ladder network with accurate resistor ratios, because the resistor ratios determine the accuracy of the conversion. The settling time is primarily determined by the nodal RC time constant at the resistor network and the current switches. This time constant depends on the g_m/C ratio of the FETs used in the current switches, where g_m is the FET transconductance and C is its input capacitance. Figure 3.62 illustrates a general circuit diagram of a 4-bit DAC using depletion FETs.

High-speed D-A conversion is needed for high-speed signal processing and high-resolution display applications. Monolithic Si based DACs have been reported with up to 500 MHz data rates with accuracy in the 8-bit range.[39] GaAs DACs, on the other hand, exhibit performances in the GHz realm with various accuracies.

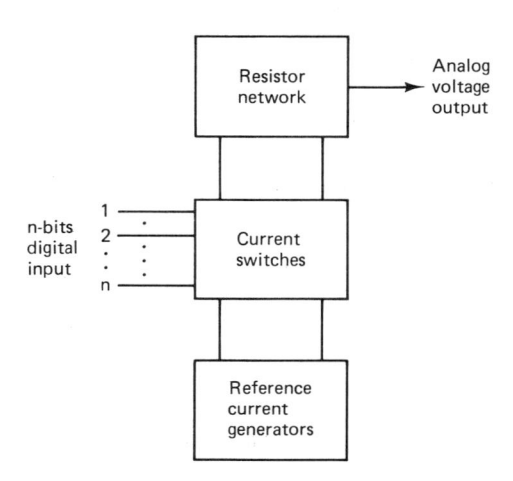

Figure 3.61 Functional block diagram of an n-bit digital-to-analog converter (DAC).

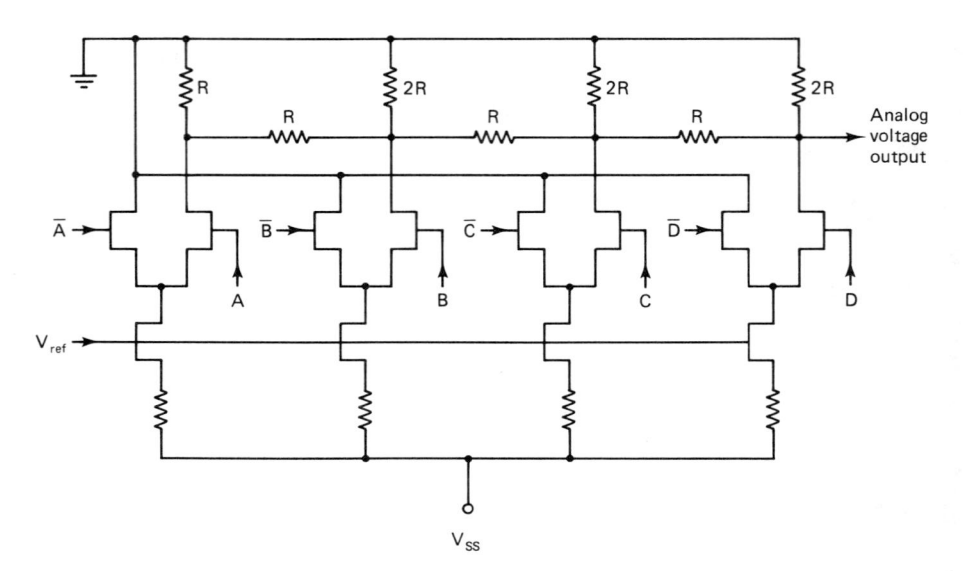

Figure 3.62 General circuit diagram of a 4-bit DAC based on an R-2R ladder architecture.

The circuit schematic of a 4-bit DAC, reported in Reference 40, is shown in Figure 3.63 with the FET widths indicated in microns. This circuit was designed using depletion mode devices with 1 μm gates. The binary-weighted current sources consist of two FETs in series. The lower FET controls the amplitude of

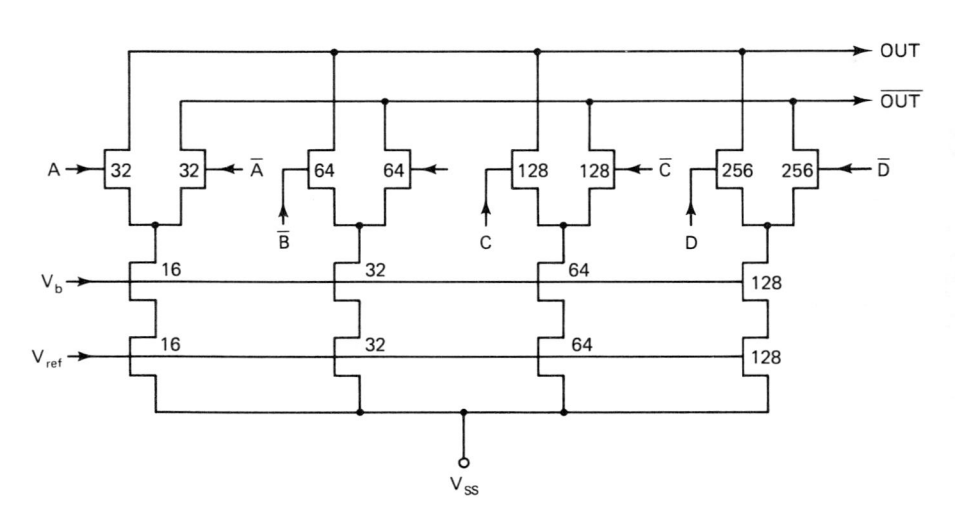

Figure 3.63 Circuit schematic of a 4-bit DAC.

Sec. 3.9   Data Conversion Circuits

the current, while the upper FET desensitizes the current-to-voltage variations at the drain by increasing the drain resistance. The switch network steers the current between OUT and $\overline{OUT}$ under the control of the input code.

With $V_{ref} = Vss$ and $V_b = Vss + 2V$, this circuit settles to within 1 percent of full scale in 1 nsec, thus accommodating 1 GHz transitions of the digital input. DC measurements result in a maximum error of 0.2 percent of the full scale, which is equivalent to an accuracy of 8-bits ± 1/2 LSB.

A higher resolution DAC, designed with depletion-mode FETs, is reported in Reference 41. This is an 8-bit circuit with on-chip current sources, multiplexed ECL compatible inputs, and a fully latched data path. The current sources are implemented using a set of cascaded binary weighted FETs with inverse-binary weighted thin-film precision resistors. A 50 ohm reverse-termination resistor is used to transform the switched current into an analog output voltage. The circuit is capable of settling to within 0.2 percent (±1/2 LSB) of the final value when the input transitions occur at a 1 GHz rate. At this level of performance, the circuit dissipates 3.5 W of power.

An even higher resolution DAC, designed with depletion-mode FETs, is reported in Reference 42. This circuit has 12-bit resolution and includes an on-chip deglitcher. However, the circuit relies on off-chip silicon binary weighted current sources to obtain the required precision. The circuit operates with ECL-compatible differential inputs and has a latched data path. Current buffer transistors isolate the sensitive common source node in the current switch from the capacitance associated with the off-chip current source. The deglitcher samples the DAC output after it has settled, thus blocking the potential glitch present on the edge transition of the DAC's analog output. A block diagram of this circuit is shown in Figure 3.64.

The dc differential nonlinearity of this circuit was measured to be ± 1/8 LSB, while its dynamic differential nonlinearity is ± 1/2 LSB for 1 GHz data rates. The circuit dissipates 5 W of power, of which 2.5 W are attributed to the deglitcher.

All GaAs DAC circuits discussed above have double output termination (i.e., OUT, $\overline{OUT}$) into 50 ohms.

3.9.2 A-D Converters

A variety of circuit organizations may be employed in the design of an ADC. Each organization has particular performance characteristics in terms of speed, accuracy, and area. The major circuit organizations are parallel (flash), serial feedback or successive-approximation, series-parallel feedback, and series-parallel feedforward.

The parallel ADC organization employs $2^n - 1$ comparators for n-bit conversion. All inputs are compared to a certain reference voltage, which is kept constant at each comparator. For a given input, the comparators below that input

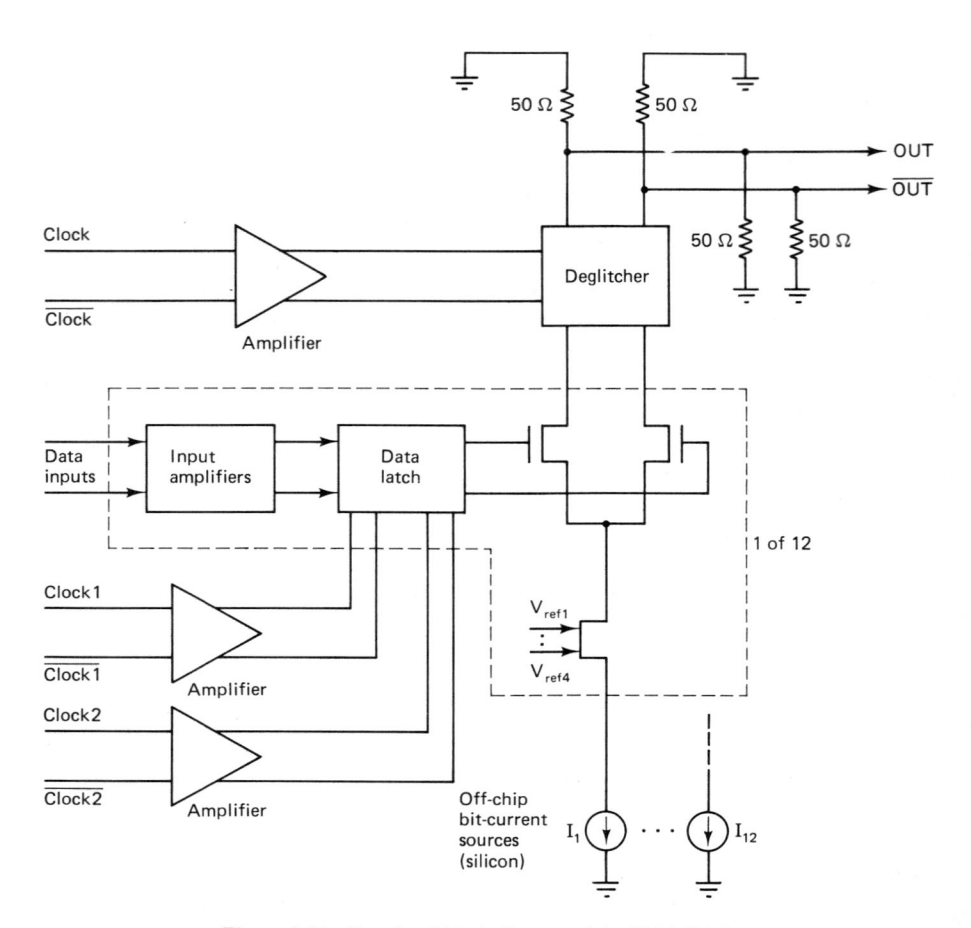

Figure 3.64 Functional block diagram of the 12-bit DAC.

level will all give an output "high" logic level and those above that level will give the complementary logic level. Following the comparators, the parallel ADC employs a decoder to convert from the linear or "thermometer" code at the comparator outputs to a standard code. For some applications, a sample and hold circuit precedes the comparators. However, in many cases, the utilization of this sample and hold circuit may be difficult due to the high input capacitance of the comparators that it has to drive. Figure 3.65 illustrates a specific example of 3-bit conversion with a quantization level of $V_{ref}/8$.

The parallel ADC organization is the fastest approach to A-D conversion, because the speed of the circuit is limited only by the response of the comparator and the propagation delay of the decoder. The comparator is probably the most critical circuit in the converter, because its speed and accuracy determine the

Sec. 3.9   Data Conversion Circuits   165

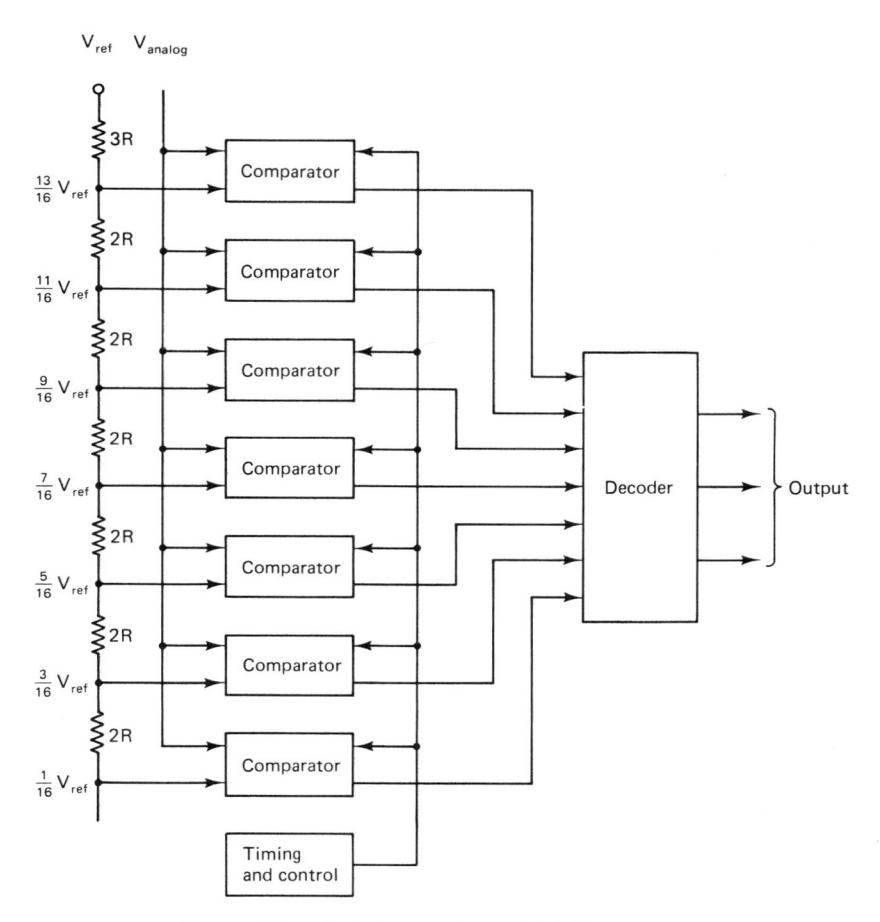

Figure 3.65 Block diagram of a parallel A-D converter.

performance of the converter. The comparator speed has two parameters—namely, setup time and delay time. The sum of these parameters, called decision time, determines the time required by the comparator to determine a valid logic output level for a given difference in the magnitude of the inputs. The resolution of the comparator is also very important because it affects the resolution of the ADC. The comparator resolution is described by two parameters: offset and hysteresis, both of which can be measured from the comparator transfer curve, as illustrated in Figure 3.66. Both hysteresis and offset determine the smallest possible voltage difference which can be reliably sensed or resolved, and therefore the resolution that can be obtained from the ADC. Offset matching between comparators is a major design concern in configuring a flash ADC. The offset plus

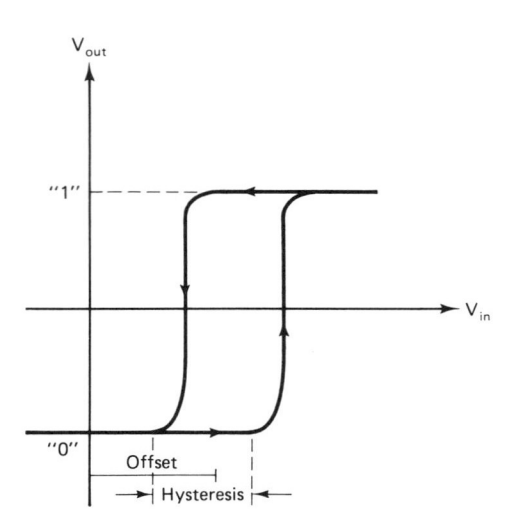

Figure 3.66 Comparator transfer curve.

hysteresis between each comparator have to match within 1 LSB in order to avoid errors like missing codes.

A circuit diagram of a comparator circuit based on depletion-mode devices is shown in Figure 3.67. This circuit has two modes of operation: track and hold. The track mode is active when the clock is "low." During this time, the flip-flop comprised of the two cross-coupled inverters is inactive, the analog input voltages are applied, and the difference is amplified and level-shifted. In the hold mode, the clock becomes "high" and activates the cross-coupled inverters which, combined with the source followers, create an amplifier with positive feedback to produce high gain to help resolve small differences of the input voltages. The comparator decision time is inversely proportional to the gain bandwidth product of the comparator during the track mode. The gain bandwidth product, on the other hand, depends on the transconductance-to-input capacitance ratio of the FETs used.

A BFL comparator operating along the principles discussed above is reported in Reference 43. This circuit was designed as a building block for a 4-bit ADC. This comparator circuit was demonstrated to operate with a clock frequency of 1.8 GHz, having a resolution on the order of 50 mV. A higher performance comparator operating on similar principles and implemented also in BFL is reported in Reference 44. This comparator uses a calibration circuit, driven by a different signal than the comparator clock, that samples the output at designated intervals and generates a compensating signal at the inverting input of the comparator. This internally generated signal combines with an externally provided calibration signal at the noninverting input of the comparator to maintain a dy-

Sec. 3.9 Data Conversion Circuits   **167**

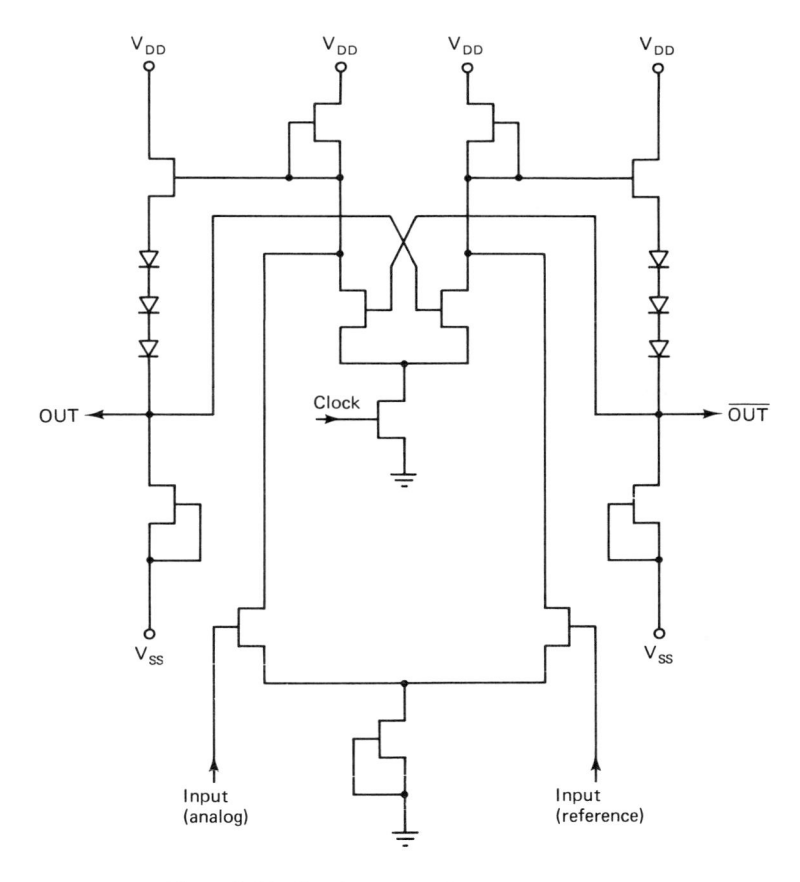

Figure 3.67 Circuit diagram of a voltage comparator.

namic residual voltage at a reference capacitor. At 1 GHz clock rate, the comparator demonstrates resolutions of 1 to 5 mV, quite suitable for utilization in 6- to 10-bit ADC circuits.

A 4-bit flash ADC circuit designed with enhancement mode devices is reported in Reference 45. This circuit utilizes a comparator structure similar to the one presented in Reference 43. NOR gates are used to decode the comparator output and provide the output binary code. The operation of the circuit was demonstrated with a 2 GHz clock rate and differential nonlinearity of ± 1/4 LSB. The typical power consumption of this circuit is 150 mV, its low value attributed to the utilization of E-MESFETs. The performance of this circuit shows promising results for designing flash ADC in the 2- to 6-bit range and for some applications probably to 8-bits. However, the large number of comparators involved will be the limiting factor because of the effect on the ADC accuracy due to their offset and hysteresis.

The successive-approximation A-D conversion technique is based on the functional block diagram in Figure 3.68. The comparator receives the buffered input signal and the output from the DAC. At the start of the conversion, all the bits in the shift register are set to 0 except the MSB. Having the MSB set to 1, the DAC provides a voltage output that represents half of the full scale. If the output of the comparator indicates that the input signal is larger than the DAC output, the MSB remains set; if not, the MSB is reset to 0. At the same time, the second most significant bit in the register is set to 1 and the overall process is repeated until all n bits have been determined.

The successive-approximation is the simplest A-D conversion organization. However, since the output bits are determined sequentially, this configuration offers the slowest conversion time for a given technology. The accuracy of the circuit is determined mainly by the accuracy of the DAC, and in fact the best accuracy that can be achieved with this circuit cannot exceed the accuracy of the DAC.

A variation of the successive-approximation configuration is the series-parallel feedback organization. This organization offers a higher conversion rate than the successive-approximation by using more than one comparator to deter-

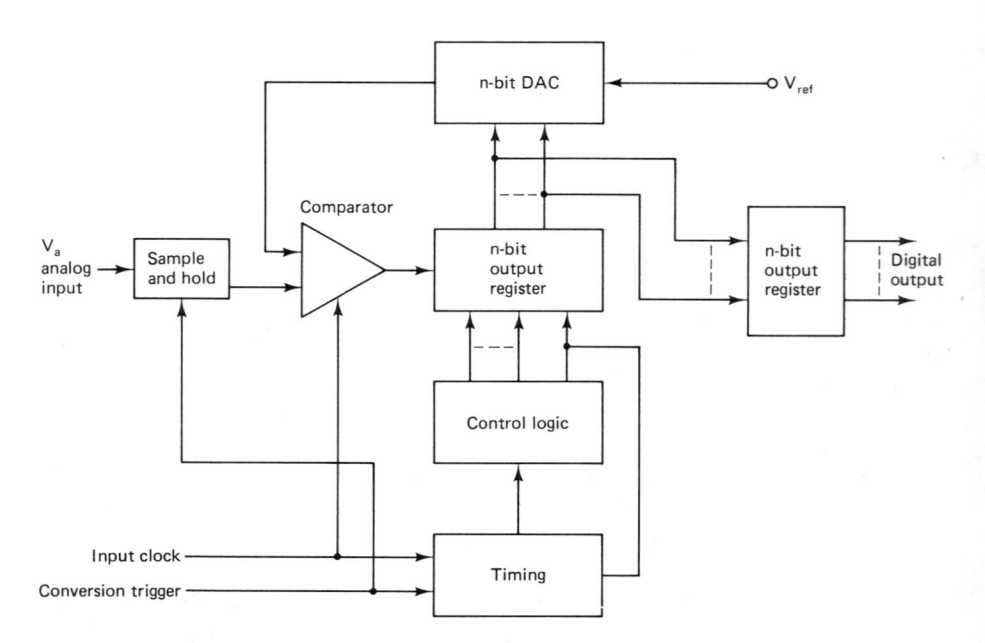

Figure 3.68 Functional block diagram of a successive-approximation ADC.

Sec. 3.9 Data Conversion Circuits

mine more than one bit per clock cycle, therefore requiring fewer cycles to complete the conversion. However, the utilization of more than one comparator increases the complexity of the ADC circuit and also its nonlinearity error. Figure 3.69 illustrates a functional block diagram of a series-parallel ADC configuration.

A high-performance ADC configuration is the series-parallel feedforward organization. It employs a bank of $2^m - 1$ comparators that determine the m most significant bits of an n-bit A-D conversion. This bank is followed by one or more similar comparator banks that determine the remaining $(n-m)$ bits of the final output. Between successive banks of parallel comparators, either the analog signal or the reference voltage, used as the input to the next bank of comparators, is shifted so that the analog equivalent of the partial digital output is subtracted from the original analog input. This causes the latter comparator bank to encode the analog remainder, thus generating the next group of bits in the digital output. This ADC configuration has a significant speed advantage over the series-parallel feedback organization, but it requires the analog signal to be delayed by a time span equal to the conversion time of each stage. The analog signal is delayed by using a sample and hold circuit, which can be a source of error in its own right as well as an added complexity.

A functional block diagram of an ADC using this organization is shown in Figure 3.70. A single comparator per stage is used for this illustration; this is the simplest version of this organization. The series-parallel feedforward ADC configuration is often called a *pipelined A-D conversion scheme*.[46]

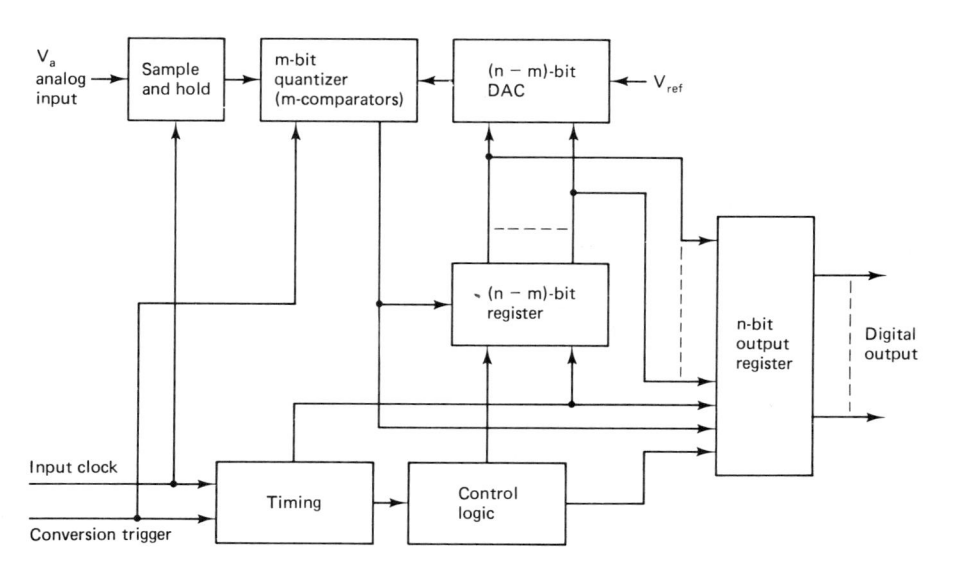

Figure 3.69 Functional block diagram of a series-parallel feedback ADC.

Figure 3.70 Functional block diagram of a feedforward ADC.

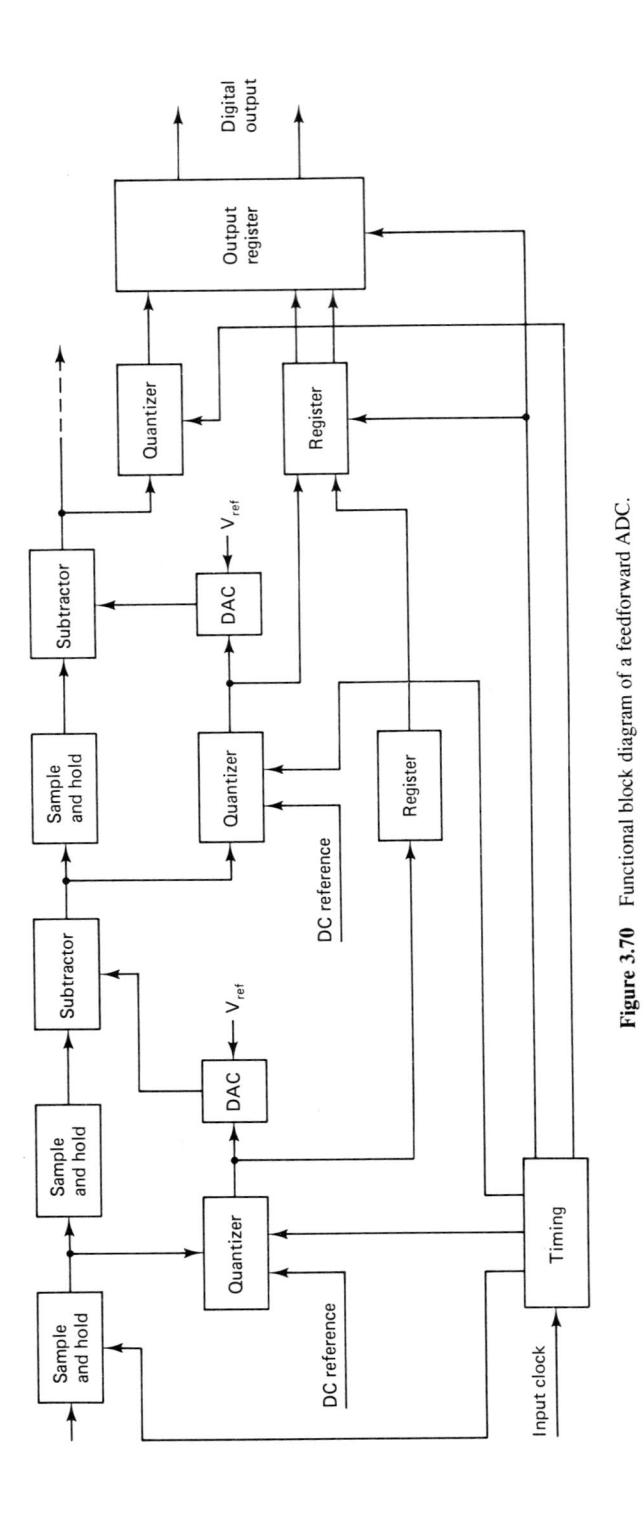

3.10 I/O DRIVERS

The design of I/O interface circuits is a critical task in GaAs technology, because most GaAs logic families use internal logic levels that require translation to and from common I/O standards established by Si technologies. In addition, protection against electrostatic discharge (ESD) is carried out in the I/O stages. If ESD propagates into the internal circuitry unfiltered, it can cause a malfunction or even destroy the circuit.

I/O buffers are often neglected during the design of a circuit and in fact they are, in many cases, the weak part of a GaAs circuit in terms of performance. Although GaAs foundries design standard I/O drivers optimized for their technology and suitable for various interface requirements, in special design situations it may be mandatory to design buffers tailored to a specific implementation. In general, however, it is recommended to carefully select appropriate I/O drivers from a vendor/foundry circuit library that has been fully characterized in terms of parameters that are of interest to the overall circuit and system design. The most critical of these parameters are bandwidth, compatibility with interfacing circuits that are designed using Si technologies, power dissipation, sensitivity to temperature and voltage variations, and stability over the required range of operating conditions.

The primary consideration in interfacing GaAs circuits to TTL and CMOS circuits is the slow rise times of the latter logic families. A slow input transition from a TTL or CMOS gate can cause a fast GaAs input buffer to oscillate or amplify noise during the transition. To minimize sensitivity to slow risetimes, input buffers must have high gain. The main consideration for interfacing to ECL circuits is the worst-case input swing requirements. The low-amplitude voltage swing delivered by an ECL circuit requires a high-gain amplifier at the GaAs input buffer, with an accurate threshold reference to ensure reliable operation. On the other hand, GaAs output buffers must control their output voltage swing to avoid saturating the ECL input drivers, since ECL circuits are sensitive to excessive input swings.

Most GaAs circuits are designed to be ECL-compatible which implies that their I/O drivers can operate with ECL logic levels. To achieve compatibility, the input buffer includes a differential amplifier, with one or more gain stages, which provide the voltage swing required by GaAs. This amplifier usually employs an on-board reference generator that is relatively insensitive to temperature and voltage supply variations. The design of the input differential amplifier depends on the logic family used for the design of the GaAs circuit. A typical design of an input driver amplifier using depletion mode FETs is shown in Figure 3.71. The input signal is capacitively coupled to the device $M1$, and only a small voltage appears at the gate of $M1$ because of voltage charge-sharing between $C1$ and the gate capacitance of $M1$. To maximize the gain of the amplifier, $M1$ can be biased-closed to pinch-off by using the diode $D1$ and the bias across the current source $M2$. However, the actual bias point of $M1$ depends on the ratio of the for-

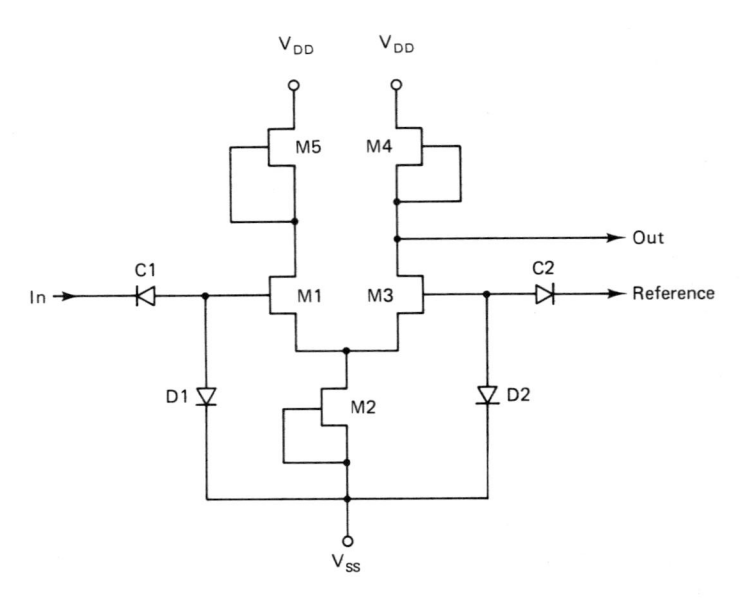

Figure 3.71 Differential amplifier for ECL-compatible GaAs input buffer.

ward-biased current of *D1* and the leakage current of *C1*. Since this voltage depends largely on operating conditions, the reference voltage is derived in a similar fashion, using *C2* and *D2*.

High gain is a very important characteristic of an input buffer, because it reduces sensitivity to slow input slew rates and allows the creation and management of large amplitude clock signals—a difficult task in high performance circuits. Also, stability of temperature and supply voltage variations is essential for reliable and error-free operation of the circuit. All input drivers must include ESD protection, which becomes more essential for small-feature, and therefore low-breakdown voltage, circuits. An ESD protection network is shown in Figure 3.72. This network acts like a low pass filter which prevents sudden spikes in the input signal from reaching the logic circuit. The low pass filter action results from the combination of the resistor *R1* with the capacitance of the reversed-biased diodes *D1* and *D2*. In addition, the two diodes protect the circuit against voltages above certain levels. If the input signal rises above $V_{DD} + V_{D(on)}$ or below $V_{SS} - V_{D(on)}$, where $V_{D(on)}$ is the threshold of the diodes, the diodes become forward-biased and route the signal to either V_{DD} or V_{SS}. Both these diodes are usually fabricated so that their reverse breakdown voltage is lower than the breakdown voltage of the gate of the active logic to assure reliable protection.

ECL-compatible GaAs output buffers can be designed using an open source output driver, which typically drives 50 ohms terminated to −2.0 V. The output buffer offers variable output impedance. Figure 3.73 illustrates the design of a

Sec. 3.10 I/O Drivers 173

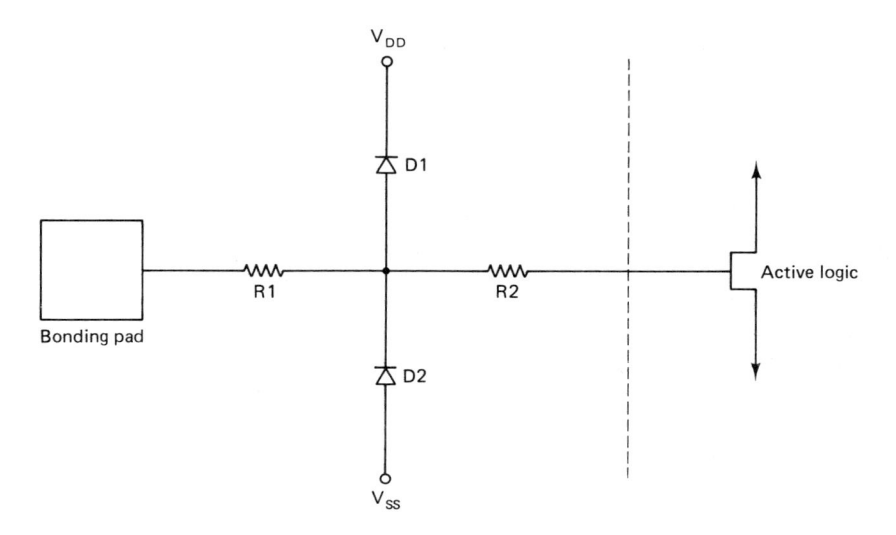

Figure 3.72 Electrostatic discharge protection network.

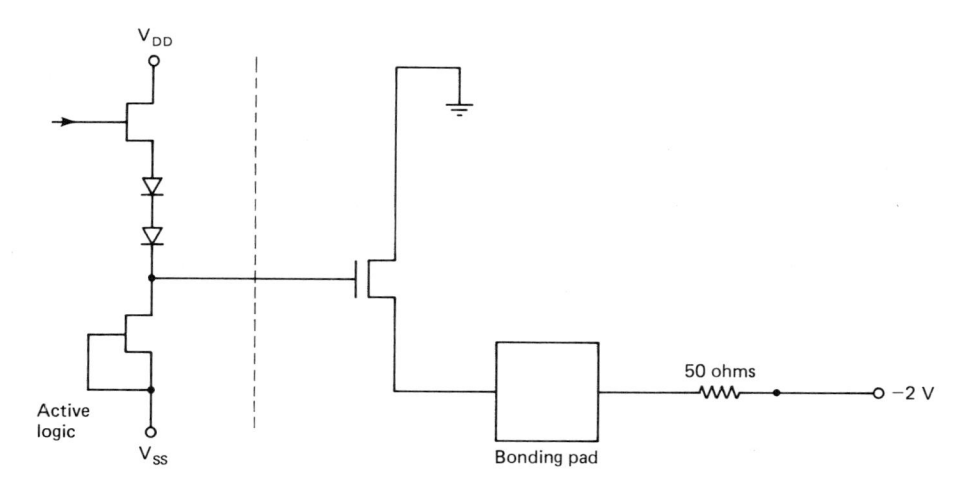

Figure 3.73 ECL-compatible output driver.

typical ECL-compatible output buffer. When the output is high, the output impedance can range from 20 to 50 ohms, while the impedance is larger than 100 ohms when the output is low. The output impedance of the driver is a function of the size of the driver FET. The size of this device also affects speed, power dissipation, area, and noise. Therefore, the design of the output driver must take into consideration all these constraints and the designer must make appropriate trade-offs so the design is optimized for a specific application or class of applications. Output drivers with 200- to 400-μm wide devices are not unusual.

To facilitate interfacing to CMOS or TTL circuits, an external pull-up device connected to a positive supply can be added, thus doubling the size of the driver circuit shown in Figure 3.73. More advanced output drivers compatible with TTL and CMOS circuits use on-board push-pull stages.

Despite the relatively large size of the device in the output driver, ESD protection may still be needed, especially for positive-going transitions (to prevent excessive gate-source reverse bias) so that output current can be limited, thus preventing damage of the output FET.

3.11 MISCELLANEOUS CIRCUITS

In some high-speed circuit applications it is desirable to detect the transition of clocking or control signals. This transition or edge detection can be achieved with a relatively simple logic structure. Figure 3.74 illustrates a "high" to "low" edge detector (a), and a "low" to "high" edge detector (b). The logic principle of operation for these circuits is that during and immediately after the transition, the ungated signal coincides with the previous level of the gated signal and produces a short pulse indicating the transition. As soon as the new level is propagated through the inverter (gated path), a level is established at the output. Although the operating principle of this circuit is simple, its operation is sensitive to input signal slew rate because of the difference in the delay of the two inputs to the NOR (NAND) gate. To verify proper operation, this circuit must be simulated at the device level including parasitic capacitances at gate inputs.

A circuit operating on the same principle as the edge detector is the pulse stretcher, illustrated in Figure 3.75. This circuit will maintain the "low" level at the output, after the input has switched to "low," for as long as it takes for the

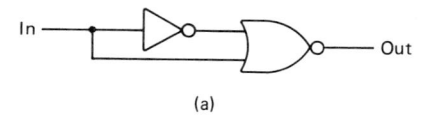

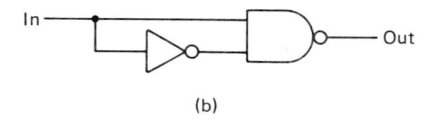

Figure 3.74 Edge detector circuits: (a) "high" to "low" and (b) "low" to "high."

Sec. 3.11 Miscellaneous Circuits

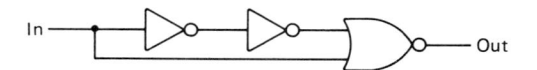

Figure 3.75 Pulse stretcher circuit.

input signal to propagate through the gated path. Since the two inputs to the NOR gate have the same levels at steady state but are established at different times after the transition at the input, this circuit will filter short glitches that can occur at the input. Similar design considerations, as in the edge detection circuits, apply in the case of the pulse stretcher circuit.

A circuit used in low noise phase-locked loops and as part of wide-band frequency synthesizers is a digital phase/frequency discriminator. This is an edge-triggered circuit composed of four flip-flops with common reset. Figure 3.76 illustrates the logic design of the phase/frequency discriminator.

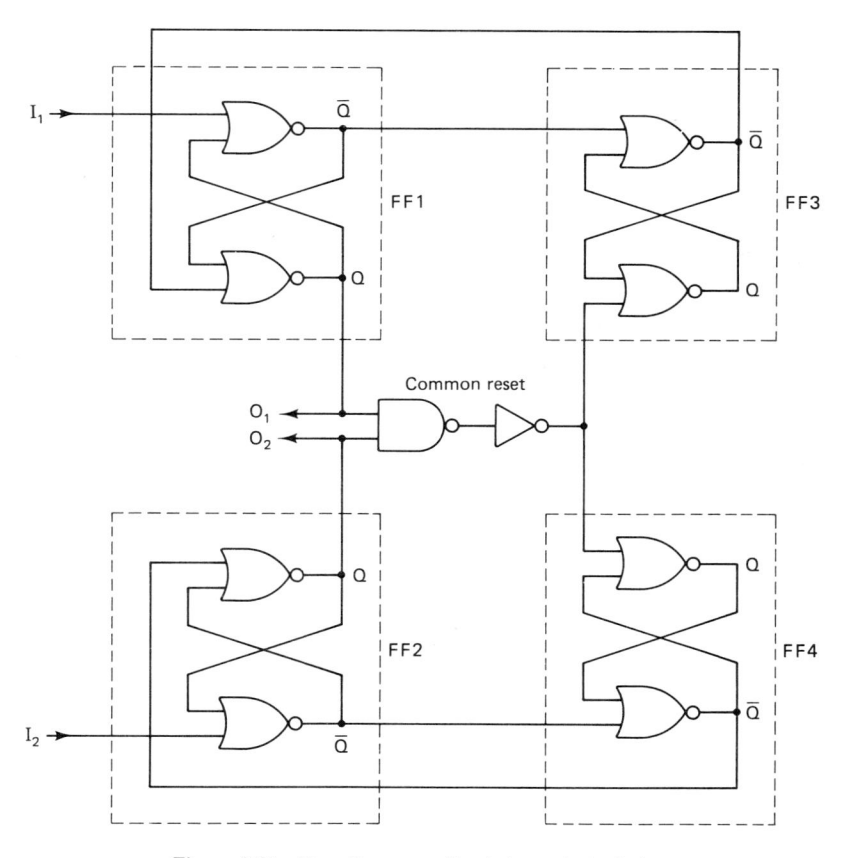

Figure 3.76 Phase/frequency discriminator logic design.

Assuming that flip-flops 3 and 4 are in their set state (i.e., $Q = 1$, $\overline{Q} = 0$), flip-flops 1 and 2 are in their reset state (i.e., $Q = 0$, $\overline{Q} = 1$), and both inputs I_1 and I_2 are "low," then when one of the inputs, say I_1, becomes "high," flip-flop 1 will be set, causing output O_1 to be "high." At this time, the outputs of the other flip-flops do not change and the circuit will maintain this state irrespective of any other transitions of the signal at input I_1. Next, if input I_2 becomes "high," flip-flop 2 is set, causing output O_2 to be at logic one. At this point, both O_1 and O_2 are at logic one, causing the output of the reset gate to be one (i.e., NAND-INVERT), thus forcing the outputs of flip-flops 3 and 4 to reset. This action subsequently forces the outputs of flip-flops 1 and 2 to be low, because the $\overline{Q}$ feedback from flip-flops 3 and 4 acts as a reset of flip-flops 1 and 2. This pulse termination at O_1 and O_2 ccurs whether or not either or both I_1 and I_2 have switched to "low." If either input switches "low" prior to termination of the output pulses, the corresponding flip-flop would be reset, forcing its associated flip-flop to be set.

The discriminator produces output pulses that can drive a differential amplifier to produce a linear voltage versus phase discriminant of the phase relationship between the appropriate edges of the two input pulse sequences. Another important characteristic of this circuit is that the output is a function of the frequency difference of the input pulse sequences. This is a natural consequence since, after a common reset, the higher frequency input will set its associated flip-flop first and the lower frequency input will be the one that will consistently trigger the common reset. As a result of this action, the output of the flip-flop associated with the low frequency will be a sequence of minimum width pulses while the output of the flip-flop associated with the high frequency will be a sequence of varying width pulses.

A phase/frequency discriminator circuit with good linearity and low-noise performance was developed by Hughes.[47] This circuit, implemented in BFL, performs its function with input pulse sequences of up to 480 MHz while dissipating 500 mW of power.

A circuit particularly useful in digital communication between multichannel systems is the cross-point switch. The function of the switch is to allow any input to be connected to any output that is not busy, which is defined as a nonblocking connection. In addition, the switch allows any number of outputs to be connected to one input, which is defined as broadcasting. Finally, the switch has memory in the sense that it allows a cross-point to be set and maintained in this set position until this cross-point is addressed again. Figure 3.77 shows the functional block diagram of a 4×4 cross-point switch. The multiplexers are used to channel the inputs to the appropriate output. The address $A_1 A_2$ selects the input and the address $A_3 A_4$ selects the output. The control signal $\overline{CP}$ closes (logic 1) or opens (logic 0) a cross-point, while the control signal WE enables the latches to be updated. Switches of different sizes can be designed by adding address bits and

Sec. 3.11   Miscellaneous Circuits   177

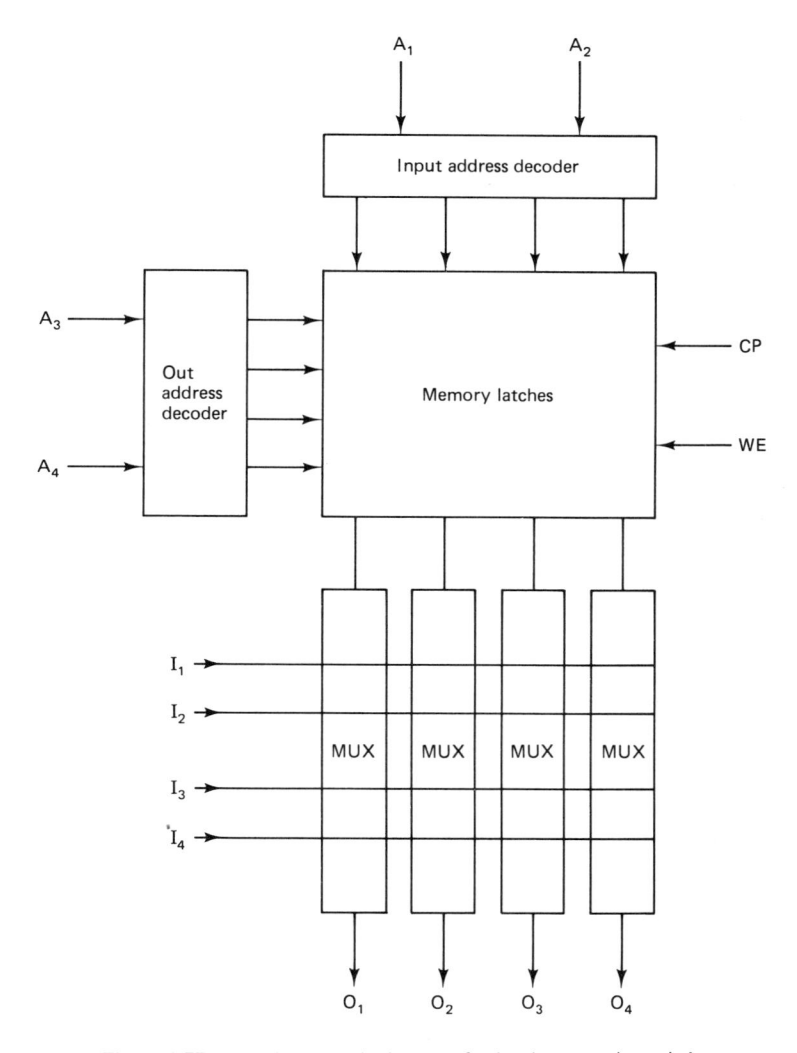

Figure 3.77 Functional block diagram of a 4 × 4 cross-point switch.

latches to store them. An n-bit address is required for a $2^n \times 2^n$ cross-point switch.

A 4×4 cross-point switch designed by IBM is reported to operate error-free with 1 GHz input signals.[48] This switch is designed using 1 μm DCFL technology with source follower gates. Source followers are added to the DCFL gates to increase the gate output voltage swing and thus the noise margin. The 4 × 4 switch design is based exclusively on NOR gates and inverters.

PROJECTS

3.1. Provide the transistor-level description for the flip-flops in Figures 3.1, 3.2, 3.3 using buffered FET logic.

3.2. Repeat Project 3.1, using direct coupled FET logic.

3.3. Using a logic simulator of your choice, can you verify the timing diagram shown in Figure 3.7? Discuss the gate delay propagation assignments you have to make to verify this timing diagram.

3.4. Design a frequency divider-by-two in BFL, SDFL, and DCFL. Provide transistor level diagrams of these circuits so that all of them have the same number of gate delays in their input and output sections.

3.5. Design a mod-4 counter. Provide its logic diagram and counting sequence. If the number of active components (transistors, diodes) is used as the criterion for efficient implementation, what logic family would you use to implement this counter most efficiently?

3.6. Design a 16-bit linear feedback shift register with maximum-length period. The all-zero code must be a valid code for this LFSR.

3.7. Design a circuit that performs the function:

$$A_{i+1} = A_i + B_i, \quad i = 0, 1, \ldots$$

where A_i and B_i are 4-bit sign-magnitude operands. Discuss every decision you make in your design.

3.8. Design a multiplier that multiplies an unsigned number that is available in parallel with an unsigned number that enters the multiplier in bit-serial fashion. Assume 4-bit operands and an 8-bit product. Provide logic simulation results to demonstrate the functional integrity of your design. What are the design differences of this multiplier (called parallel-serial) compared to the array multiplier (Figure 3.45) and the bit-serial multiplier (Figure 3.48)?

3.9. Suggest a logic circuit combination that allows a RAM to operate as a FIFO. Can any RAM circuit operate as a FIFO? Provide a discussion that justifies your answers.

3.10. What is the effect of the device widths (W) on the performance of the DAC circuit in Figure 3.62? Assume $L = 1\ \mu$ and $R = 2$ K ohms and perform circuit simulations to determine this effect.

REFERENCES

1. M. Racchi, B. Gabillard, "GaAs Digital Dynamic ICs for Applications up to 10 GHz," *IEEE J. Solid-State Circuits*, vol. SC-18, p. 369, June 1983.
2. R. Joly, C. Liechti, M. Namjoo, "A GaAs MSI Pseudo-Random Bit-Sequence Generator and Error Detector Operating at 2 GBits/S Data Rate," *Proc. Intl. GaAs IC Symposium*, p. 33, 1982.
3. M. R. Namordi, et al., "A 2.2 GHz Transmission Gate GaAs Shift Register," *ISSCC Digest of Technical Papers*, p. 218, February 1985.

4. F. Lee, P. Miller, "4-GHz Counters Bring Synthesizers Up to Speed," *Microwaves & RF*, Article 2, June 1984.
5. M. Rocchi, "GaAs Dynamic Frequency Dividers for Applications up to 10 GHz," *GaAs IC Symposium Digest*, San Diego, 1981.
6. J. Wakerly, *Error Detecting Codes, Self Checking Circuits and Applications*, North Holland: Elsevier, 1978.
7. C. Meyer, W. Tuchman, "Pseudorandom Codes Can Be Cracked," *Electronic Design*, 23, p. 74, 1972.
8. W. W. Peterson, E. J. Weldon, *Error Correcting Codes*, Cambridge, Mass.: The MIT Press, Second Edition, 1975.
9. E. H. Perea, et al., "A GaAs Low-Power Normally-On 4 Bit Ripple Carry Adder," *IEEE J. Solid-State Circuits*, vol. SC-18, no. 3, p. 365, June 1983.
10. T. P. Ngu, M. Gloanec, G. Nuzillat, "High-Density Submicron Gate GaAs-MESFET Technology," *Proc. 7th European Specialist Workshop Active Microwave Semicond.*, p. 11, Greece, 1981.
11. Y. Nakayama et al., "An LSI GaAs DCFL Using Self-Aligned MESFET Technology," *Proc. GaAs IC Symposium*, p. 6, 1982.
12. R. C. Eden et al., "Multi-Level Logic Gate Implementation in GaAs ICs Using SDFL," *ISSCC Digest of Technical Papers*, p. 122, February 1980.
13. R. Yamamoto, "Design and Fabrication of Depletion GaAs LSI High-Speed 32-Bit Adder," *IEEE J. Solid-State Circuits*, vol. SC-18, p. 591, October 1983.
14. N. Toyoda et al., "4×4 Bit GaAs DCFL Parallel Multiplier," *Proc. IEDM-82*, p. 598, 1982.
15. Y. Kato et al., "GaAs MSI with JFET DCFL," *Proc. GaAs IC Symposium*, p. 182, 1983.
16. A. R. Schlier, "A High-Speed 4×4 Bit Parallel Multiplier Using Selectively Doped Heterostructure Transistors," *Proc. GaAs ICs Symposium*, p. 91, 1985.
17. Y. Nakayama et al., "A GaAs 16×16 Bit Parallel Multiplier," *IEEE J. Solid-State Circuits*, vol. SC-18, no. 5, p. 599, October 1983.
18. F. S. Lee et al., "A High-Speed LSI GaAs 8×8 Bit Parallel Multiplier," *IEEE J. Solid-State Circuits*, vol. SC-17, no. 4, p. 638, August 1982.
19. A. D. Booth, "A Signed Binary Multiplication Technique," *Quarterly Journal of Mechanics and Applied Mathematics*, vol. 2, part 2, p. 236, 1951.
20. L. P. Rubinfield, "A Proof of the Modified Booth's Algorithm for Multiplication," *IEEE Trans. on Computers*, vol. C-24, p. 1014, October 1975.
21. R. L. Van Tuyl et al., "GaAs MESFET Logic With 4-GHz Clock Rate," *IEEE J. Solid State Circuits*, vol. SC-12, no. 5, p. 485, October 1977.
22. G. D. McCormack, A. G. Rode, E. W. Strid, "A GaAs MSI 8-bit Multiplexer and Demultiplexer," *Proc. GaAs IC Symposium*, p. 25, 1982.
23. P. G. Flahive et al., "A GaAs DCFL Chip Set for Multiplex and Demultiplex Applications at Gigabit/sec Data Rates," *Proc. GaAs IC Symposium*, p. 7, 1984.
24. S. J. Lee et al., "Ultra-Low Power, High Speed GaAs 256-Bit Static RAM," *Proc. GaAs IC Symposium*, p. 74, 1983.

25. G. L. Troeger, J. K. Notthoff, "A Radion-Hard Low-Power GaAs Static RAM Using E-JFET DCFL," *Proc. GaAs IC Symposium*, p. 78, 1983.
26. M. Ino et al., "GaAs 1 Kb Static RAM with E/D MESFET DCFL," *Proc. GaAs IC Symposium*, p. 2, 1982.
27. F. Katano et al., "Fully Decoded GaAs Static 1 Kb Static RAM Using Closely Spaced Electrode FETs," *Proc. IEDM-82*, p. 336, 1982.
28. N. Yokoyama et al., "A GaAs 1 K Static RAM Using Tungsten Silicide Gate Self-Aligned Technology," *IEEE J. Solid State Circuits*, vol. SC-18, no. 5, p. 520, 1983.
29. W. V. McLevige, C. T. M. Chang, "An ECL-Compatible GaAs E/D MESFET 1K-Bit Static RAM," *Proc. GaAs IC Symposium*, p. 203, 1985.
30. N. Toyoda et al., "A 256 × 4-Bit GaAs Static RAM," *Proc. GaAs IC Symposium*, p. 86, 1983.
31. T. Mizoguchi et al., "A GaAs 4K Bit Static RAM with Normally-ON and -OFF Combination Circuit," *Proc. GaAs IC Symposium*, p. 117, 1984.
32. M. Hirayama et al., "A GaAs 4Kb SRAM with Direct Coupled FET Logic," *Proc. Intl. Symposium Solid State Circuits*, p. 46, 1984.
33. N. Yokoyama et al., "A 3 ns GaAs 4 K × 1 SRAM," *Proc. Intl. Symposium Solid State Circuits*, p. 44, 1984.
34. A. Fielder, "A GaAs 256×4 Static Self-Timed Random-Access Memory," *Proc. GaAs IC Symp*, p. 89, 1986.
35. K. Nishiuchi et al., "A Subnanosecond HEMT 1 Kb SRAM," *ISSCC Digest of Technical Papers*, p. 48, February 1984.
36. N. H. Sheng et al., "A High-Speed 1 K-Bit High Electron Mobility Transistor Static RAM," *Proc. GaAs IC Symposium*, p. 97, 1986.
37. N. Kanopoulos, J. J. Hallenbeck, "A First-In, First-Out Memory for Signal Processing Applications," *IEEE Trans. on Circuits and Systems*, vol. CAS-33, p. 556, May 1986.
38. E. H. Perea, G. Nuzillat, C. Arnodo, "MESFET PLAs for GaAs LSI and VLSI Integrated Circuits," *Proc. GaAs IC Symposium*, p. 104, 1982.
39. K. Maio, et. al, "A 500 MHz 8-Bit D/A Converter," *IEEE J. Solid State Circuits*, vol. SC-20, no. 6, p. 1133, December 1985.
40. G. S. Larue, "A GHz GaAs Digital to Analog Converter," *Proc. GaAs IC Symposium*, p. 70, 1983.
41. F. G. Weiss, "A 1 Gs/s 8-Bit GaAs DAC with On-Chip Current Sources," *Proc. GaAs IC Symposium*, p. 217, 1986.
42. K. H. Hsieh, T. A. Knotts, G. L. Baldwin, "A GaAs 12-Bit Digital-to-Analog Converter," *Proc. GaAs IC Symposium*, p. 187, 1985.
43. D. Meignant, M. Binet, "A High Performance 1.8 GHz Strobed Comparator for A/D Converter," *Proc. GaAs IC Symposium*, p. 66, 1983.
44. K. Fawcett et al., "High-Speed, High Accuracy, Self-Calibrating GaAs MESFET Comparator for A/D Converters," *Proc. GaAs IC Symposium*, p. 213, 1986.
45. T. Ducourant et al., "3 GHz, 150 mW, 4-Bit GaAs Analog To Digital Converter," *Proc. GaAs IC Symposium*, p. 209, 1986.

46. K. D. Graaf, K. Fawcett, "GaAs Technology for Analog-to-Digital Conversion," *Proc. GaAs IC Symposium*, p. 205, 1986.
47. I. Shahriary et al., "GaAs Monolithic Digital Phase/Frequency Discriminator," *Proc. GaAs IC Symposium*, p. 183, 1985.
48. C. J. Anderson, "GaAs MESFET 4×4 Crosspoint Switch," *Proc. GaAs IC Symposium*, p. 155, 1986.

4

GaAs DIGITAL INTEGRATED CIRCUIT DESIGN PRINCIPLES

GaAs technology is attractive for the design and implementation of high-speed digital integrated circuits (ICs), mainly because of the inherent properties of the material—high electron mobility, high peak electron velocity, and low intrinsic carrier concentration (which yields semi-insulating substrates). This latter property reduces device and interconnection capacitances, which is a requirement for high-speed operation at reduced-power dissipation.

Effort to develop GaAs circuits started in the early 1960s; the objective was to realize bipolar transistors in a GaAs substrate. These efforts did not produce encouraging results, due mainly to low-quality material and the difficulties of introducing dopants by diffusion, coupled with the thermal conversion of the semi-insulating substrates. Thermal conversion means the material becomes electrically conductive when treated in high temperatures. When the thermal conversion problem was solved and ion implantation replaced diffusion for introducing dopants into the material, greater progress in GaAs circuit implementation began. MESFETs emerged as devices suitable for realizing ICs, and they were used to produce the small-scale integration (SSI) and medium-scale integration (MSI) circuits that are commercially available today. Large-scale integration (LSI) complexities have been exhibited by prototype circuits, and commercial standard parts of this complexity are currently being fabricated. The advantages of higher levels of integration are understood, and they are highly desirable by system designers. However, for each GaAs logic family, there appears to be an optimum level of integration that acts as an upper boundary to speed improvement, that the technology can offer to systems design.[1] Beyond this limit, system speed may actually degrade.

Several issues should be considered when designing GaAs digital ICs. Interconnection of devices becomes crucial because gate delays may be comparable to delays in the interconnections. Although on-chip speed is very high, this performance advantage can be lost at the system level by the interchip connections, which are significantly slower. Therefore, a different approach to on-chip architectures is required than with Si-based VLSI ICs. Exploitation of on-chip gate speed will be optimum when the functionality of the chip is increased for a given system application, resulting in the implementation of the intended system function with as few chips as possible. Furthermore, optimum utilization of the advantages offered by the GaAs technology requires a holistic approach to system design: The architectural design of the system should be considered as a whole, including algorithms, architectures, and device performance. The IC designer also should consider in particular which chip architecture can achieve the performance required without exceeding the allocated power budget. Design should allow for interfacing with other technologies. Fast rise times generated by GaAs circuits and small logic swings make their interface with Si ICs difficult and the design of on-chip I/O drivers crucial.

One of the most important issues in the development of GaAs ICs is functional verification testing. Since GaAs ICs outperform test equipment capabilities, at-speed testing is often a problem. Furthermore, if no provisions are taken during design for testing the prototypes, functional verification can be a formidable task. As the levels of integration increase, the design of digital GaAs ICs must include testing as a design requirement of equal importance to performance or power dissipation.

This chapter presents IC design methodologies and tools used to perform various designs. It also offers some practical guidelines for making certain design decisions and resolving design problems. An example of a chip design from concept to functional verification testing is also presented. Finally, specific architectures that are attractive for GaAs IC implementation are discussed in some detail.

4.1 CHIP DESIGN METHODOLOGIES

Three generic design methodologies apply to the development of high performance GaAs digital ICs. These design methodologies are: custom design, standard cell design, and gate array design. All three approaches have been used to design digital GaAs chips and the selection of one approach over the other depends on the performance, schedule, and budget requirements the design engineer has to fulfill.

This section presents the characteristics of each design methodology and the trade-offs to be made for selecting each design approach.

4.1.1 Custom Design

Custom design is the first and best-known methodology for GaAs IC design. This approach is called *custom* because the designer has the freedom to design the intended structure for the circuit using transistors and diodes of any size, placed at any position allowed by the design rules, and interconnected in any way that does not violate the design rules. Following this approach, the designer can take maximum advantage of the technology, thus producing a circuit that meets the required speed, power, noise margin, and I/O driving specifications while maintaining the best possible utilization of GaAs area. Minimizing the area occupied by the circuit is important because the chip die size affects the process yield. Utilizing the fewest devices for performing a required function results in small area die as well as low power dissipation. Sizing each device independently yields faster circuits than in cases where sizing restrictions apply.

On the other hand, custom design is a laborious process, requiring good knowledge of circuit design and understanding of the GaAs fabrication process and device characteristics. Since the design is essentially performed at the transistor level, the design time is the longest compared to any other design approach. The final design has to be verified exhaustively before fabrication because there is no utilization of circuits known to be functioning correctly. This process adds time to the design cycle and, if not done properly, can result in the release for fabrication of chips containing design errors. The implication of long design times is, of course, high development cost, which is an important parameter in a competitive semiconductor market.

Custom design is used mainly for SSI and MSI circuits where the number of transistors yields acceptable design time and a manageable design verification process. As the level of integration increases, custom design is used to design a few basic circuit blocks and then these blocks are repeatedly used to complete the overall design. A memory circuit is a good example of this. The storage cell can be custom designed and then repeated thousands of times to yield the storage section of the memory circuit.

As both level of integration and demand for GaAs chips increase, there are requirements for rapid turnaround, and the custom design approach may not be acceptable—especially for circuits to be manufactured in relatively small quantities. For this type of circuit, standard cells or gate array approaches will be more attractive, mainly because they offer shorter turnaround times at lower cost. Custom techniques, however, will continue to be employed in the design of memory and microprocessor components where the chip quantities produced can justify the development cost. Finally, custom design will always be used for special circuits where performance and reliability outweigh any other parameters.

4.1.2 Standard Cells

The standard cell approach is the best way to achieve maximum speed performance short of a custom chip design, without paying the penalty of long turnaround times and high development costs. This approach to the IC design is similar to board design using off-the-shelf standard parts. The standard cells are functional building blocks of various complexities that are custom designed at the transistor level, and their function and performance have been verified through implementation and testing. The functions implemented by the standard cells are chosen so that their applicability covers a wide spectrum of applications and their implementation is area-efficient in the technology selected. The performance characteristics and the electrical parameters of each cell are known to the designer. In most cases, the same function is available in several cells that have the same logic design but different performance and, thus, different area and power requirements. In most cases, standard cells are laid out so that all have a standard dimension (i.e., either standard height or standard width). This facilitates efficient layouts because the cells can be abutted at the standard dimension when they are connected. To achieve this layout effectively, in addition to the standard dimension, standard cells place the power and often all input and output signal lines at standard distances from the edge of the standard dimension designated as alignment edge. If the input and output signals are available on the side of the nonstandard dimension, then interconnection of the cells requires external connections for these signals, although the cells are abutted at the side of the power lines. Additional flexibility is provided when the input and output signals are available on both sides of the same dimension of the cell. In this case, the cell can be rotated 180 degrees to facilitate interconnection with other cells without disturbing the abutting of the power lines. An example of the standard cell concept is shown in Figure 4.1.

The individual standard cells are essentially custom designs of all mask levels, but their layout and function remain constant over time. Therefore, their layout can be stored in a database, usually called a standard cell library, and can be recalled for repeated use. A number of standard cells can be interconnected to realize a complete functional entity on a chip. Since the function and performance of the cells have been verified, the designer need only verify the correctness of the interconnection and the performance of the cells after they have been interconnected. Although the designer knows the speed and driving capabilities of each cell along with its input and output capacitances, if the effect of interconnection is not taken into account properly, the performance of interconnected cells can violate design specifications while the individual cells meet those specifications.

The standard cell approach provides a low-risk, rapid-turnaround method for designing high-performance ICs. The penalty one pays, however, compared to custom design, is in area utilization and maximum performance allowed by the technology. Standard cell designs occupy more area than custom designs, for two reasons. The first reason is that one of its dimensions is kept constant independent of the circuit complexity in a cell and this can lead to suboptimal utilization of GaAs area. The cell structure also affects the interconnect patterns; this is another area-consuming factor attributed to the constant layout of the cells. The

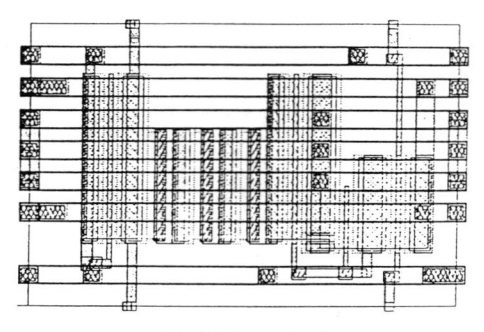

(a) A high-speed cell

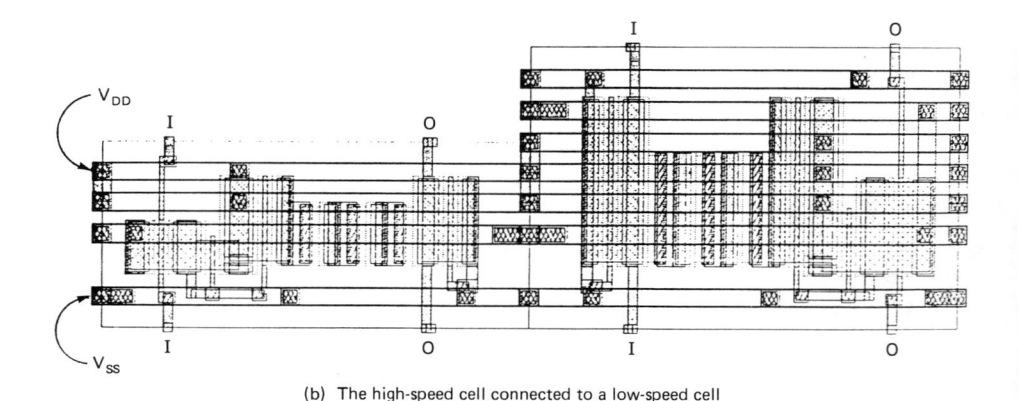

(b) The high-speed cell connected to a low-speed cell

Figure 4.1 The standard cell concept. (a) Cells are of fixed "height." Some exceptions are made for high-speed cells. In any case, connection of cells results in an abutment of power lines. (b) Also note the availability of I/O on both sides of the cells.

second reason is that quite frequently the designer has to use a cell but not all of its functions and therefore creates more devices than are actually needed to perform the function intended. Extra devices naturally imply extra area. Assume, for example, that a designer uses a standard cell library where all flip-flops available have a reset capability. It is quite possible that in some sections of the designer's chip reset is not required and in fact not used. However, since the designer has only one choice, he uses the flip-flop cell available and properly connects the reset signal so this function is always disabled. In custom design, on the other hand, the logic that performs the reset function would have been omitted.

Designing with standard cells makes the technology available to a wide spectrum of designers who are not necessarily experts in GaAs device operation. This is a major advantage for logic and system designers who intend to design special-purpose ICs to impact the performance of systems. This approach is particularly attractive in the design of high throughput signal processors where there is often a high-speed requirement for the circuits used and the computational structures are very regular and can be implemented using only a few different cells replicated over the chip.

Currently, several GaAs foundries offer commercial standard cell libraries for designing integrated circuits. The most well-known of these foundries are Tri-Quint Semiconductor (the first full-service GaAs foundry), GigaBit Logic, Harris Microwave Semiconductor, and Vitesse Electronics. Other foundries offer this capability only for military and aerospace applications. As GaAs matures and fabrication process yield increases, the utilization of standard cells will continue to expand. This is because of the short turnaround design times, which lower the component's price and give system developers a competitive edge by allowing for timely introduction of a product to market. On the other hand, for standard components that will be produced in large quantities (i.e., memory and microprocessor circuits), as well as components for military applications with special performance requirements, custom design will continue to be a viable approach.

Some applications call for standard cells and custom design to be combined to yield a chip designed according to certain specifications. The combination of the two approaches is often called semicustom design. In this case, standard cells are used where possible while other portions of the circuit are designed using the custom approach in order to achieve necessary high performance or area efficiency not possible with currently available standard cells. The custom-designed portion of the circuit does not follow the standard dimension principle of the standard cells, but it is "handcrafted" into the circuitry that is developed with the standard cells. This approach resolves design bottlenecks due to performance or area requirements that cannot be met by the standard cells, while maintaining the benefits of rapid turnaround and low design risk, since the custom-designed portions of the circuit are usually small.

4.1.3 Gate Arrays

The gate array approach became popular with Si technologies and it is being widely used for low-cost rapid turnaround development of ICs with no stringent performance requirements. This trend is also being used for GaAs technologies. Almost every GaAs foundry at present offers some kind of gate array capability and service.

Gate arrays are die containing collections of prefabricated transistors and diodes placed in fixed positions on the die, along with suitable general purpose I/O structures and associated bonding pads. The transistors and diodes are usually configured as arrays of active "islands" with regions around the islands provided for interconnections. This type of gate array is called a configurable gate array. An improvement to this approach is to design each active island as a cell containing a small number of transistors and diodes that can be connected to form any one of a set of basic gates or even more complex circuits such as flip-flops. This type of gate array is called a cell-configurable array. A design on a gate array, often called the array personalization, entails the interconnection of the premade, fixed-position transistors and diodes that manifest the intended function. The designer in this case defines only the interconnection pattern, which determines the logic function the gate array will perform. By varying the interconnect pattern, a gate array can be configured to a virtually unlimited number of circuits. Cost and turnaround time when using this design approach are significantly reduced. Since the gate array structure is fixed, the array is prefabricated up to the metalization steps; therefore, only these steps need to be performed when a design is being fabricated.

The trade-offs the designer has to make when using a gate array concern mainly circuit performance and I/O structure, assuming of course that the design can be accommodated by an available gate array. Since the transistor sizes and placement are fixed, the designer may not be able to meet performance specifications for that design using a gate array. Also, trade-offs may have to be made about the I/O structure intended for the design if the gate array does not offer as many I/O ports as the design may require. It should be noted here that only the number of the available I/O ports is significant and not their type (i.e., input or output). Any available port in the array can usually be personalized to be either an input or output port.

GaAs gate arrays are available in both depletion-mode and enhancement/depletion-mode technologies. The gate arrays offered by different foundries have a variety of sizes in terms of the number of equivalent gates, die size, and number of I/Os available for signals, as well as different performance in terms of gate speed and nominal chip power dissipation. A depletion-mode gate array that allows the implementation of both digital and analog functions is offered by Tri-Quint Semiconductor. This is a reconfigurable cell array, called the Quick Chip, and it contains 28 core cells and 24 I/O pads. The array core cells support two UFL or BFL gates or one flip-flop configuration. The typical gate delay for the Quick Chip is 190 psec (fan-in = fan-out = 2) at a nominal chip power dissipation of 2 W. Personalization of this array for digital circuits requires the fabrication of only three mask layers. A variety of circuits implemented using this gate array have exhibited measured performance at 1.4 GHz.[2]

Honeywell has reported the design of an SDFL cell reconfigurable array. The array contains 432 cells, of which 292 can be used for logic and 32 for I/O signal buffers. The remaining 108 cells are used for 64-bits of static RAM that is available on this array. The basic gate configuration that the cells are optimized to implement is NOR or OR/NAND. A number of circuits were implemented with this gate array, and ring oscillator results show a nominal gate delay of 250 psec at 1 W of chip power dissipation.[3]

The largest gate arrays reported, as may be expected, are designed in DCFL. In Reference 4, a 2 K gate array with 66 I/O signal buffers is reported to have a nominal delay of 215 psec for loaded gates at a maximum chip power dissipation of 1 W. Each cell in this array can be personalized as an inverter or a 2- or 3-

input NOR gate. This array was used successfully to implement an 8×8 array multiplier that exhibited a multiplication time of 8.5 nsec with 400 mW of power dissipation.

A yet-larger DCFL gate array of 3 K gates is reported in Reference 5. This gate array offers 64 signal I/Os, and its cells can be personalized into NOR gates of 2 or 3 inputs. The loaded gate delay of this array was measured to be 285 psec while the total array power dissipation is 2.5 W plus 20 mW per ECL output driver used.

Gate arrays from other manufacturers, still in the prototype stage, are reported in References 6, 7, and 8. It is expected that more GaAs foundries will offer gate array design services and the design trend toward gate arrays that began in Si technologies will have a parallel in GaAs.

4.2 CHIP DESIGN SEQUENCE AND COMPUTER-AIDED DESIGN TOOLS

Regardless of the design approach chosen for developing a GaAs IC, the designer must take a sequence of steps from the development of the design concept to the final pattern generation tape that is used for mask fabrication. The work involved in the tasks of the design sequence can be supported by a number of Computer-Aided Design (CAD) tools the designer can use to maximize the probability of a successful design with the first fabrication run. Absence of a particular tool does not necessarily imply failure, but it increases the probability of it, especially when the size of the design tends to be too complicated for manual analysis. On the other hand, the availability of CAD tools does not guarantee success unless the tools are judiciously used and the results they provide are correctly interpreted.

CAD tools are computer programs that are driven with input files containing information that reflects important parameters and technological features characteristic of the design level and the technology supported by the tools. Therefore, it is very important for the designer to provide accurate input to these tools in order to receive meaningful results. One way, and probably the best way, of ensuring the accuracy of the input data is to compare test results of fabricated circuits with results predicted by the tools during different stages of the design. This action should also be complemented by an understanding of the underlying assumptions on which the tools operate and possible differences between these assumptions and the electrical principles of circuit operation.

4.2.1 Schematic Capture

The initial task in designing an integrated circuit is to capture the overall logic function by interconnecting logic primitives such as gates, flip-flops, registers, etc. A schematic capture editor can be used to perform this task in a CAD environment. The schematic capture of the design can be independent of the design methodology used to implement the design, its purpose being the logical representation of the functions to be performed. Many commercial schematic capture systems, however, are supported by libraries of logic primitives that correspond to standard cell or gate array primitives, which cause the schematic capture of the design to be linked to its physical implementation in terms of libraries. If the custom design approach is used, this correspondence does not exist at this point—you have only the logic representation of primitives and their interconnection.

During schematic capture, the user provides the names of the signals and nodes in the schematic. It is preferable that the designer provides meaningful names whenever possible. The reason is that if the user does not name nodes and signals, most schematic editors will produce these names automatically but not necessarily in a meaningful sense as far as the user is concerned. In a complex design, this can cause confusion, especially when certain nodes need to be monitored during logic simulation. Although name lists can be traced to identify signal names given by the schematic editor, such a process often can be tedious.

The schematic capture of the design is performed using a graphics device and a menu of logic primitives available in the CAD system. The user selects the logic primitives from the menu and interconnects them manually using the graphics capability of the system, or automatically by specifying which nodes need to be connected. Schematic capture editors have different capabilities and are supported by logic primitive libraries of various sizes. The way the designer performs this task depends largely on the flexibility of the schematic capture editor and its utilities. Most editors, however, offer in their menu a set of minimum functions such as zoom-in or zoom-out for viewing the schematic, selecting a component, connecting a line, and plotting a copy of the schematic. All of these functions are performed using the graphics capability of the CAD system, which allows the designer to execute these functions from the editor's menu using a "puck" or "mouse" and, on occasion, by typing a few simple characters.

The database created by the design capture editor usually is formatted by a net-list extractor program that allows the description of the logic primitives and their interconnection to be read by a logic simulator.

4.2.2 Logic Simulation

The purpose of logic simulation is to verify that the circuit to be implemented on a chip performs its intended logic function correctly. The logic simulator operates at the gate level and the inputs are circuit descriptions composed of logic gate primitives such as NOR, NAND, AND, OR, INVERT or more complex primitives such as flip-flops. The logic simulation can predict the output of a circuit statically (i.e., in a truth table form) or dynamically, as a function of time. In both cases, however, the logic simulator may not account for circuit characteristics dependent on the technology used, device geometry, bias, or loading. In fact, some of the assumptions made by logic simulators can mislead the user if more than the intended information is retrieved from the output.

For example, logic simulators assume that a gate's input capacitance is linear, that all gates switch when their inputs change (i.e., constant switching threshold), and in some cases, that all gates have unit propagation delay. These assumptions are not correct, because gate capacitance is nonlinear, depending on the bias of the gate, and subsequently the state of the gate affects the gate delay and loading characteristics. Furthermore, the interconnection adds a linear-capacitance component to the nonlinear gate capacitance, which adds complexity in simulating accurately the behavior of this structure. The assumption of linear capacitance is made to facilitate simulation. Gate switching threshold depends on the gate fan-in ratio as well as the state of the inputs. Variation of the switching threshold implies variation of the gate delay times. Some logic simulators assume that all gates have the same delay equal to a time unit, and that is an oversimplified approximation. More flexible simulators allow the user to define the delay of each gate used but they treat this delay as constant independent of input conditions—not a valid assumption in the general case because the gate delay can be affected by a possible glitch when the inputs change simultaneously or almost simultaneously. These assumptions made by logic simulators will introduce design errors if the user relies on the logic simulation output to estimate circuit performance in terms of switching speed and driving capabilities of the circuit under consideration. However, the designer can rely on the output of the logic simulator in deciding if the circuit performs its intended function correctly in a boolean sense. In terms of performance, the simulation output should be taken as an *indication* of the circuit performance, not as an accurate prediction.

Logic simulation is a design step necessary to ensure correct logic operation for a wide range of input logic variations. The need for using a logic simulator becomes greater as the number of states the circuit can assume increases and when asynchronous logic is used. The latter is difficult to tackle manually and be able to guarantee error-free logic operation, even for relatively small circuits. If a logic simulator is used to obtain performance measures, then the designer should accurately specify circuit gate delays, avoid cases of extreme fan-in and fan-out ratios, and design with substantial margin (i.e., 30–40%) over the design specification for the circuit.

A list of commonly used, commercially available logic simulators is given in Table 4.1. Since this is a highly competitive environment, there is rapid change in the performance and capability of each simulator due to continuous development efforts by the different vendors. Therefore, the reader should always consult with manufacturers for up-to-date information on the current version of their simulators.

The discussion on logic simulators shows that these tools apply to generic logic circuits independent of the technology used to implement these circuits. This naturally introduces a weakness in the link between logic representation and physical structure and, as already mentioned, the results of logic simulation should be used cautiously when the designer strives to achieve certain circuit performance. In addition, certain circuit structures may not be able to be represented in a logic simulation environment. For example, many logic simulators cannot

TABLE 4.1 Logic Simulators

Simulator	Source
VERILOG	Gateway, Lowell, MA
HILO	GenRad, Concord, MA
CADAT	HHB-Softron, Mahwah, NJ
SILOS	Silos, Incline Village, NV
LASAR	Teradyne, Boston, MA
TEGAS	GE/CALMA, Austin, TX
LOGCAP	Phoenix Data Systems, Albany, NY
LOGSIM	ERADCOM/RCA, Camden, NJ
HELIX	Silvar-Lisco, Menlo Park, CA

handle the bidirectional nature of pass devices in dynamic circuits or accurately describe the operation of push-pull buffers. On the other hand, due to circuit simplification resulting from the assumptions made, logic simulators are suitable for simulating the logic behavior of large circuits in a timely fashion.

Finally, a logic simulator, as with any simulator, will provide output that is as good and complete as is the input that drives the simulator. The user must accurately describe the circuit under design in terms of connectivity, individual gate delays, clocking scheme, and input vector space, in order to increase the probability of a logically correct designed circuit.

4.2.3 Circuit Simulation

Circuit simulation is performed at the transistor level by taking into consideration device characteristics dependent on the technology of implementation as well as active (i.e., FETs, diodes) and passive (i.e., resistors, capacitors) device sizes. The basis for this type of simulation is the solution of a set of equations relating circuit voltages, currents, and resistances, as discussed in Chapters 1 and 2. Since device sizes and technology factors are taken into consideration, there is a strong link between the simulated and the physical structure, and consequently there is high accuracy in the simulation output that presents the circuit node behavior as a function of time.

This simulation output can be used to derive the maximum and minimum propagation times as a function of bias, device geometry, temperature, and loading (i.e., fanout). In addition, rise and fall times also can be estimated with a circuit simulator. The propagation delay times often specified in logic simulators are obtained from circuit simulations. The penalty for the high accuracy, however, is longer simulation time. The simulation time is typically proportional to n^r, where

n is the number of transistors in the circuit, and r is an empirically determined exponent that takes values between 1 and 2. It is unrealistic to use this type of simulation for large circuits because of the excessive simulation time that may be required.

The most widely used circuit simulator is the SPICE circuit simulator, which originated at the University of California at Berkeley and has been incorporated into numerous CAD tool sets.[9] Different versions of the SPICE simulator can improve the software performance, aid user interface, and correct certain convergence problems of the initial versions of SPICE.

The input parameters specified by the user of a SPICE simulation include device models indicating the electrical characteristics of the devices, parameters characterizing the device geometry, and input conditions indicating the circuit bias, clocking, and input transistors. One of the most important aspects of an accurate simulation concerns the device models. Since some of the device parameters include intrinsic properties of GaAs and the fabrication process, the user must have access to this information for a reliable simulation. The problem is that this information is frequently proprietary. Also, even when these parameters theoretically can be obtained, the user should introduce corrections based on results obtained from measured parameters of devices fabricated with the process under consideration. Circuit simulation integrity can be claimed only when the simulation output does not differ from measured circuit outputs by more than an acceptable margin.

What is acceptable in this case depends on the circuit design accuracy required and the design margin that can be afforded. As a rule of thumb, this margin should be at least 5 percent. The model parameters are constant in the circuit equations solved by the simulator, while the device geometry characteristics and the input conditions are variables. The designer obtains the desired circuit response by varying the device sizes for a set of input conditions (i.e., bias) in order to satisfy certain design requirements such as power dissipation. However, the circuit designed using information from a circuit simulator output will be as good as the device models used to characterize electrically the behavior of the circuit components. Therefore, the user should not put too much faith in the simulator results without adequately questioning the device models used and the suitability of these models for the application at hand.

4.2.4 Layout

Layout is the process of translating a design from its logic and circuit representation to its physical representation. The way layout is performed depends on the design methodology used for implementing an integrated circuit and the capability of the layout system. Specifically, when the designer uses the custom design approach, layout is performed for all circuits included in the design on a per-gate basis, and all gates are manually interconnected. The designer in this case has maximum freedom concerning the layout style and the arrangement of gates and interconnects. This freedom allows the designer to optimize the utilization of area devoted to active logic and, most importantly, to interconnections. However, since this is a manual process, custom layout is a time-consuming process and often accounts for more than 50 percent of the total design time of an integrated circuit. Since each gate is laid out individually, the designer has to pay close attention to the layout rules during the entire process of custom layout. The layout rules are established based on the fabrication process used and the capabilities of the available lithography system. The layout rules define limits for the mechanical dimensions of, and spacing between, different features on the various layers in order to ensure maximum circuit reliability for a given process technology. Some general layout rules are shown in Figure 4.2. These rules vary according to the fabrication process used. Figure 4.3 illustrates the layout of a 3-input NOR gate. Figure 4.3a shows the circuit diagram, and Figures 4.3b through g show the transistor sizes and layers 2 through 7 of the seven-layer layout. Figure 4.3h shows the final composite.

If the standard cell design approach is used, the designer concentrates on placing the cells and interconnecting them in a way that minimizes overall chip area, assuming that the selected cells provide the performance required by the chip design specification. Layout is not performed at the gate level; rather, the arrangement of the interconnections is the most important aspect of layout when using standard cells. The designer is not free to compact the layout of the individual cells, only their interconnections. This is the main reason that standard cell layout is generally less area-efficient than custom layout. On the other hand, the layout process using standard cells may be performed in a fraction of the time required for custom layout. Attention to design rules is paid primarily to the interconnects, since the lay out of the cells are predefined and is known to be correct.

Efficiency and time required to lay out a chip design using standard cells are also affected by the capability of the CAD tools used. Placement of the cells and routing of the interconnections can be performed either manually or automatically. The trade-off in most cases is area utilization efficiency versus layout time. Manual placement and routing achieve area efficient layouts at the expense of longer layout times. For relatively small circuits with high performance requirements, manual placement and routing are preferred to minimize the chip area, thus

Rule	Range	Units
Minimum interconnection line width (metal 1)	1.0-2.0	μm
Minimum interconnection line width (metal 2)	2.0-3.0	μm
Minimum interconnection line spacing (metal 1)	1.5-2.0	μm
Minimum interconnection line spacing (metal 2)	2.0-3.0	μm
Minimum via size	$1.0 \times 1.0 - 2.0 \times 2.0$	μm
Minimum FET gate length	0.8-1.0	μm
Minimum diode tab	0.8-1.0	μm

Figure 4.2 A sample of general layout rules.

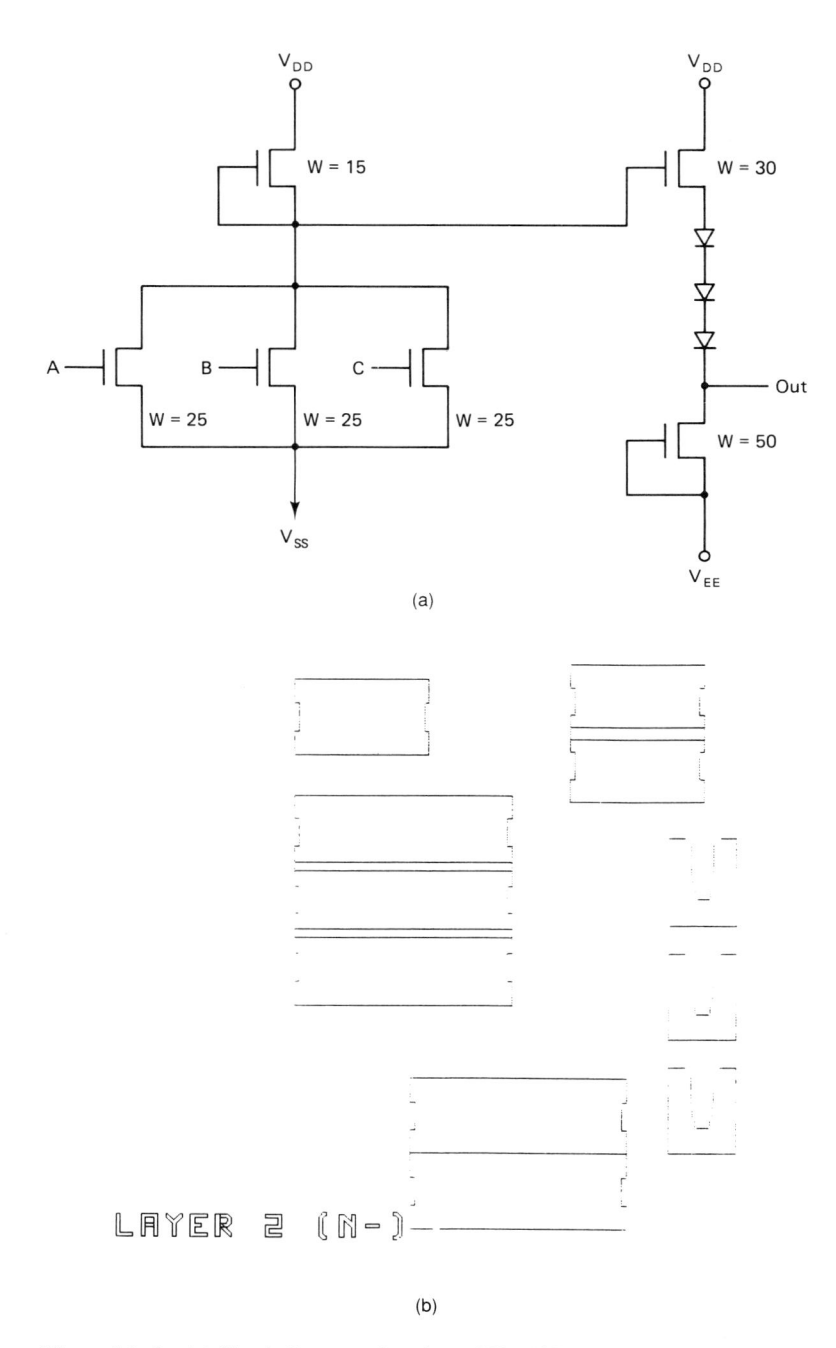

Sec. 4.2 Chip Design Sequence and Computer-Aided Design Tools   195

Figure 4.3a-b (a) Circuit diagram a three-input BFL NOR gate. All devices have gate length $L = 1$; and gate widths W are as shown. (b) Layout of layer 2 (N$^-$).

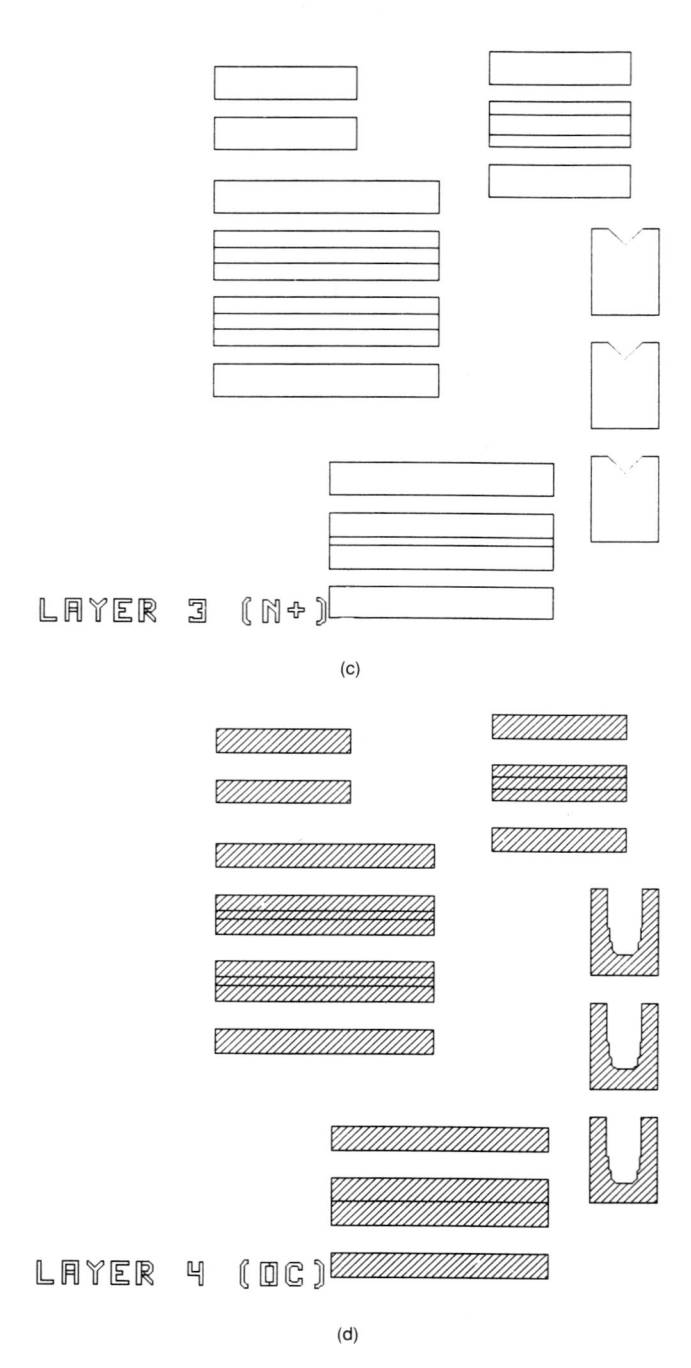

196 GaAs Digital Integrated Circuit Design Principles Chap. 4

(c)

Figure 4.3c-d (c) Layout of layer 3 (N^+. (d) Layout of layer 4 (OC).

Sec. 4.2 Chip Design Sequence and Computer-Aided Design Tools 197

(e)

(f)

Figure 4.3e-f (e) Layout of layer 5 (SMET). (f) Layout of layer 6 (via).

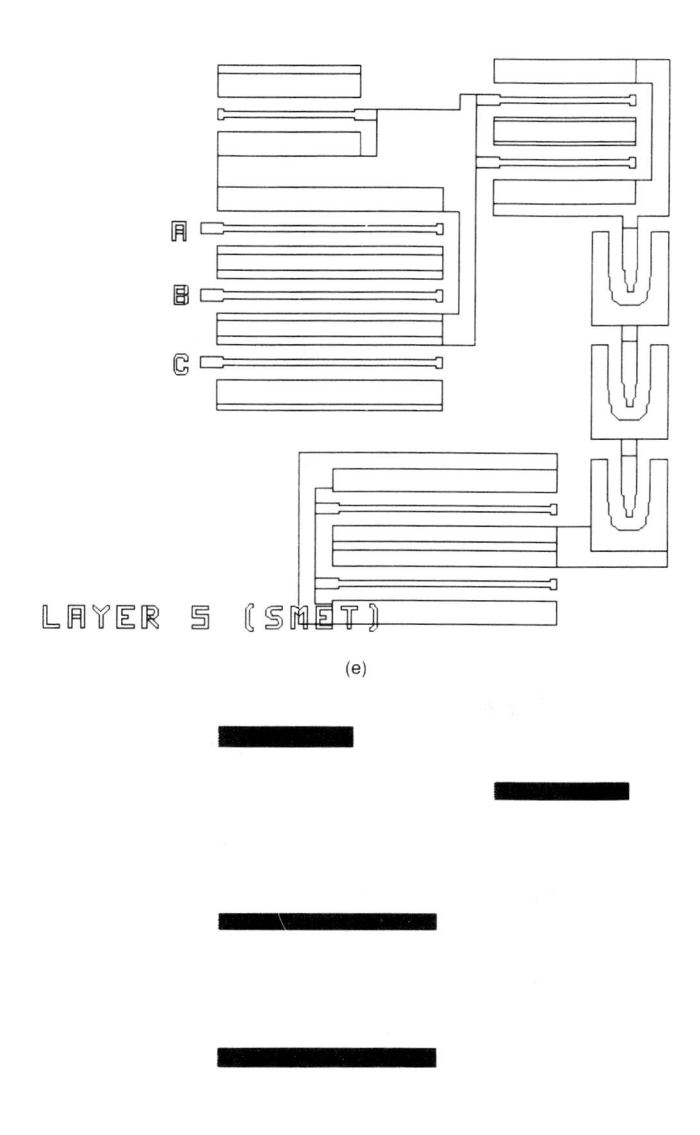

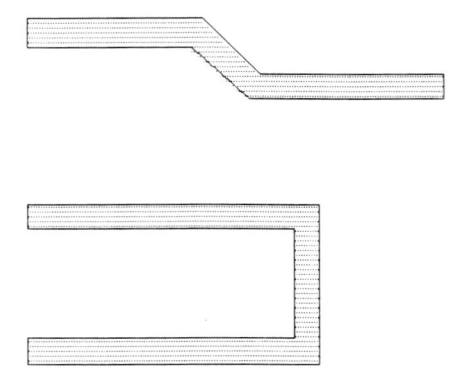

198   GaAs Digital Integrated Circuit Design Principles   Chap. 4

LAYER 7 (MET2)

(g)

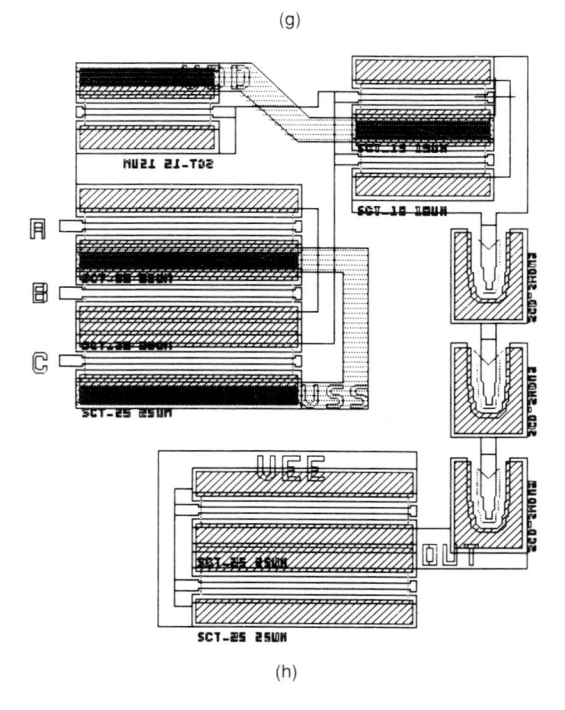

(h)

Figure 4.3g-h (g) Layout of layer 7 (MET2). (h) Composite layout of the three-input NOR gate.

Sec. 4.2   Chip Design Sequence and Computer-Aided Design Tools   199

increasing yield and decreasing production cost, also at the expense of relatively longer layout times. The difference in layout time between manual and automatic placement and routing becomes significant only when the circuit complexity rises to LSI levels and above. Figure 4.4 illustrates the logic, and circuit diagrams and the layout of a NOR gate standard cell. Cells can be placed directly next to each other ensuring the connection of power, clock, and signal lines, since these lines are placed at fixed intervals on the standard dimension (i.e., height or width) of the cell. Many cell designs also can be placed after rotation by 180 degrees without any routing penalty. This is possible when the cells provide input and output connections on two opposite sides.

When designing with gate arrays, layout can be custom or one can use cells. However, the meaning of *custom* in this case refers only to interconnecting the devices of the array so they perform a desired function. Both the device sizes and

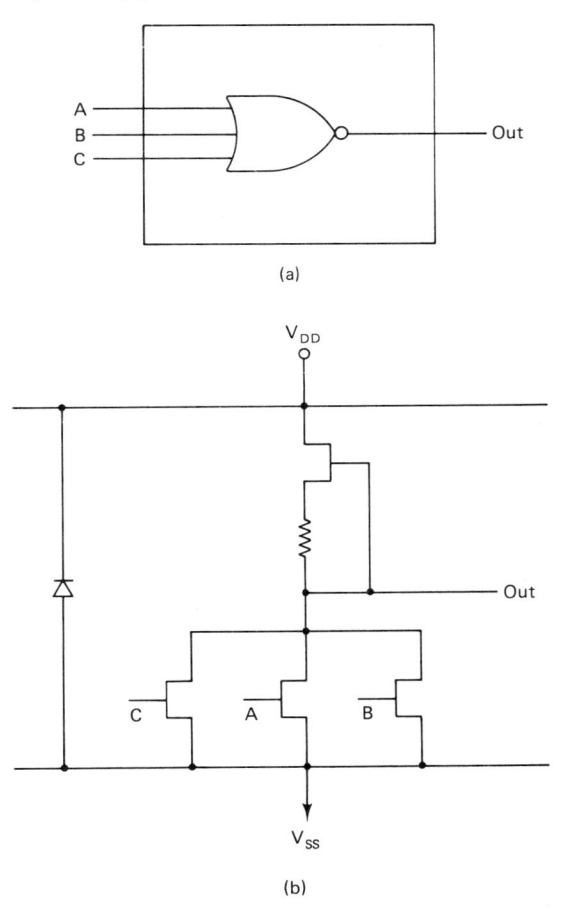

Figure 4.4a-b   Three-input CDFL NOR standard cell: (a) logic diagram; (b) circuit diagram. (*Figure continues.*)

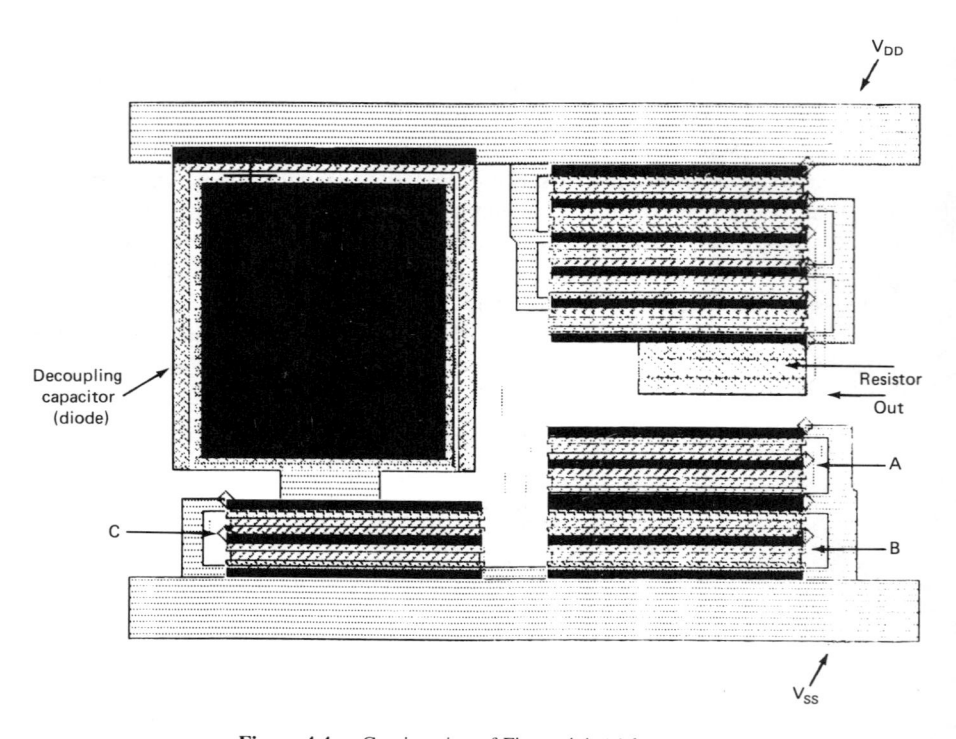

Figure 4.4c Continuation of Figure 4.4. (c) layout.

their placement are fixed, and so are the routing channels the designer can use for interconnections. The layout process entails the proper interconnection of the devices via the available routes so that a required function can be performed as specified. A *cell approach* refers to using a configurable array where each cell is personalized to some primitive logic function and the cells are connected through the available routes so that the required function in the chip design specification is achieved. Although the placement of the devices in the array is fixed, the mapping of the functional partition of the design onto the array is decided by the designer. This is a very important decision because it affects the utilization efficiency of the array (i.e., percentage of array devices used in the design) and the overall routing.

Regardless of the approach used, the primary objective during layout is efficient utilization of chip area. However, consideration should be given to all aspects of circuit performance during layout. Frequently, trade-offs must be made among optimum area utilization, performance, yield, and turnaround time. Such trade-offs involve a careful examination of circuit design, device parameters, and layout rules. Optimum exploitation of the performance offered by GaAs technology can be achieved only when the layout guarantees minimization of parasitic capacitances. In addition, a very important consideration is layout simplicity. To reduce the possibility of errors in mask fabrication and increase the accuracy of parameter extraction and, therefore, circuit performance calculations, the layout should be as simple as possible to accomplish the function in the chip design specification. Complex layout configurations should be used only as a last resort for achieving specific characteristics that are otherwise unattainable.

A multitude of available CAD layout systems support one or more chip design methodologies. The main function of these tools is to define the required shapes to form devices and interconnects, or define the appropriate cells from a standard cell database (i.e., library) and provide their interconnection. Most of the CAD layout tools are hierarchical and interactive, providing immediate feedback to the designer in graphical form of the actions taken in laying out a circuit. The user interface in most of these tools is menu driven. Some tools provide concurrent layout rule checking, so as the user specifies active areas (i.e., FETs, diodes) and interconnections, the tool identifies possible violations of the layout rules. The layout rules can be updated in the technology file used by the tool so that they reflect the process to be used to fabricate the design.

4.2.5 Design Rule Checking

When the layout is completed, it must conform to the geometric design rules determined by photolithography requirements and the fabrication process to be used for implementing the design. The design rule checking is performed with a program called design rule checker. This program checks for possible violations of minimum dimensions of structures allowed in each layer and also for minimum spacing violations of both unrelated and related structures. For example, as shown in Figure 4.2, a design rule defines the minimum gate length of a transistor, while another rule defines the minimum metal interconnection line width. A design rule defines the minimum spacing of two metal lines carrying unrelated signals (same mask layer), while another rule defines the minimum spacing between the edge of a contact and the edge of the metal line that that contact is connecting to another metal line or to the active area (different mask layers).

The design rule checker should detect layout violations of these rules in order to enhance the probability for successful circuit implementation. If design rules are violated and the violations are undetected, the resulting circuit will either malfunction or will have very low reliability. This is because the geometry of devices and interconnections is directly related to their electrical characteristics and also to the capability of the fabrication processing technology. Violating the minimum sizes and distances dictated by the design rules will result in structures that have lower electrical tolerances than those required by the technology for reliable operation (i.e., current density) or require photolithography of higher resolution than the fabrication process can provide.

Design rule checking is a computer-time-intensive process but it is a very important design step and never should be omitted. Even when standard cells or gate array approaches are used, design rule checking should be performed to guarantee that the interconnection is free from design rule violations. Computer time can be shortened for this operation if a hierarchical design rule checker is available. These design rule checkers use the hierarchical nature of a design to reduce the number of design portions that have to be individually checked. Many variations of design rule checkers exist, but typical approaches can be found in References 10, 11, and 12.

4.2.6 Net List Extraction

When a standard cell or gate array approach is used, the layout can be produced by placing and routing the logic cells that constitute the circuit. The designer in this case needs to ensure that the circuit produced by the layout is identical to the circuit captured by the design capture editor and simulated at the logic and circuit levels. A net list extractor can identify the interconnection of the logic blocks by the signals propagating between connecting nodes. This identified pattern can then be compared to the interconnection pattern (net list) provided by the schematic capture editor.

This procedure allows the designer to verify the connectivity of the layout and also the utilization of the proper logic cells. This program does not provide any information about the electrical characteristics of the interconnections and their effect on the circuit operation. Furthermore, the designer should have access to a program that can verify the equality (or lack thereof) of two graphs; namely, the net list produced by the schematic capture editor and the one produced by the net list extractor. A typical program that performs this function is WOMBAT.[13]

4.2.7 Switch Level Simulation

An alternative to net list extraction for ensuring correspondence between the physical layout and the circuit initially simulated is to perform switch level simulation. In fact, this approach applies to any methodology the designer may follow and guarantees that both the logic and the connectivity of the laid-out circuit correspond to the circuit initially captured and simulated.

A switch level simulator is a three-state simulator using 1, 0, and X for undetermined states. The input to the simulator is the circuit that is already laid out, where all transistors are treated like switches and the connectivity of the nodes is preserved. For each set of inputs, the circuit is assigned a steady state that changes as the inputs change. The state transition is evaluated by using a relaxation technique on the nodes in the circuit.

A switch level simulator can reveal logic and interconnection problems due to layout errors. Since the simulation at this level includes the I/O pads, the designer has a global picture of the logic behavior of the circuit after layout has been completed. However, this simulation does not account for any layout effects on the circuit's electrical characteristics and performance and, therefore, the results obtained do not guarantee correct operation. They can be used only to ensure that the circuit in the layout performs the logic function intended.

Not all circuits may be simulated this way. Whether such simulation is appropriate depends on the input configuration for the transistors and the signal flow convention used by the program.

4.3.8 Circuit Extraction

Circuit extraction ensures logic and electrical performance correctness of the circuit after layout. A circuit extraction program examines the interrelationship of mask layers to infer the existence of transistors and other components. In addition to device connectivity, the circuit extractor reports parasitic capacitances and resistances at all nodes. This information is extremely valuable to the designer, because it can be used to perform circuit simulations on various sections of the circuit with actual values of parasitics derived from the layout. Circuit simulation including these parasitics is the closest prediction or indication of how the fabricated circuit will behave. Depending on the values of parasitic capacitance and resistance produced by the circuit extractor, the designer may have to change either the layout or the device sizes used if the circuit simulation including these parasitics produces results that do not meet circuit design requirements.

The need for circuit extraction increases as performance requirements for a circuit become higher, thus allowing for smaller design margins. Given the relatively high cost of design and implementation of GaAs integrated circuits, this design step should be performed to ensure compatibility of design requirements and circuit performance after layout.

Various approaches have been used to implement circuit extraction tools. Details of these approaches and the software design of tools can be found in References 11 and 14.

4.3 SELECTION OF CIRCUITS TO MEET DESIGN REQUIREMENTS

The designer of an integrated circuit has to meet certain design requirements concerning the performance of the circuit—speed and power dissipation, size of the chip—which depend on the availability of packages and affect the fabrication yield. The optimum design of an integrated circuit is the one that meets the speed requirements while dissipating the minimum possible power, without exceeding the power dissipation requirements, and occupies the minimum possible area without exceeding the limits on die size. Speed, power, and area, however, are all interdependent design parameters, and therefore, certain trade-offs must be made to meet certain design requirements.

The speed of a circuit is determined by the time required to charge and discharge the device and interconnect parasitic capacitances. This time is proportional to the parasitic capacitances and the logic voltage swing and inversely proportional to the average drain current that charges and discharges the parasitic capacitances. The power dissipated by the circuit is proportional to the drain current and the logic voltage swing. The size of the die for a given number of devices is proportional to the device sizes and the amount of interconnect. Expressions that relate propagation delay and power dissipation to capacitance, logic voltage swing, and drain current are given in Chapter 2.

The logic voltage swing depends on the pinch-off voltage, which is determined by the fabrication control and thus cannot be affected by the designer. The bias of the circuit, on the other hand, should guarantee a logic voltage swing large enough for adequate noise margin. For a given logic voltage swing, the speed of a circuit can be increased by decreasing the parasitic capacitances and/or by increasing the drain current. The drain current of a device is directly proportional to its width, but so are the parasitic capacitances associated with the device. The fringing capacitances at the ends of the device and the interconnection capacitances are independent of the device width, and these capacitances become a larger portion of the total parasitic capacitances as the device width decreases. Since the drain current of the device decreases with decreasing width, the charging time of the parasitic capacitances increases, thus decreasing the circuit speed. At the same time, however, the circuit power dissipation also decreases, as does the die size. Therefore, the designer has to determine device widths so that all requirements are met. At the same time, effort must be made to minimize the parasitic capacitances attributed to interconnects, which are independent of the device width. For a given minimum gate length and circuit bias, the designer can manipulate only the width of the active devices and the geometry of the interconnects in order to achieve circuit design requirements. It is generally assumed that the circuit designer has no control over the fabrication process parameters for a given technology and uses these parameters as constants in modeling circuit behavior.

When custom design methodology is used to design an IC, it is important that the circuit characteristics be determined by modeling the circuit with all parasitic capacitances included. Although one can use analytical expressions to determine the device sizes required to obtain desired values for output voltage (i.e., Equation 2.1), propagation delay (i.e., Equation 2.11), and power dissipation (i.e., Equation 2.12), the designer almost exclusively uses circuit simulation to solve these equations, in both a static as well as a dynamic sense. Therefore, the circuit description used as input to the circuit simulator becomes critical and essentially determines the accuracy of the obtained results, assuming trustworthy device models. To illustrate this argument, let us consider the design of a NOR gate using BFL. Figure 4.5 shows the circuit diagram of the gate, including all relevant parasitic capacitances. Figure 4.6 shows circuit simulation results with and without the parasitic capacitances. As shown, failure to account for parasitic capacitances, especially on nodes driving the gate of a device, will result in optimistic results and the actual circuit will either exhibit lower performance or it will malfunction.

Sec. 4.3   Selection of Circuits to Meet Design Requirements   205

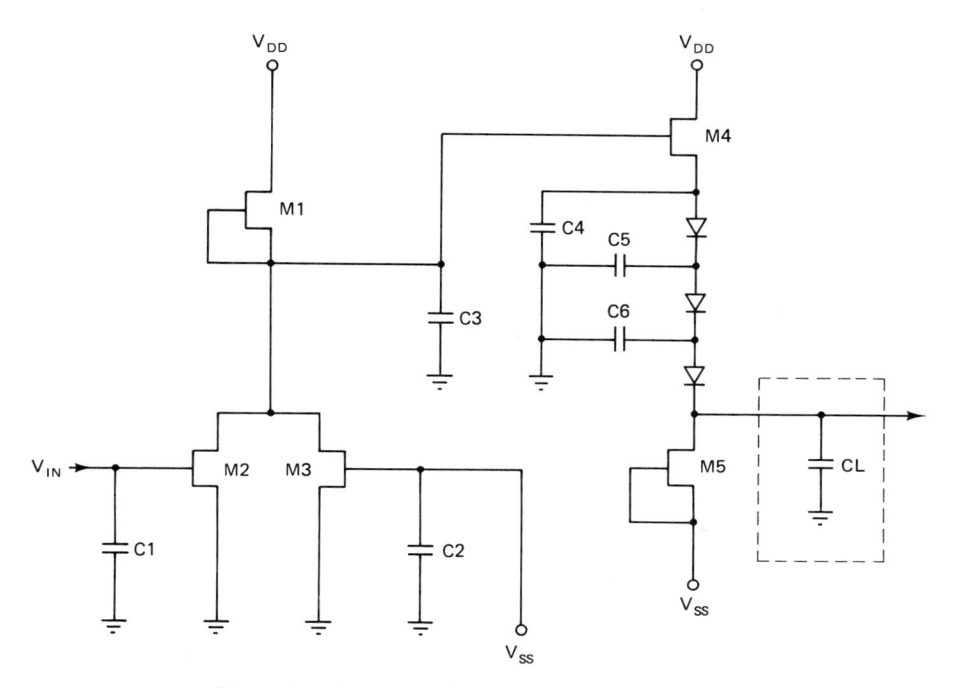

Figure 4.5 A two-input NOR with parasitic capacitances.

The device width determines the current supplied by the device; it also affects its parasitic capacitances as mentioned above. When devices are interconnected to form circuits, the designer must determine the device widths to meet design requirements. The ratio of device widths is also important, as becomes evident when we revisit the expressions from Chapter 2 mentioned above. Consider the NOR gate in Figure 4.5. The current supplied by the transistor $M1$ is proportional to its width, and it must be sufficient to charge $C2$ at a required speed when both $M2$ and $M3$ are off. When either $M2$ or $M3$ is on, it must be capable of sinking the current supplied by $M1$, which depends on the width of the "on" transistor. Therefore, the output voltage level and the time to establish this level are determined by the width ratio of the devices in the circuit, and this ratio should be determined when the circuit is modeled including all relevant parasitic capacitances.

When the designer uses the standard cell approach, the circuit selection procedure followed differs from the custom design case, because the circuit is formed by interconnecting basic circuits rather than transistors and diodes. In this case, the device sizes have been determined during the design of the cells, and the IC designer has to select the appropriate cells for satisfying the circuit design requirements. Each cell used as a logic function has certain performance characteristics including speed, driving capability, and power dissipation. Therefore, the

$V_{DD} = 4$ V

$V_{DD} = 4$ V

W = 20

C3

C4

②

V_{IN}

W = 32   W = 32

C5

C1

C6

C2

⑦

V_{OUT}

W = 40   CL

C1 = 40 fF
C2 = 40 fF
C3 = 6 fF
C4 = 9 fF
C5 = 4 fF
C6 = 4 fF
CL = 50 fF

$V_{SS} = -4$ V

(a)

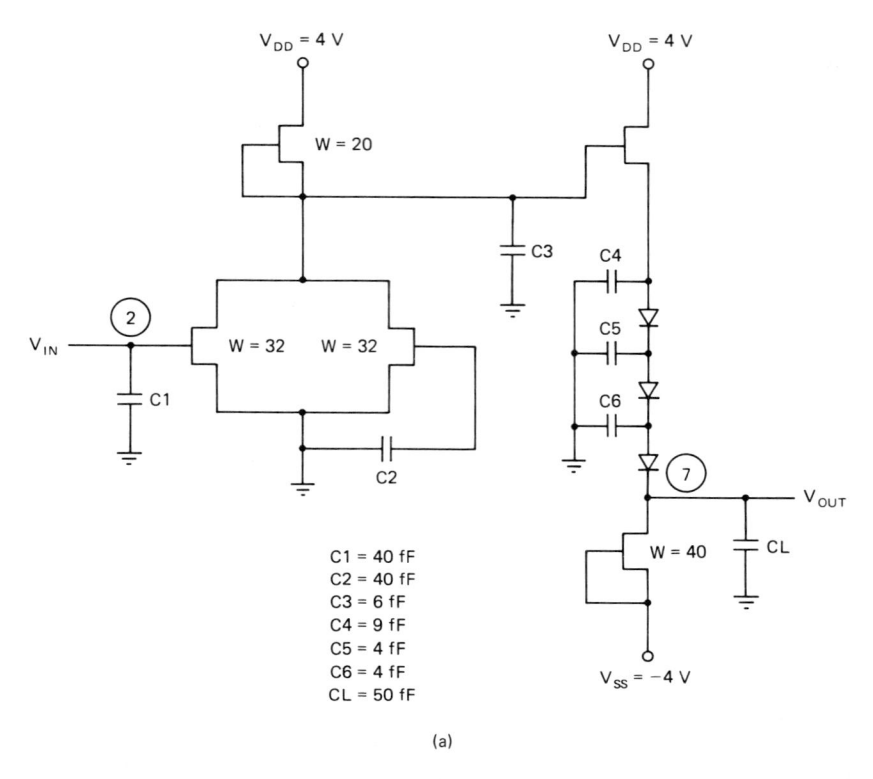

Figure 4.6 Circuit diagram of a two-input BFL NOR gate including parasitic capacitances. The device gate length is 1 mm; gate widths are set to the values shown. (b) Circuit simulation output without including any capacitances (c) Circuit simulation output including capacitances C1 through C6 but not CL. (d) Circuit simulation output including CL only. (e) Circuit simulation output including all capacitances.

emphasis in this case is shifted to the cell interface. When using this design methodology, circuit modeling must include accurate estimates of parasitic capacitances on the cell interface or at the cell output. Let us assume that the NOR gate in Figure 4.5 is a standard cell and it is to be connected to cells X, Y, Z as shown in Figure 4.7. If the designer wants to find out if the node "output" behaves as required in this circuit configuration, the NOR gate should be simulated including the capacitive load CL at the "output" node corresponding to the sum of the input capacitances of cells X, Y, Z (i.e., these capacitances are characteristics of the cells and are known to the designer), plus any estimated capacitance attributed to the interconnect between node "output" and cells X, Y, Z. Figure 4.8 illustrates the output response of the NOR gate for different values of CL. As the simulation results indicate, omitting CL in circuit analysis will result in erroneous results.

Sec. 4.3 Selection of Circuits to Meet Design Requirements   207

```
 *: V(7)
 +: V(2)
X
      TIME        V(7)
 (*+)------------ -2.000d+00      -5.000d-01       1.000d+00       2.500d+00   4.000d+00
                             - - - - - - - - - - - - - - - - - - - - - - - - - - - - - -
  0.    d+00   1.467d+00  +              .              .       *       .              .
  1.500d-10   1.467d+00  +              .              .       *       .              .
  3.000d-10   1.467d+00  +              .              .       *       .              .
  4.500d-10   1.467d+00  +              .              .       *       .              .
  6.000d-10   1.467d+00  +              .              .       *       .              .
  7.500d-10   1.467d+00  +              .              .       *       .              .
  9.000d-10   1.467d+00  +              .              .       *       .              .
  1.050d-09   1.467d+00  .+             .              .       *       .              .
  1.200d-09   1.467d+00  .        +     .              .       *       .              .
  1.350d-09   1.467d+00  .            + .              .       *       .              .
  1.500d-09   1.467d+00  .              +              .       *       .              .
  1.650d-09   1.362d+00  .              .       +      .    *          .              .
  1.800d-09   1.030d+00  .              .              +       *       .              .
  1.950d-09   6.591d-01  .              .              .   * +.        .              .
  2.100d-09   2.898d-01  .              .              .  *    +       .              .
  2.250d-09  -7.817d-02  .              .              .*      +       .              .
  2.400d-09  -4.449d-01  .              .            .*        +       .              .
  2.550d-09  -8.092d-01  .              .      *     .         +       .              .
  2.700d-09  -1.137d+00  .              .   *        .         +       .              .
  2.850d-09  -1.365d+00  .              *            .         +       .              .
  3.000d-09  -1.491d+00  .          *                .         +       .              .
  3.150d-09  -1.535d+00  .        *                  .      +          .              .
  3.300d-09  -1.519d+00  .        *                  .          +      .              .
  3.450d-09  -4.993d-01  .              .    *+      .                 .              .
  3.600d-09   6.713d-01  .              +    .              *          .              .
  3.750d-09   1.140d+00  .          +        .              .  *       .              .
  3.900d-09   1.328d+00  .   +               .              .      *   .              .
  4.050d-09   1.410d+00  +              .              .       *       .              .
  4.200d-09   1.442d+00  +              .              .       *       .              .
  4.350d-09   1.456d+00  +              .              .       *       .              .
  4.500d-09   1.463d+00  +              .              .       *       .              .
  4.650d-09   1.465d+00  +              .              .       *       .              .
  4.800d-09   1.467d+00  +              .              .       *       .              .
  4.950d-09   1.467d+00  +              .              .       *       .              .
  5.100d-09   1.467d+00  +              .              .       *       .              .
  5.250d-09   1.467d+00  +              .              .       *       .              .
  5.400d-09   1.467d+00  +              .              .       *       .              .
  5.550d-09   1.467d+00  +              .              .       *       .              .
  5.700d-09   1.467d+00  +              .              .       *       .              .
  5.850d-09   1.467d+00  +              .              .       *       .              .
  6.000d-09   1.467d+00  +              .              .       *       .              .
                             - - - - - - - - - - - - - - - - - - - - - - - - - - - - - -
```

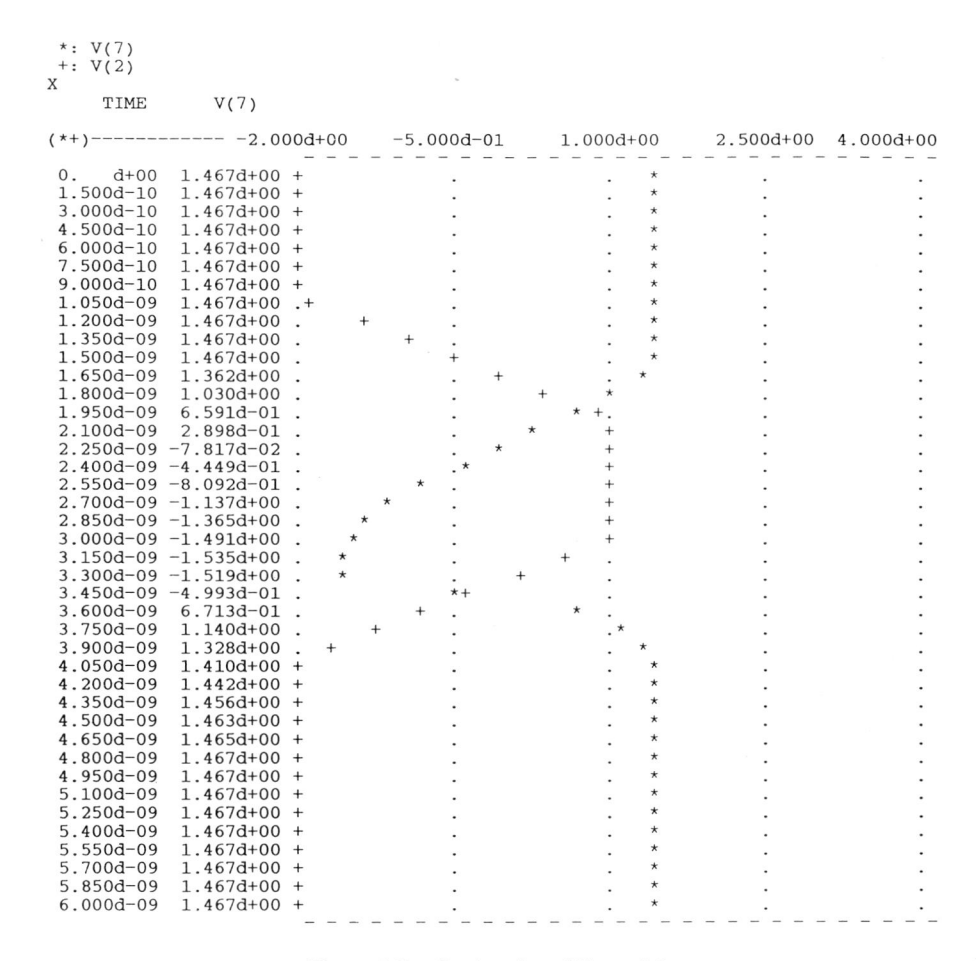

Figure 4.6b Continuation of Figure 4.6

Now the question becomes how to begin choosing the cells to implement a design that has been completed at the functional and logic levels. Designers follow different approaches to this problem, but one common procedure uses the following steps:

1. Determine the highest speed required for a signal to propagate on some path in the circuit.
2. Determine the longest path (i.e., the path including the largest number of serially connected cells) on which this signal has to propagate.
3. Start with the cell at the end of the path.
4. Choose an appropriate cell based on speed requirements and estimated CL.
5. Proceed backward on the path by repeating step 4 until all cells are selected.

```
*: V(7)
+: V(2)
X
      TIME        V(7)
(*+)------------ -2.000d+00     -5.000d-01      1.000d+00      2.500d+00   4.000d+00

 0.      d+00   1.467d+00 +              .              .    *         .          .
 1.500d-10      1.467d+00 +              .              .    *         .          .
 3.000d-10      1.467d+00 +              .              .    *         .          .
 4.500d-10      1.467d+00 +              .              .    *         .          .
 6.000d-10      1.467d+00 +              .              .    *         .          .
 7.500d-10      1.467d+00 +              .              .    *         .          .
 9.000d-10      1.467d+00 +              .              .    *         .          .
 1.050d-09      1.467d+00 .+             .              .    *         .          .
 1.200d-09      1.467d+00 .         +    .              .    *         .          .
 1.350d-09      1.467d+00 .              +              .    *         .          .
 1.500d-09      1.466d+00 .              .    +         .    *         .          .
 1.650d-09      1.282d+00 .              .              +    *         .          .
 1.800d-09      8.005d-01 .              .              +  *  .        .          .
 1.950d-09      3.051d-01 .              .         *    +.            .          .
 2.100d-09     -1.875d-01 .              .         *    .    +         .          .
 2.250d-09     -6.631d-01 .              .    *    .              +    .          .
 2.400d-09     -1.075d+00 .              .    *    .              +    .          .
 2.550d-09     -1.345d+00 .         *    .              .              +          .
 2.700d-09     -1.479d+00 .         *    .              .              +          .
 2.850d-09     -1.546d+00 .    *         .              .              +          .
 3.000d-09     -1.578d+00 .    *         .              .              +          .
 3.150d-09     -1.576d+00 .    *         .              .         +    .          .
 3.300d-09     -1.532d+00 .    *         .              +              .          .
 3.450d-09     -4.066d-01 .              .    .X       .              .          .
 3.600d-09      7.602d-01 .              .    +        .    *         .          .
 3.750d-09      1.182d+00 .              +              .         *    .          .
 3.900d-09      1.352d+00 .    +         .              .         *    .          .
 4.050d-09      1.422d+00 +              .              .         *    .          .
 4.200d-09      1.448d+00 +              .              .              *          .
 4.350d-09      1.459d+00 +              .              .              *          .
 4.500d-09      1.464d+00 +              .              .              *          .
 4.650d-09      1.466d+00 +              .              .              *          .
 4.800d-09      1.467d+00 +              .              .              *          .
 4.950d-09      1.467d+00 +              .              .              *          .
 5.100d-09      1.467d+00 +              .              .              *          .
 5.250d-09      1.467d+00 +              .              .              *          .
 5.400d-09      1.467d+00 +              .              .              *          .
 5.550d-09      1.467d+00 +              .              .              *          .
 5.700d-09      1.467d+00 +              .              .              *          .
 5.850d-09      1.467d+00 +              .              .              *          .
 6.000d-09      1.467d+00 +              .              .              *          .
```

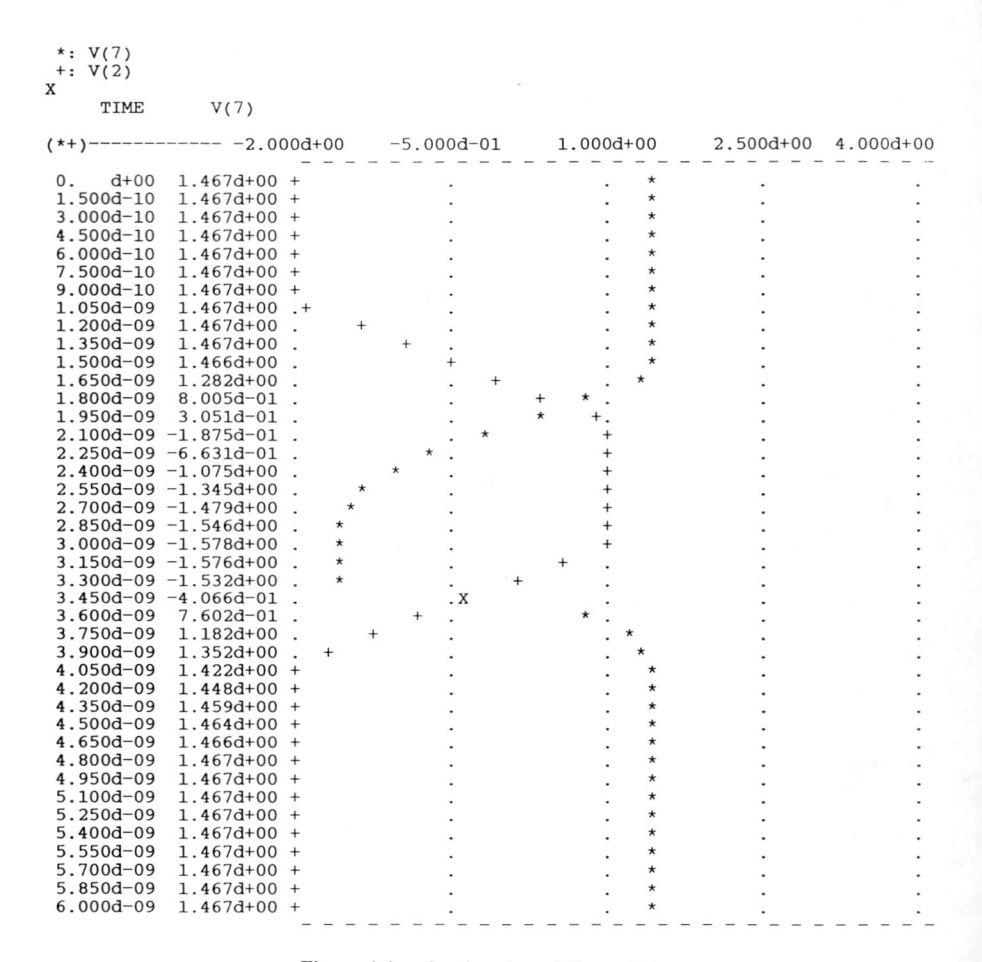

Figure 4.6c Continuation of Figure 4.6

Keep in mind that the cell selection and placement should be such that during layout the parasitic capacitance is minimized. This is a critical design issue because it allows the designer to meet the design performance requirements with cells having smaller device sizes, thus occupying less area and dissipating less power. In the event that there are high fanout nodes in the path, the designer should not try to drive these nodes with cells that have larger devices because this approach shifts the capacitance to the cells' inputs. A better approach for driving high fanouts is to split the load into a tree and use more than one cell to drive it. Scaling the device sizes of a cell upward should be used primarily to counter the effect of parasitic capacitance on the signal propagation between cells.

Sec. 4.3   Selection of Circuits to Meet Design Requirements   209

```
*:  V(7)
+:  V(2)
X
      TIME       V(7)
 (*+)------------ -2.000d+00      -5.000d-01      1.000d+00       2.500d+00    4.000d+00
    0.    d+00   1.467d+00 +          .               .      *         .           .
 1.500d-10      1.467d+00 +          .               .      *         .           .
 3.000d-10      1.467d+00 +          .               .      *         .           .
 4.500d-10      1.467d+00 +          .               .      *         .           .
 6.000d-10      1.467d+00 +          .               .      *         .           .
 7.500d-10      1.467d+00 +          .               .      *         .           .
 9.000d-10      1.467d+00 +          .               .      *         .           .
 1.050d-09      1.467d+00 .+         .               .      *         .           .
 1.200d-09      1.467d+00 .      +   .               .      *         .           .
 1.350d-09      1.467d+00 .          .   +           .      *         .           .
 1.500d-09      1.463d+00 .          .               +      *         .           .
 1.650d-09     -1.133d+00 .          .   *           .   +  .         .           .
 1.800d-09     -1.523d+00 .    *     .               .      .  +      .           .
 1.950d-09     -1.588d+00 .    *     .               .      +.        .           .
 2.100d-09     -1.602d+00 .    *     .               .      .   +     .           .
 2.250d-09     -1.602d+00 .    *     .               .      .   +     .           .
 2.400d-09     -1.602d+00 .    *     .               .      .   +     .           .
 2.550d-09     -1.602d+00 .    *     .               .      .   +     .           .
 2.700d-09     -1.602d+00 .    *     .               .      .   +     .           .
 2.850d-09     -1.602d+00 .    *     .               .      .   +     .           .
 3.000d-09     -1.602d+00 .    *     .               .      .   +     .           .
 3.150d-09     -1.550d+00 .    *     .               .   +  .         .           .
 3.300d-09     -1.438d+00 .      *   .               .   +  .         .           .
 3.450d-09      1.455d+00 .          .   .+          .      *         .           .
 3.600d-09      1.467d+00 .          .       +       .      *         .           .
 3.750d-09      1.467d+00 .        +            .    .      *         .           .
 3.900d-09      1.467d+00 .  +      .               .      *         .           .
 4.050d-09      1.467d+00 +          .               .      *         .           .
 4.200d-09      1.467d+00 +          .               .      *         .           .
 4.350d-09      1.467d+00 +          .               .      *         .           .
 4.500d-09      1.467d+00 +          .               .      *         .           .
 4.650d-09      1.467d+00 +          .               .      *         .           .
 4.800d-09      1.467d+00 +          .               .      *         .           .
 4.950d-09      1.467d+00 +          .               .      *         .           .
 5.100d-09      1.467d+00 +          .               .      *         .           .
 5.250d-09      1.467d+00 +          .               .      *         .           .
 5.400d-09      1.467d+00 +          .               .      *         .           .
 5.550d-09      1.467d+00 +          .               .      *         .           .
 5.700d-09      1.467d+00 +          .               .      *         .           .
 5.850d-09      1.467d+00 +          .               .      *         .           .
 6.000d-09      1.467d+00 +          .               .      *         .           .
```

Figure 4.6d Continuation of Figure 4.6.

In both custom design and standard cell design approaches, the selection of device sizes or cells to meet design requirements is not final. The designer has to verify that the circuit performs as predicted by the circuit simulation after layout is completed. This is a necessary step because, during circuit simulation, the values of parasitic capacitances are estimates that can differ from what is actually produced by the layout. After the layout is completed, a circuit extraction program provides the parasitic capacitance values at points of interest, and a new circuit simulation should be performed to ensure proper operation. This is the point in the design where the final selection of devices or cells is performed, or a decision to change the layout to reduce parasitic capacitances is made.

```
 *: V(7)
 +: V(2)
X
     TIME        V(7)
 (*+)------------ -2.000d+00      -5.000d-01      1.000d+00       2.500d+00    4.000d+00
                                 - - - - - - - - - - - - - - - - - - - - - - - - - - -
 0.    d+00   1.467d+00 +                          .           *       .            .
 1.500d-10  1.467d+00 +                          .           *       .            .
 3.000d-10  1.467d+00 +                          .           *       .            .
 4.500d-10  1.467d+00 +                          .           *       .            .
 6.000d-10  1.467d+00 +                          .           *       .            .
 7.500d-10  1.467d+00 +                          .           *       .            .
 9.000d-10  1.467d+00 +                          .           *       .            .
 1.050d-09  1.467d+00  .+                        .           *       .            .
 1.200d-09  1.467d+00  .         +               .           *       .            .
 1.350d-09  1.467d+00  .              +          .           *       .            .
 1.500d-09  1.465d+00  .                   +     .           *       .            .
 1.650d-09  8.849d-01  .                        +       *.           .            .
 1.800d-09 -5.567d-01  .                 *.         +                .            .
 1.950d-09 -1.532d+00  .      *                  .      +.           .            .
 2.100d-09 -1.602d+00  .      *                  .      +            .            .
 2.250d-09 -1.602d+00  .      *                  .      +            .            .
 2.400d-09 -1.602d+00  .      *                  .      +            .            .
 2.550d-09 -1.602d+00  .      *                  .      +            .            .
 2.700d-09 -1.602d+00  .      *                  .      +            .            .
 2.850d-09 -1.602d+00  .      *                  .      +            .            .
 3.000d-09 -1.602d+00  .      *                  .      +            .            .
 3.150d-09 -1.560d+00  .      *                  .          +        .            .
 3.300d-09 -1.460d+00  .        *                 .          +       .            .
 3.450d-09  1.158d+00  .                        .+          .*       .            .
 3.600d-09  1.375d+00  .                   +     .           *       .            .
 3.750d-09  1.458d+00  .              +          .           *       .            .
 3.900d-09  1.468d+00  .    +                    .           *       .            .
 4.050d-09  1.468d+00 +                          .           *       .            .
 4.200d-09  1.468d+00 +                          .           *       .            .
 4.350d-09  1.467d+00 +                          .           *       .            .
 4.500d-09  1.467d+00 +                          .           *       .            .
 4.650d-09  1.468d+00 +                          .           *       .            .
 4.800d-09  1.467d+00 +                          .           *       .            .
 4.950d-09  1.467d+00 +                          .           *       .            .
 5.100d-09  1.467d+00 +                          .           *       .            .
 5.250d-09  1.467d+00 +                          .           *       .            .
 5.400d-09  1.468d+00 +                          .           *       .            .
 5.550d-09  1.467d+00 +                          .           *       .            .
 5.700d-09  1.467d+00 +                          .           *       .            .
 5.850d-09  1.468d+00 +                          .           *       .            .
 6.000d-09  1.468d+00 +                          .           *       .            .
                                 - - - - - - - - - - - - - - - - - - - - - - - - - - -
```

Figure 4.6e Continuation of Figure 4.6

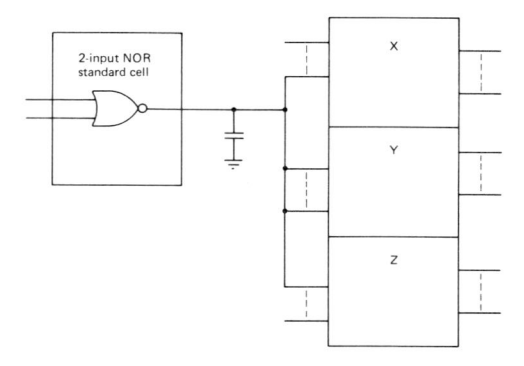

Figure 4.7 Standard cell interconnection.

Sec. 4.4   Areas for Special Consideration   211

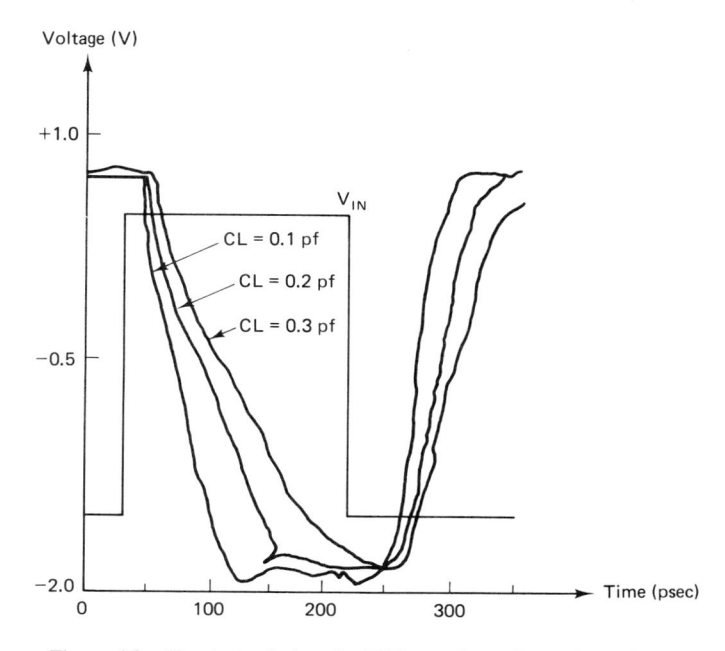

Figure 4.8   Circuit simulation of a NOR gate for various values of CL.

4.4 AREAS FOR SPECIAL CONSIDERATION

The main consideration during the design of the circuit is the minimization of parasitic capacitances throughout the chip, and the layout should be dictated by this consideration. However, some other global design aspects deserve special attention because they impact the functional as well as the electrical performance of the circuit.

Power distribution is one of the sensitive design issues because *IR* drops on the supply lines limit the dc noise margin of the circuit. Given the high speed of GaAs, the di/dt factor (inductive energy storage) can be high during switching and therefore the supply inductance should also be minimized to maintain low induced noise. In addition, the current densities on the supply lines should be kept within safe limits so that electromigration is avoided. A common approach to these problems is the use of multiple pads and power distribution networks on a chip. It is also recommended that each power distribution line carry approximately the same load, and that different distribution lines be used for the I/O and the logic on the chip.

When designing a circuit, it is generally assumed that the heat generated is uniformly distributed over the circuit and that the temperature rise affects the behavior of devices in a uniform manner (e.g., Equation 1.26, Chapter 1). This assumption can be violated if there is a cluster of large-size devices dissipating high power at some part of the circuit. The clustering of such devices can cause a local temperature rise much greater than expected, leading to possible noise margin deterioration of the devices in the cluster. Formation of high-power dissipation clusters should be avoided as much as possible.

The most critical area in the design of sequential circuits is clock distribution. This criticality arises from the requirement for unskewed clock phases in a dual-phase clock system—which can be violated if the two clock signals are loaded with significantly different capacitive loads and therefore experience different delays. A small contributing factor, but also significant, is the length of interconnect between the buffers producing the clock phases and the circuits using the clock. The routing of the clock signals should ensure that the two phases have routes of approximately equal length. Overlapping of dual-phase clocks can cause circuit malfunction (as was pointed out in Chapter 3) for flip-flops where the feedback is controlled by a single clock phase. Clock phase overlap, equal to the delay of the logic, can cause oscillation or even loss of count in a counter-circuit. Smaller amounts of overlap can create glitches at the circuit outputs that may propagate further into the circuit, causing logic errors. If clock phase overlap is unintentionally introduced during the design, the designer will become aware of it, because the logic simulator will not be able, in most cases, to determine the output states of circuits driven by this clock. This is due to the simulator assumption of infinite gate bandwidth, which results in an inability to handle glitches of this type. Upon receiving an indeterminate output on a simulated sequential circuit, the designer should check for possible clock skew. If there is a need to measure the amount of skew and its effect on circuit behavior, then a circuit simulation at the device level, including parasitics, should be performed.

Race conditions are quite possible when a single-phase clock is used in a circuit, and the designer should try to identify the circuit areas sensitive to race conditions by conducting simulations using a whole range of clock rates. Most commonly, race conditions occur in circuit arrangements such as the one shown in Figure 4.9. The clock is used to enable both the output of memory element-1 as well as the input of memory element-2. A race condition will exist if the time the clock is valid is less than the time required for the output of memory element-1 to become valid, plus the time to propagate through the logic, plus the time required by memory element-2 to sample the data. Memory elements in this case can be

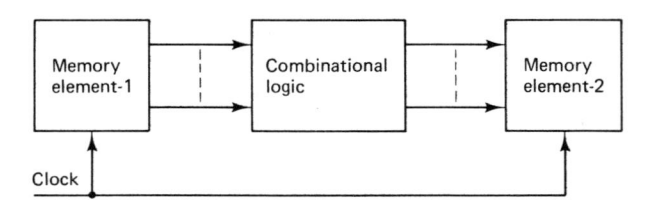

Figure 4.9 Configuration where race conditions can occur.

Sec. 4.4 Areas for Special Consideration   **213**

flip-flops, latches, or registers. If a race condition exists, then memory element-2 may contain erroneous data part of the time.

There are two approaches one can use to eliminate this race problem. One way is to reduce the path delay in the combinational logic section by splitting the combinational logic into two or more parts and by latching the intermediate results. Figure 4.10 illustrates an implementation of this approach. Although this approach may not affect the throughput of the circuit, it affects the computational delay, and this may not be acceptable in some applications. In addition, this implementation requires additional logic and also increases the loading on the clock distribution network. Another approach is to introduce a delay on the clock line that enables memory element-2. If this delay is made equal to the delay required to propagate through the combinational logic, then the problem is resolved. Figure 4.11 illustrates this scheme. The trade-off in this case is again circuit overhead. In addition, there is a risk in trying to match the delay of the combinational logic path, especially if this section of logic includes inputs other than the memory element-1 outputs. Following either approach is a design decision; however, in both cases circuit simulation with different clock rates should be performed to ensure that the race problem is resolved.

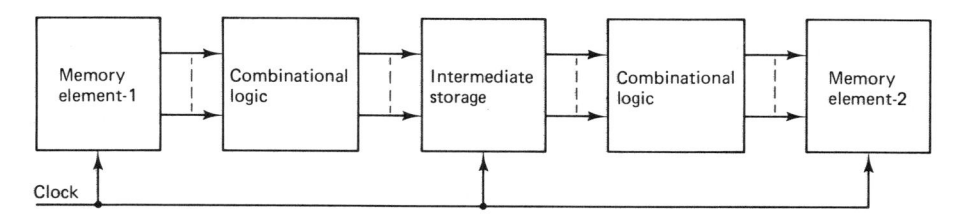

Figure 4.10 Approach to resolving the race condition problem.

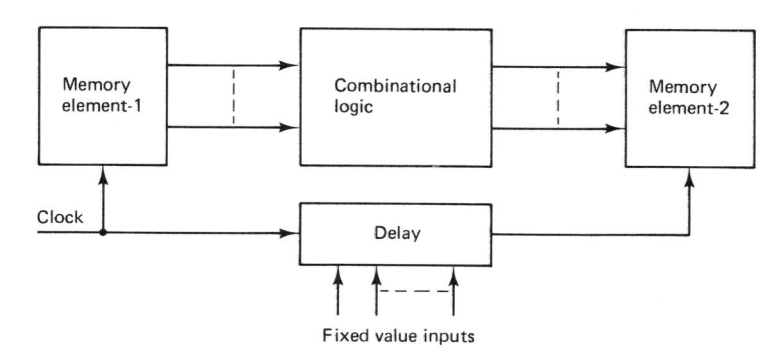

Figure 4.11 Alternative approach to resolving the race condition problem.

4.5 INTERFACE WITH Si TECHNOLOGIES

An integrated circuit eventually will be used in a system design and, in most cases, it will have to be placed on a circuit board with mixed logic families and/or different semiconductor technologies. This is the point at which interfacing becomes very important because it essentially determines the usefulness and the applicability range of an integrated circuit.

In the case of GaAs, I/O interface is often considered the weakest factor in the performance of these ICs. The reason for this assessment is that most GaAs logic families operate with logic levels that differ from those commonly used by Si logic families, and logic transitions in GaAs circuits are, in general, much faster than those in Si circuits. The logic swing and the speed of logic transition (i.e., rise and fall times) are important performance factors. The smaller the logic swing, the faster the circuit and the lower the power dissipation, but also the lower the noise margin. Large voltage swings of an input signal can cause excessive drain current in the input driver and adversely affect the reliability of a circuit. Certain logic families are particularly sensitive to the high value of the logic swing (i.e., ECL). Slow logic transitions of an input signal can cause oscillation or amplify noise when it drives a GaAs input buffer.

The worst-case values for the logic swing required by different logic families are shown in Figure 4.12. As this figure illustrates, interfacing GaAs circuits to Si circuits requires a translation of logic levels at the I/O. To be flexible in receiving signals from different Si logic families, the input section of a GaAs circuit should be capable of (a) limiting the input current and over-voltage (i.e.,

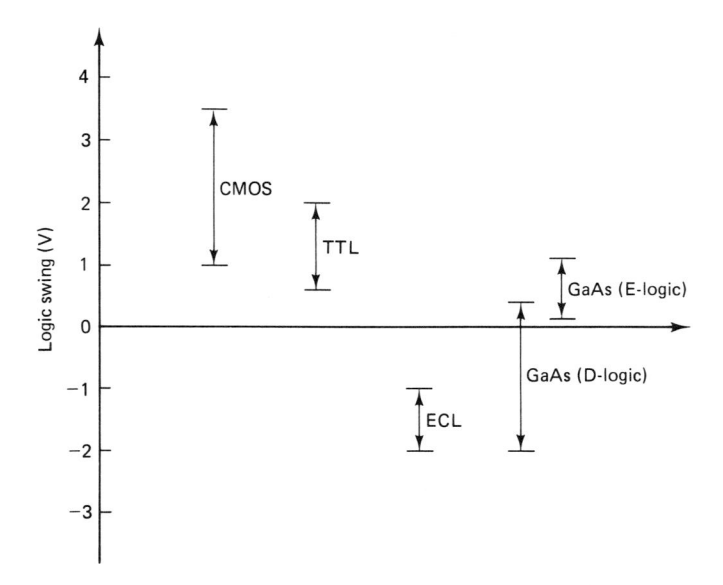

Figure 4.12 Worst-case logic swings for Si and GaAs logic families.

electrostatic discharge protection), (b) providing downward logic voltage level translation for proper MESFET operation at the internal logic section, and (c) adjusting the input logic level to compensate for threshold variations as a function of temperature when interfacing with ECL. Assuming depletion-mode logic, Figure 4.13 shows an approach at the input of a circuit that can be used to provide these capabilities.[15] The input signal is passed through a FET, which operates as a saturated resistor. Although the actual input impedance varies with the frequency, from approximately 5K Ω at dc to 500 Ω at 1 GHz, the input FET will go into current saturation for currents typically greater than 15 mA, not allowing the input current to exceed this level even with input voltage overdrive. The clamp diodes DCL-1 and DCL-2 following the saturation resistor are used to limit the peak-to-peak value of the input waveform peak-to-peak value to prevent damage. This capability allows input waveforms with large swing to be used as inputs to a GaAs circuit. Figure 4.14 shows the action of the clamp diodes when the input waveform is a sine wave, which is often used for clocking GaAs circuits. Given the bias of the diodes at V_{ICH} and V_{ICL}, the voltage $V(1)$ will be as follows:

$$V(1) = V_{ICH} + V_D \text{ for } V_{IN} > V_{ICH} + V_D$$
$$V(1) = V_{ICL} - V_D \text{ for } V_{IN} < V_{ICL} - V_D$$
$$V(1) = V_{IN} \text{ for } V_{ICH} - V_D < V_{IN} < V_{ICH} + V_D$$

The action of these diodes is important because they can effectively "square-up" a high-frequency input sine wave and at the same time preserve or even improve edge rates, as shown in Figure 4.14. These diodes switch at roughly 1000 GHz[15] and their low on-resistance results in a relatively low field, across the gate of the logic circuit following the input, in case of an electrostatic discharge. Input clamping is not needed when normal ECL level logic signals are applied to the input. The action of the diodes can be disabled in this case by connecting V_{ICH} to V_{DD} and V_{ICL} to V_{SS}. Although now the diodes do not have any influence on the shape of the input waveform, they still provide a discharge path for input voltage transients.

Another design requirement in interfacing with high-performance ECL circuits is immunity to logic threshold sensitivity due to temperature and voltage variations. Unlike TTL and MOS logic families, ECL I/O voltage levels shift considerably with temperature variations and can cause noise margin deterioration if this effect is not compensated. For example, if due to temperature rise the output "low" voltage level rises by 100 mV, the same increase should occur in the output "high" voltage level to maintain the required voltage swing and hence the required noise margin. Ideally, the corresponding input voltage levels should change by the same amount. A circuit proposed in Reference 15 to track threshold variations and compensate for differences is shown in Figure 4.15. The outputs of both ECL and GaAs inverters are connected back to their respective inputs. This autozeroed scheme causes each gate to settle to its logic switching threshold. The capacitors at the input of the inverters are used to eliminate

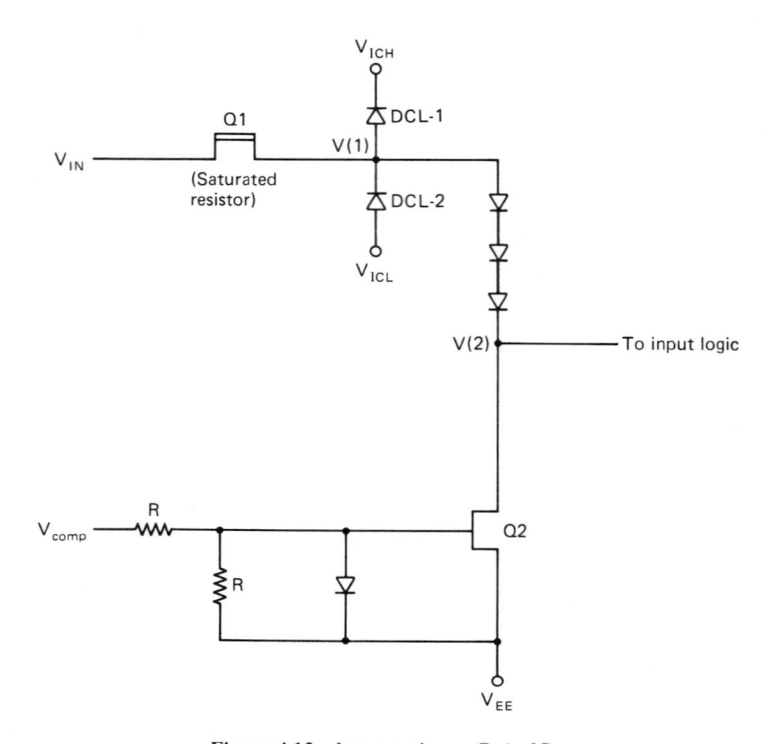

Figure 4.13 Input section to GaAs IC.

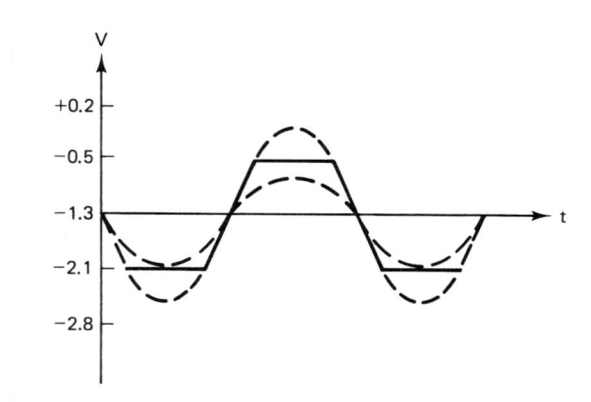

Figure 4.14 Clamping action of the input diodes DCL-1 and DCL-2.

possible oscillations that can occur if the physical length of the feedback connection is not kept short. The difference between thresholds is obtained using a high-gain op-amp. The output of the op-amps, V_{comp}, is used as the V_{comp} input of the GaAs circuit shown in Figure 4.13. In steady state, V_{comp} is the voltage

Sec. 4.5 Interface with Si Technologies

required to make the GaAs switching threshold equal to the ECL switching threshold. Any shift in the ECL gate threshold will result in a feedback adjustment of the V_{comp} which, in conjunction with the bias circuit in Figure 4.13 (i.e., pull down current source $Q2$), will realign the GaAs logic threshold to equal again the ECL threshold. This circuit can be used on a board with GaAs and ECL mixed logic to supply the V_{comp} input to all GaAs circuits, thus maintaining the required system noise immunity.

The output of a GaAs circuit can directly drive a TTL or CMOS circuit by using a FET driver biased at 0 V and having the output terminated at 5 V through a single load resistor R_L. Figure 4.16 illustrates this approach. An appropriate value for R_L is in the range of 500 Ω to 2 KΩ with a preference for high values if a large number of TTL inputs are connected to the output. It should be noted that this scheme provides logically inverted output levels with respect to input levels. The same circuit configuration can be used to provide the interface for ECL or GaAs circuits. In this case, however, the output is terminated at a negative supply V_T and R_L equals the impedance of the interconnection line, Z_o. Figure 4.17 illustrates this circuit. One problem when interfacing to ECL is the output "high" level, which should be kept lower than approximately −0.4 V because

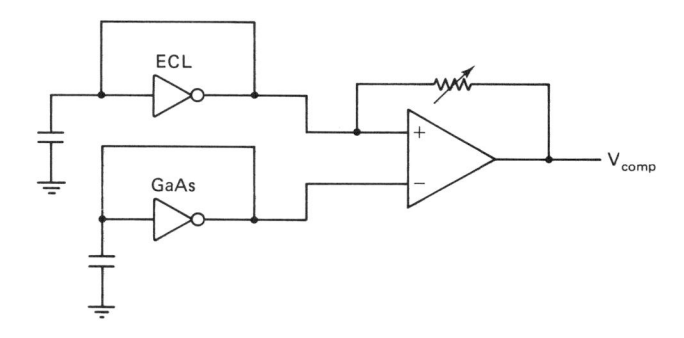

Figure 4.15 Tracking circuit for threshold shift caused by temperature and voltage variations.

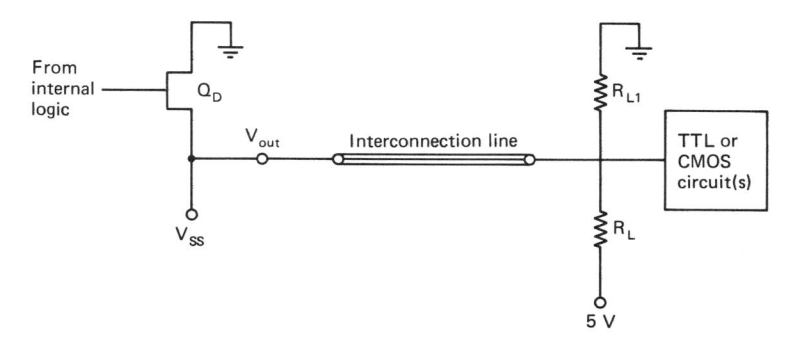

Figure 4.16 GaAs interface to TTL/CMOS circuits.

higher values of V_{out} will tend to drive the input bipolar transistor of the ECL circuit into saturation, thus degrading its ac performance. The "high" value of V_{out} for given values of V_T and R_L depends on the *on* resistance of the output transistor driver Q_D. The smaller the *on* resistance of Q_D, the higher the V_{out}. One way to increase the *on* resistance of Q_D is to limit the "high" level coming from the internal logic that drives Q_D. This can be accomplished using an output clamping diode DCL-1, as shown in Figure 4.17. The bias V_{OCH} can be used to hard limit the high level gate input to Q_D, thus eliminating the possibility of ECL input saturation. The V_{OCH} can be set to V_{DD} to disable the action of DCL-1, in case the interfacing logic is GaAs. Another clamping diode DCL-2 can be used to limit the "low" level gate input to Q_D. This action can aid ac performance, particularly if large negative V_T termination values are employed. If the value of V_{OCL} is set to V_{SS}, the clamping action of DCL-2 is disabled.

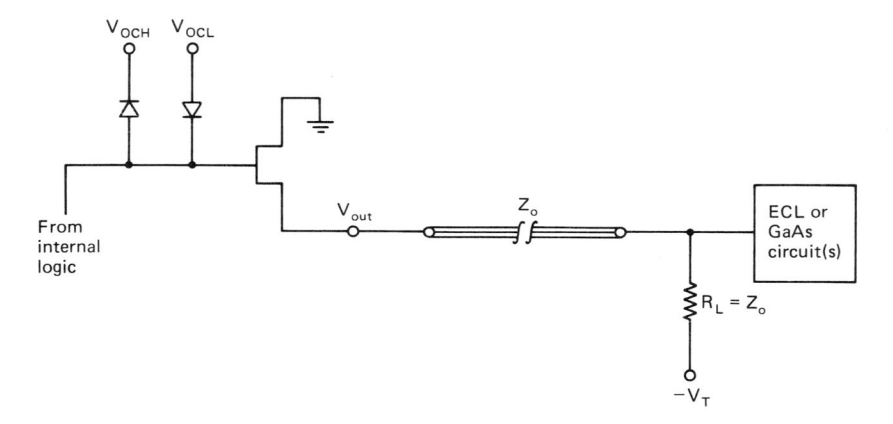

Figure 4.17 GaAs interface to ECL/GaAs circuits.

4.6 A CHIP DESIGN EXAMPLE

A popular design technique for achieving fault tolerance is Triple Modular Redundancy (TMR).[16] TMR can be configured in either software or in hardware. A hardware TMR configuration uses what is known as a voter circuit. Figure 4.18 presents an example of a generic TMR circuit. Circuits $C1$, $C2$, and $C3$ perform the same function with identical data. Although these circuits are not required to have identical designs, they are required to perform their function in step (i.e., identical I/O performance). Under normal operation, all outputs are, of course, identical. In case an error occurs in one of the circuits, this configuration is able to detect the error, isolate the erroneous circuit, and still produce the expected (correct) results. It is clear that the capability of tolerating errors increases the

Sec. 4.6 A Chip Design Example

reliability of the circuit, and this is the main reason that TMR is commonly used in circuit applications with high reliability requirements such as flight controls and space-borne processors. The circuit that performs the error detection and data reconfiguration is the voter circuit in Figure 4.18. The function of the voter is very important because it determines the overall performance of the TMR configuration. In addition, the performance of the voter is very important because it adds delay to the circuit's normal function, as well as power dissipation and size. The design of a voter circuit will be discussed as an example of a GaAs chip design.

The design of the voter is based on the assumption that only one of the circuits $C1$, $C2$, or $C3$ can fail at a given time and an error on the output of one of the circuits does not affect the output of another. Thus, the voter can perform error detection using a majority voting function. Based on this function, the characteristic Boolean expression for outputs $F1$, $F2$, and $F3$ of the voter (Figure 4.17) are as follows:

$$F1 = A_1 \overline{A_2} \, \overline{A_3} + \overline{A_1} A_2 A_3$$
$$F2 = A_2 \overline{A_1} \, \overline{A_3} + \overline{A_2} A_1 A_3$$
$$F3 = A_3 \overline{A_1} \, \overline{A_2} + \overline{A_3} A_1 A_2$$

When $F_i = 1$, then A_i is in error, for $i = 1, 2, 3$. The error reconfiguration (i.e., substitution of an erroneous output with an error-free output) can be performed by defining the voter $O1$, $O2$, and $O3$ outputs as follows:

$$O1 = F1A_1 + F2A_2$$
$$O2 = F2A_2 + F3A_3$$
$$O3 = F3A_3 + F1A_1$$

These expressions determine the logic behavior of the circuit and can be used as the starting point for the chip design. Additional information is needed, however, before the design can begin. This information is related to system requirements and actually determines the chip design specification. In this case, it was specified that the chip must be capable of operating on single-bit inputs, it must perform the error detection function in ≤ 1 nsec, and it must be able to

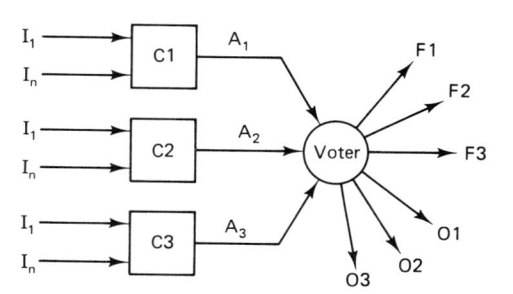

Figure 4.18 A TMR circuit configuration.

reconfigure the data in no more than 3 nsec. Finally, the chip should not dissipate more than 1 W of power, and it should be able to interface directly with ECL components.

These design requirements cover a wide range of applications in which TMR configurations are used. Based on these requirements, the designer can decide what technology to use. However, this decision is also affected by other factors which usually do not appear clearly in the design specification list. Such factors are associated mainly with cost. Specifically, the maturity of technology, yield, and fabrication costs are always factors to consider.

In the case of the voter circuit, because of the speed requirements, technology maturity, and versatility in fanout driving, the decision was made to implement the circuit in BFL. Based on this decision and the logic definition of the chip function, the logic design can begin. It is important to recognize that the logic design is affected by the technology selected for implementation, the reason being that certain types of gates are more efficiently implemented in one technology than others, as discussed in Chapter 2. The selection of BFL favors the utilization of NOR gates or AND-NOR complex gates. Then, the logic design of the voter circuit that implements the logic expressions describing its operation can be performed. Figure 4.19 illustrates the logic configuration of the voter circuit.

The input data is latched at the input ports (I_1, I_2, I_3) during CLK "high" and is valid during NCLK "high." The input latches are flip-flops clocked with a two-phase clock. Fault detection takes place in the upper half of the circuit, and the faulty line is identified with a "high" signal at one of the $F1$, $F2$, $F3$ outputs, which indicate errors at the I_1, I_2, and I_3 inputs, respectively. The error indication signals $F1$, $F2$, and $F3$ will become valid shortly after the data is valid at the output of the input latches, which is on $\overline{NCLK}$ "high."

The data reconfiguration takes place in the lower half of the circuit. This operation is accomplished using the latched input data and the error indication signals $Q4$, $Q9$, and $Q10$ (outputs of NOR gates 10, 11, and 12, respectively, in Figure 4.19). Each output port $O1$, $O2$, and $O3$ can receive data from its corresponding input, I_1, I_2, and I_3, or an alternative input. Direct transfer of data from input to output occurs when there is no fault detection, while the alternate path is used in case a fault is detected. The alternate path is allowed to replace the direct path under the control of the error indication signals. The output is enabled with the clock and it is valid on NCLK "high." Since the output data is not latched, the circuit is not immune to glitches that can occur due to different delays between the data path and the path that produces the error indication signals used to control the data reconfiguration network. The possibility of glitches at the output is minimized by adding three inverters to the data path so that the delays on the paths of the signals that determine the output are equal.

At this point, a logic simulation is performed to ensure that the circuit operates as it is intended. If the logic simulator allows the assignment of different gate delays to the logic gates used, the designer can take advantage of this capability to obtain an estimate of the performance required for the different sections

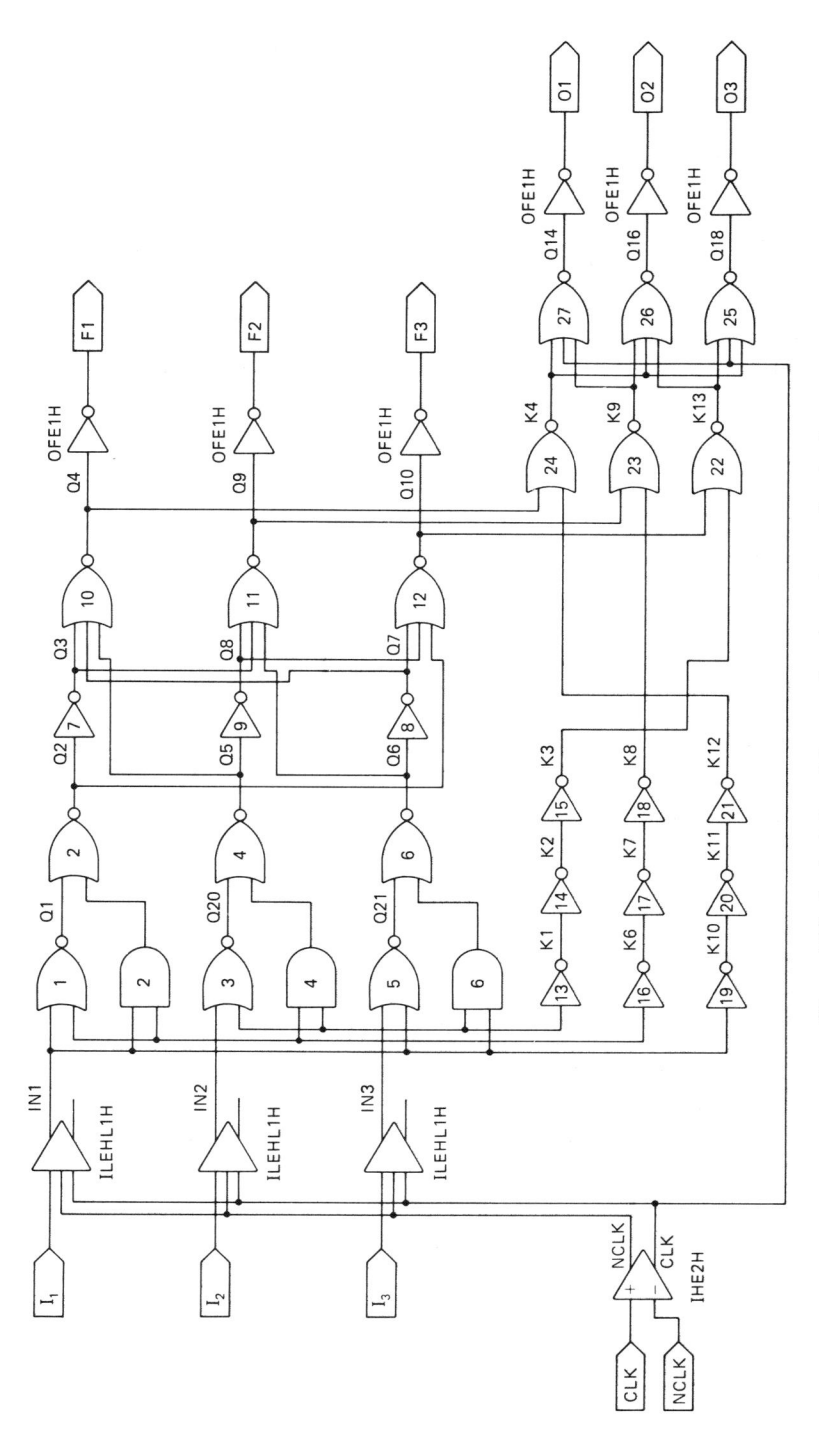

Figure 4.19 Logic design of the error detection and correction circuit.

of the circuit in order to meet the chip design specifications. The logic simulation output for the circuit in Figure 4.19 is shown in Figure 4.20. As can be seen, the logic response of the circuit is in agreement with the expression describing its operation. Based on the steps taken so far, some observations can be made about this circuit. First of all, the amount of logic in the circuit does not make great demands on the chip area because there is a small number of gates and all of them are efficiently implemented in BFL. There are no high fanout points (hence, no requirement for large drivers) and the interconnection network is rather regular (feed-forward, no global feedback). Based on these observations and the desire to maximize the probability of a successful design with the shortest possible design time, the decision to use the standard cell approach for implementing the voter chip is a reasonable one.

The cell selection is based on the performance and power dissipation requirements of the overall design, as well as the driving capability required by

Cycle mode	CNIII LCNNN KL123 K	Display processor OOOFF UUUAA TTTUU 123LL TT YY 12	F A U L T Y 3
Step #	IIIII	OOOOO	O
pwr pindef	xxxxx		
pindef	xxxxx	xxxxx	x
1	10000	11100	0
2	01000	00000	0
3	10100	11110	0
4	01100	00010	0
5	10010	11101	0
6	01010	00001	0
7	10110	11100	1
8	01110	11100	1
9	10001	11100	1
10	01001	00000	1
11	10101	11101	0
12	01101	11101	0
13	10011	11110	0
14	01011	11110	0
15	10111	11100	0
16	01111	11100	0
17	10000	11100	0
18	01000	00000	0
19	10100	11110	0
20	01100	00010	0
21	10010	11101	0
22	01010	00001	0
23	10110	11100	1
24	01110	11100	1
25	10001	11100	1
26	01001	00000	1
27	10101	11101	0
28	01101	11101	0

Figure 4.20 Logic simulation output of the error detection and data reconfiguration chip.

Sec. 4.6   A Chip Design Example   **223**

each cell in terms of fanout, starting with the cells having the highest fanout. In our case, these cells are the latches that sample the inputs and they are selected first, followed by the cells driven by these latches, and so on. Upon completion of cell selection, a logic simulation is performed again with gate delays equal to those of the selected cells (this information is provided to the designer by the foundry and accompanies every standard cell library). The output of the simulation reveals correct operation at the required speed with no race condition. However, it is important to remember this result is an approximation of the behavior of the actual circuit. A timing diagram corresponding to the logic simulation output for different input scenarios is shown in Figure 4.21. The selection of the cells provides enough performance margin to counter any variance in simulation findings that may result in violation of design requirements. If the margin is small, the designer should perform a circuit simulation of the critical path at this point with assumed parasitic loads.

The next step is to place and route the cells. Figure 4.22 illustrates the cell placement, while Figure 4.23 illustrates the routing of the cells. Note the special consideration given to power distribution. Three routes between cell columns are devoted to power distribution for every cell column (6 cells) and perimeter power

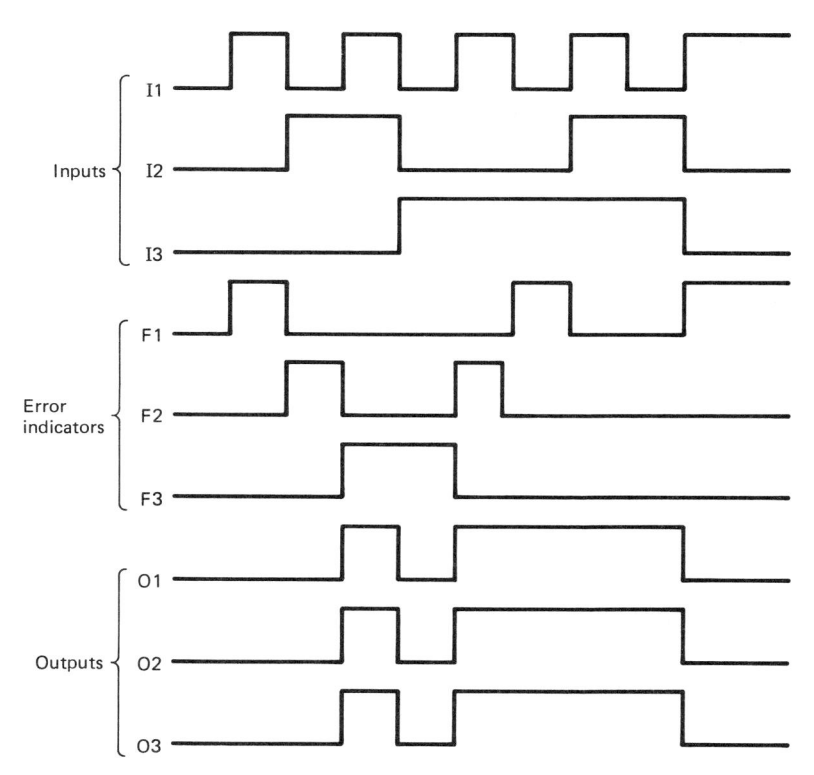

Figure 4.21 Timing diagram of the operation of the voter circuit.

buses are used for the pads. There are multiple power supply and GND bonding pads. Upon completion of the cell routing, a design rule checking program is used to ensure correctness of the layout, and a parameter extraction program is used to determine the actual loading on different nodes with the interconnections included. As Figure 4.24 illustrates, the nodes with the highest fanout and relatively long routes represent the heaviest load. Given the extracted parameters, circuit simulations are performed on different paths, starting with the critical path, to determine if the circuit still meets the design specifications. In addition to performance, these circuit simulations provide the most accurate estimate of the power to be dissipated by the circuit. When all the obtained results meet design specifications, the design is released for mask fabrication and subsequently fabrication of the chip.

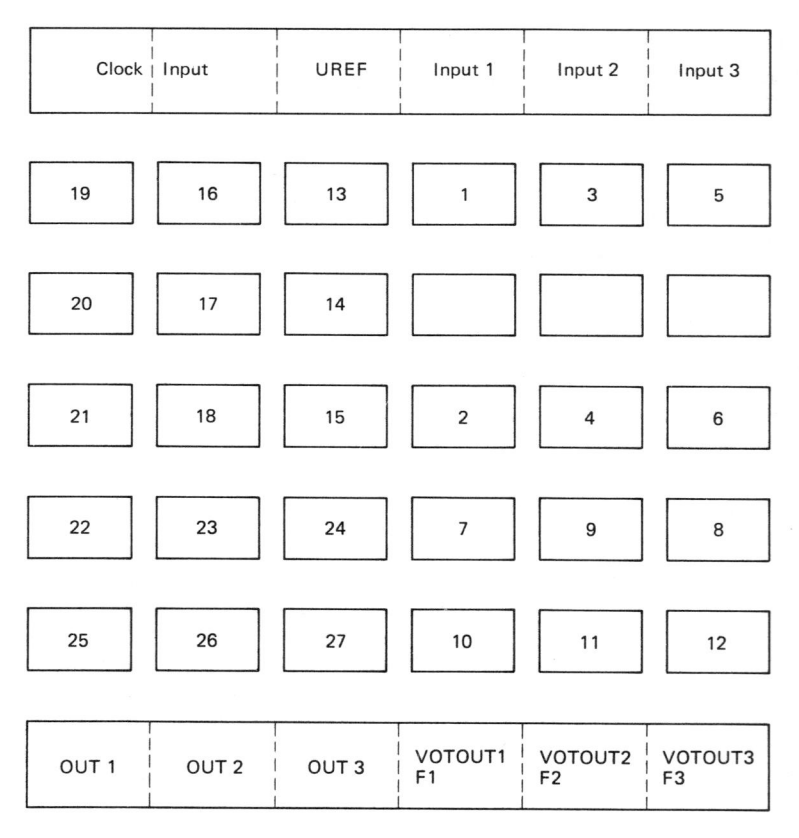

Clock	Input	UREF	Input 1	Input 2	Input 3
19	16	13	1	3	5
20	17	14			
21	18	15	2	4	6
22	23	24	7	9	8
25	26	27	10	11	12
OUT 1	OUT 2	OUT 3	VOTOUT1 F1	VOTOUT2 F2	VOTOUT3 F3

Figure 4.22 Cell placement map (the cell numbers correspond to the numbers of the logic gates in Figure 4.19).

Sec. 4.6 A Chip Design Example

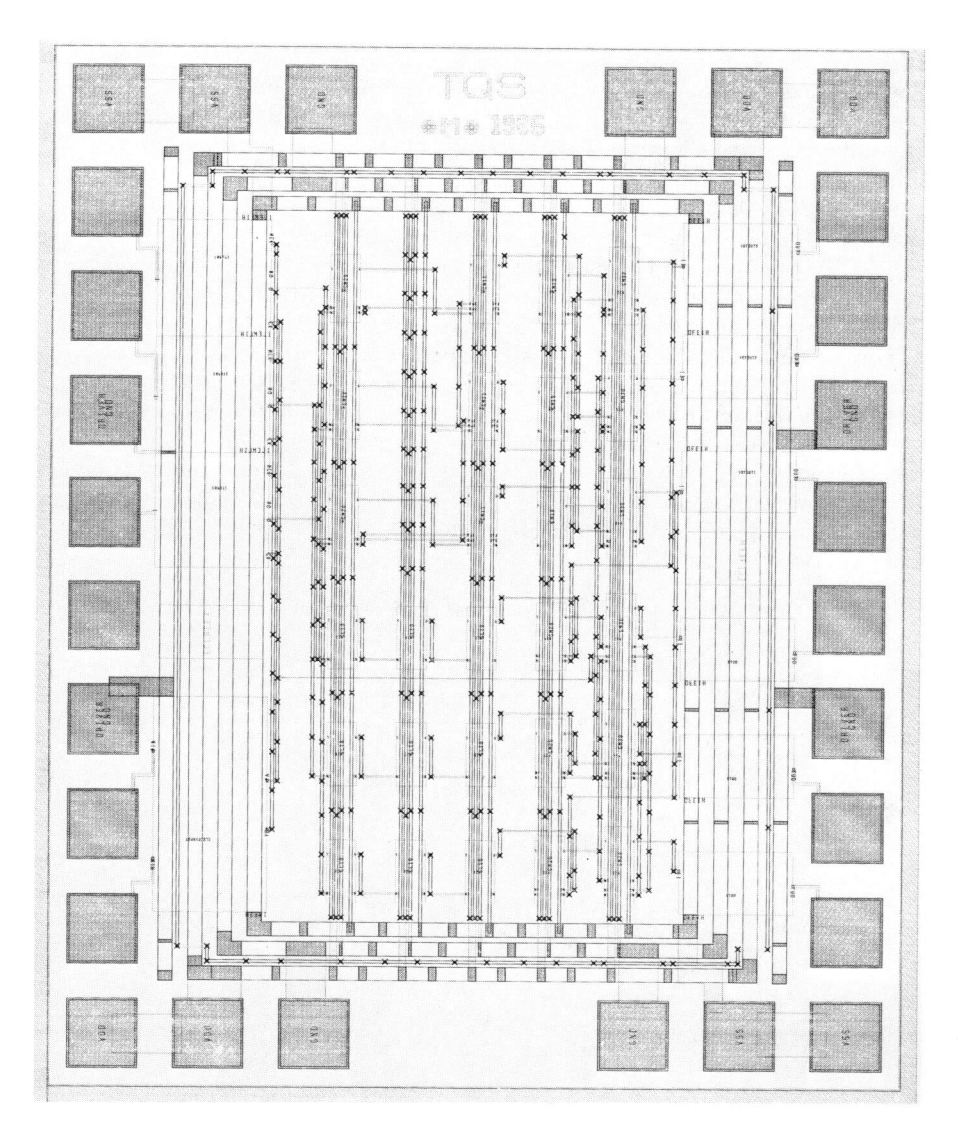

Figure 4.23 Cell routing of the voter chip, including the I/O pads.

A very important step in the design that has not been mentioned thus far is the development of tests for the chip. Testing, design for testability, and test set development will be treated in detail in Chapter 6.

After the chip is fabricated, measurements are performed on standard test devices contained in "process control monitor" (PCM) sites on the wafer to determine the electrical characteristics of the fabrication "run" and the fabrication defect-free die. The identified "good" die are cut and packaged, and the chip can now be tested for functional verification and performance validation.

The voter chip was implemented using BFL with 1 μm minimum gate length devices. The circuit operates at $V_{DD} = 4$ V, $V_{SS} = -4$ V, and ground, and provides ECL-compatible outputs. The output logic voltage swing was measured to be 1.25 V (-0.65 to -1.85), sufficient for adequate noise margins. Measurement of PCM test devices on wafers indicated that the device pinch-off voltage was $V_p = -1.445$ V, with a saturation current $I_{DSS} = 43$ mA, transconductance $g_m = 42.7$ mS, and gate breakdown voltage of 40 V. Less than 1 percent variation of these parameter values was observed over the test sites on the wafer, except the pinch-off voltage, where the variation was 2 percent. Testing of the packaged chip indicated that the data reconfiguration path which provides the chip data outputs operates correctly with inputs having a bandwidth of up to 350 MHz, while the error detection path that provides the error indication outputs operates correctly with inputs having a bandwidth of up to 1002 MHz. The chip power dissipation at the maximum frequency of operation and with the above mentioned bias conditions is 760 mW.[17] A microphotograph of the chip is shown in Figure 4.25.

Node name	Capacitance (fF)	Vias	Airbridge metal (μm)	First-layer metal (μm)
IN3	1.52e + 02	10	1183	462
IN1	1.44e + 02	13	637	665
IN2	1.41e + 02	12	959	490
Q4	1.20e + 02	7	707	469
Q9	1.18e + 02	7	665	476
Q10	1.13e + 02	7	616	469
CLK	9.77e + 01	7	826	266
Q16	8.01e + 01	4	154	462
Q14	8.01e + 01	4	154	462
NCLK	6.40e + 01	4	854	28
Q2	4.67e + 01	4	427	112
K4	3.97e + 01	4	357	98
Q7	3.75e + 01	4	385	70
Q6	3.20e + 01	4	217	112
Q5	3.20e + 01	4	217	112
K13	2.88e + 01	4	231	84
Q21	2.75e + 01	2	63	154
Q20	2.75e + 01	2	63	154
Q1	2.75e + 01	2	63	154
K9	2.67e + 01	4	231	70
K7	2.46e + 01	4	112	112
K2	2.46e + 01	4	112	112
K11	2.46e + 01	4	112	112
Q3	2.12e + 01	3	168	63
Q8	2.07e + 01	3	161	63
K6	1.23e + 01	2	56	56
K10	1.23e + 01	2	56	56
K1	1.23e + 01	2	56	56
K8	1.18e + 01	2	49	56
K3	1.18e + 01	2	49	56
K12	1.18e + 01	2	49	56

Figure 4.24 Parasitic capacitance extraction for the voter circuit after layout. (The node names correspond to the node names in Figure 4.19.)

Sec. 4.7 Architectures Attractive for GaAs Implementation   227

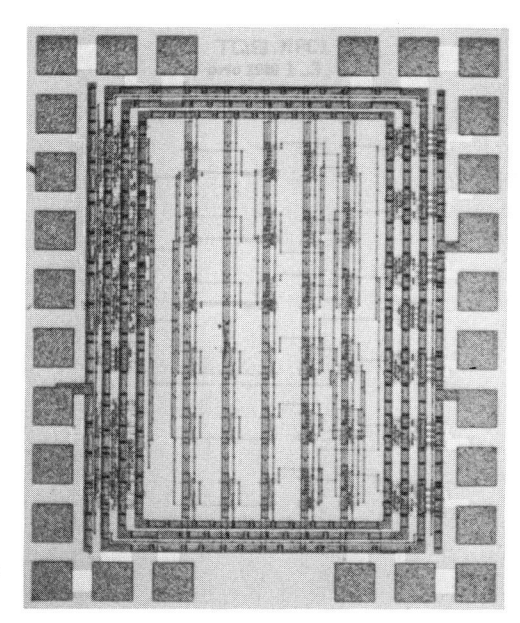

Figure 4.25 A microphotograph of the Vote-and-Reconfigure circuit.

The test measurement results indicate that the chip performs within its design specifications, as predicted by the simulation results obtained during the design phase. In case the test measurements indicate that the chip does not perform according to its design specifications, the designer will have to redesign the circuit after carefully examining the results from the CAD tools used in all design steps and the process parameters measured for the fabrication "run." A redesign is imminent when the process parameters measured agree with the parameters expected by the process engineers—yet the chip fails to meet spec. In this case, simulation models, CAD tools, and the designer's approach should be scrutinized in every step of the design.

4.7 ARCHITECTURES ATTRACTIVE FOR GaAs IMPLEMENTATION

From the discussion on GaAs technology and circuit design presented up to this point, it is clear that the gate speed advantage afforded by this technology makes it attractive for utilization in high-performance system design. However, optimum exploitation of gate speeds leading to significant impact on system performance requires a different approach to on-chip architectures than those used in VLSI Si chips. The main reason for this different approach is that the gate delays of GaAs circuits are comparable to the delays of chip-to-chip interconnections and the latter remain constant, thus limiting system performance. Therefore, decreasing on-chip gate delays by using improved technology, while at the same time requiring more chips to perform a given function, may offer no improvement to the system's performance. Also, the result may be system performance degradation, since interconnect delays rather than the components become the predominant limiting factor on performance. The underlying assumption here, of course, is the difference in the scale of integration available in Si and in GaAs technologies. Optimum use of Si circuits depends on the availability of a large number of gates on each chip, while the effect of gate speed is a second consideration. The designer in this case is usually driven to parallel architectures to obtain high performance, and less emphasis is placed on structures that exploit gate speed. On the other hand, GaAs offers very high gate speeds but significantly lower levels of integration. The on-chip architectures therefore should focus on maximizing their performance while minimizing the amount of system performance degradation due to intercomponent communication delays as well as on-chip propagation delays. A generic approach to achieving this is to maximize the fraction of the system function performed by the GaAs chip, not simply to minimize the time required for a simple logic operation such as multiplexing. One way to maximize the on-chip functionality is to increase the number of gate stages traversed by the data through appropriate selection of arithmetic and logic circuit configurations. For example, consider one of the most widely performed operations in many signal processing applications, namely, the Fast Fourier Transform (FFT). The basic operation in the FFT is:

$$A_1 = A + B$$
$$B_1 = (A - B)W$$

where A, B are the input operands, A_1, B_1 are the outputs, and W is a constant coefficient. An attempt to perform in parallel the computations required for the calculation of A_1 and B_1 (assume 16-bit operands), using arithmetic circuits operating on data words will exceed the chip size allowed by GaAs technology for acceptable yield. In this case, one can implement individual operators (i.e., multiplier, adder) on a single chip and use a chip set to perform the required computations. Although the individual components will have high throughput, the chip-set performance will be dominated by interconnect communication delays. An alternative would be to implement the whole operation on a single chip using bit-serial arithmetic. As discussed in Chapter 3, bit-serial operators can be implemented using approximately $1/n$th (n is the data wordlength) of the logic required by their word-parallel counterparts at the expense of an n-fold increase in the computational delay. However, although the logic reduction comparison is valid, the performance for the overall function is not, because the chip-set cannot maintain the on-chip clock rates over the intercomponent interconnections. Furthermore, the structure of the bit-serial operators allows for low fan-out, very short (next neighbor) on-chip interconnection, thus allowing for optimum exploitation of the gate speeds the technology can provide. Therefore, it is conceivable that the overall function can be performed in a shorter time by a single chip utilizing bit-serial operators than a chip-set employing individually faster word-parallel operators.

In general, circuit structures that are pipelined, and therefore require a minimum of decision branching, appear to offer the best approach for exploiting the inherent speed performance of GaAs technology. These structures include iterative and recursive architectures and systolic arrays.[18] Pipelining can be used both on a bit-by-bit basis as well as a word-by-word basis. The latter case, however, is highly demanding on chip area and scale of integration, both of which are limited by requirements for acceptable yield.

The approach to pipelined architectures can proliferate to memory design when certain applications are considered, where successive rather than random-access of data is required. Many signal and image processing applications fall into this category. When successive access of data is required, generation and update of the address can be performed on a clock cycle and a single "read"/"write" control line may be supplied to the chip at full operating speed, thus allowing reading and writing to be interleaved at full speed. This operation can be achieved using a conventional RAM design and two counters to control "read" and "write" sequences, respectively. The counters in this case replace the off-chip address generator normally used. Various versions of this memory configuration were investigated by the Mayo Foundation and were found to apply to a wide range of high-speed memory requirements.[19] In conventional RAM designs, address generation and updating, as well as address distribution across a page of memory, dominate the performance of the circuit because all of the information that controls the memory operation is generated off-chip on a per-memory-cycle (i.e., "read," "write") basis.

One pipelined memory configuration that can be widely used in signal processing applications is the FIFO memory described in Chapter 3. In this case, "read"/"write" is used as a trigger to enable the corresponding operation for a burst of data, and this operation is controlled by on-chip circuitry, which is also pipelined. The FIFO structure has all the characteristics required for full exploitation of the gate speeds offered by GaAs technology.

PROJECTS

4.1. You are part of the standard cell development group in a company that has access to the BFL process discussed in Chapter 1. You are asked to design a set of cells using the device models presented in Section 1.5.4, following the layout rules presented in Section 4.2.4. Three cells should be designed for each function to address different performance requirements—namely, slow, medium, and fast. The design team decided that slow, medium, and fast correspond to throughput of 500 MHz, 750 MHz, and 1 GHz, respectively, with 10 percent performance margin, which implies that the cells should exceed these requirements by 10 percent under nominal conditions.

All cells should meet the performance specifications for worst-case operation, which is described as driving a 0.05 pF load, at 75°C (nominal is 0.01 pF at 25°C). The cells should operate with $V_{DD} = 4$ V, $V_{SS} = -4$ V, and an input waveform that swings between −2 V and 1 V. The worst-case rise and fall times for the cells should be no more than 200 psec. The user should be able to use the cells by rotating them by 180° without having to route I/O signals around a cell. The user should be provided with the following information concerning each cell:

1. Functional description
2. Logic diagram (gate level)
3. Circuit diagram (transistor level)
4. Performance in different operating conditions
5. Power dissipation (worst case)
6. Noise margins (worst case)
7. Maximum fanout driving capability (assuming cells of the same design)

Design and document a standard cell library that contains the following functions:

BASIC GATES
NOR, NAND, INVERT, AND, OR (select various number of inputs)

COMPLEX GATES
AND/NOR, OR/NAND, EXOR, AND/OR, OR/AND (select various input configurations)

FLIP-FLOPS (two-phase clock, static)
D-Flip-Flop with reset, D-Flip-Flop with preset, D-Flip-Flop with both reset and preset.

All cells should have a standard dimension (vertical or horizontal), which should be indicated to the user in the cell-documentation. During placement, the user should be able to abut cells without a need for additional routing of power and clock signals.

4.2. Using cells from the cell-library developed in Project 4.1, design a comparator chip that functions as follows:

It stores two 6-bit incoming data samples, which are in sign-magnitude representation. It then compares the values of these samples and outputs the greatest sample. If the incoming samples are equal, then a flag is raised indicating this fact (a single output signal becomes high) and the sample value is a valid output. The chip must be able to accept input samples at 500 MHz rate. The two phases of the clock are externally provided.

Perform the functional and logic design of this chip. Verify your design through simulation. Place and route your cells.

At the end of the project you are required to deliver the following:

1. Logic description of the chip that includes functional block diagram and the corresponding cell net-list

Chap. 4   Projects   231

 2. Logic simulation
 3. Timing analysis that demonstrates the chip can meet the throughput requirements
 4. Layout including die size and number of I/O
 5. Estimation of power dissipation

4.3. Design the same chip as in Project 4.2 but without using the standard cells. Since this is a custom design project, in addition to the deliverables of Project 4.2, you are required also to deliver circuit simulation results, at least for the critical path.

For the same performance requirements, what are the area and power dissipation differences between the designs in Projects 4.2 and 4.3? What is the difference in design time?

4.4. Median filtering is a discrete time process in which a window of $2N + 1$ points is stepped across an input signal sequence. For each position of the window, the points inside the window are sorted according to their values, and the median value (midpoint) of the sorted set is taken as the output of the filter. To allow the filter to reach the edges of the signal, N endpoints are appended at each end of the signal. The value of the front endpoints is equal to the value of the first point of the signal; the value of the rear endpoints is equal to the value of the last point of the signal. If $Z(\eta)$ and $W(\eta)$ are the input and output sequences, respectively, then the jth filter output for a window of size $M = 2N + 1$ is:

$$W(j) = \text{Median}\,[Z(j-N),\ldots,Z(j),\ldots,Z(j+N)]$$

where the median of M elements, $Z(i)$, $i = 1,\ldots,M$, is defined as the Kth largest element for M odd and $K = \left\lceil \dfrac{M}{2} \right\rceil$.

The nonlinear nature of median filters makes them attractive for filtering signals with sharp edges or for eliminating impulse noise. Median filters are finding increasing use in high-performance image processing.

Design a monolithic median filter using a window size equal to 3 (i.e., $N = 1$) and capable of operating on signal sequences having samples with only positive values. The samples of the input sequence are described with 5 bits. An input signal becomes valid "high" one clock period before the first sample enters the filter and switches "low" immediately after the last sample enters the filter. This signal indicates the beginning and the end of the incoming sequence. The chip should operate with a two-phase clock, both phases provided externally. The filter should output a point as soon as a point is computed, and it should append the correct sample values at the endpoints of the output sequence. The filter should be capable of filtering incoming sequences providing samples at 70 MHz rate. The maximum number of I/O allowed for signals is 24 and for power 12. The die size should not exceed 70×70 mils. Although this size requirement is not a hard requirement, the success of your design depends on how close your design is to this requirement. The chip should operate with $V_{DD} = 4$ V, $V_{SS} = -4$ V, and it should be able to drive 0.5 pF loads under nominal conditions (i.e., 25°C). The same process as in Project 4.1 will be used to fabricate this chip. You are free to select a design methodology (i.e., custom, semicustom, standard cells); however, you should discuss the reason for your selection. At the end of the project, you are required to deliver the following:

1. Logic description of the chip that includes functional block diagram and the gate-level net-list.
2. Functional and logic simulation results. Select your own input sequence to demonstrate the correct functionality of your design.
3. Timing analysis that demonstrates the chip accepts input samples at 70 MHz rate.
4. Layout including die size and number of I/O.
5. Switch level simulation of enough samples from the sequence used in (2) to ensure correct translation from logic to layout.
6. Estimation of power dissipation.

4.5. Is the design specification for Project 4.4 complete? If not, provide a discussion of the parameters that must be specified so that reliable operation of the chip can be ensured. From the designer's point of view, are there any additional design parameters you need to have specified to complete your design successfully?

REFERENCES

1. L. E. Larson, J. F. Jensen, P. T. Greiling, "GaAs High-Speed Digital IC Technology: An Overview," *IEEE Computer*, p. 21, October 1986.
2. L. Pengue et al, "The Quick-Chip, a Depletion-Mode Digital/Analog Array," *Proc. GaAs ICs Symposium*, p. 27, 1984.
3. G. Lee et al, "A 432-Cell SDFL GaAs Gate Array Implementation of a Four-Bit Slice Event Counter with Programmable Threshold and Time Stamp," *Proc. GaAs IC Symposium*, p. 174, 1983.
4. N. Toyoda et al, "A 42 psec 2 K-Gate GaAs Gate Array," *ISSCC Digest of Technical Papers*, p. 206, 1985.
5. W. H. Davenport, "Macroevaluation of a GaAs 3000 Gate Array," *Proc. GaAs ICs Symposium*, p. 19, 1986.
6. H. Nakamura et al, "A 390 psec 1000-Gate Array Using GaAs Superbuffer FET Logic," *ISSCC Digest of Technical Papers*, p. 204, 1985.
7. D. Kinell, "A 320 Gate GaAs Logic Gate Array," *Proc. GaAs ICs Symposium*, p. 17, 1982.
8. A. Peczalski et al, "12×12 Multiplier Implemented on 6 K Gate Array," *Proc. Government Microcircuits Conference*, p. 177, 1986.
9. L. W. Nagel, "SPICE 2: A Computer Program to Simulate Semiconductor Circuits," Electronics Research Laboratory, Univ. of California at Berkeley, *Memorandum ERL-MS20*, May 1975.
10. H. S. Baird, "Fast Algorithms for LSI Artwork Analysis," *Proc. 14th Design Automation Conf.*, p. 303, 1977.
11. C. M. Baker, C. J. Terman, "Tools for Verifying Integrated Circuit Designs," *VLSI Design Magazine*, p. 20, 4th Quarter 1980.
12. T. G. Szymanski, C. J. Van Wyk, "Space Efficient Algorithms for VLSI Artwork Analysis," *Proc. 20th Design Automation Conf.*, p. 734, 1983.

13. R. L. Spickelmier, A. R. Newton, "Wombat: A Net Netlist Comparison Program," *Proc. Intl. Conf. on CAD*, p. 170, 1983.
14. M. Hofmann, V. Lauther, "HEX: An Instruction Driven Approach to Feature Extraction," *Proc. 20th Design Automation Conf.*, p. 331, 1983.
15. J. Haight, "Working with GaAs ICs," *Electronic Design News (EDN)*, June 1984.
16. F. P. Mather, A. Avizienis, "Reliability Analysis and Architecture of a Hybrid Redundant Digital System: Generalized Triple Modular Redundancy with Self-Repair," *Spring Joint Computer Conf., AFIPS Conf. Proc.*, vol. 36, p. 375, 1970.
17. N. Kanopoulos, "A Monolithic GaAs Error Detection and Correction Circuit for Triple Modular Redundancy Applications," *Proc. Intl. Conf. on Custom Integrated Circuits*, p. 85, 1987.
18. H. T. Kung, "Why Systolic Architectures," *Computer* 15, no. 1, p. 37, January 1982.
19. B. K. Gilbert et al, "Signal Processors Based Upon GaAs ICs: The Need for a Wholistic Design Approach," *Computer* 19, no. 10, p. 29, October 1986.

5
PACKAGING

Design of high-speed GaAs integrated circuits is the first step toward the ultimate goal—to use such circuits in very high performance systems. The next steps are packaging and high-speed testing. This chapter presents some techniques associated with high-speed integrated circuit packaging; Chapter 6 will examine testing techniques.

The methods by which integrated circuits are assembled, mounted, interconnected, provided with ground, power supplies, and signals, and protected against noise and temperature variations bear as much importance as their design. As already mentioned (and treated in greater detail in Chapter 7), at very high speeds signal lines and clock distribution networks are more complicated than they might appear at first. Stray capacitance and inductance can become limiting performance factors, while unequal conductor lengths can introduce timing errors. To maintain the high performance of an integrated circuit, its package must have short delays and minimize the ringing caused by the package loading. Package dielectrics and conductors must contribute minimal losses at very high frequency (i.e., GHz range) signals. The thermal, mechanical, and chemical requirements, which affect circuit reliability, also affect the type of packaging used.

The packaging scheme used for a high-performance system based on GaAs integrated circuits will typically have three components: chip carriers, multilayer cards interconnecting the chips, and multilayer boards interconnecting the multilayer cards.

5.1 PACKAGE DESIGN CONSIDERATIONS

The design of the chip-carrier package must carefully consider the factors that affect signal timing and propagation delays, noise immunity, and temperature effects on circuit performance. Also, the design of the package should consider the utilization of the chip at the next level of design hierarchy (i.e., board level) and should strive to minimize performance degradation when operating at that level. After all, the merit of an improved technology is assessed on the basis of the overall impact of this technology on the performance of systems—not just on bare chips.

5.1.1 Package Size

The time required for a signal to propagate over the substrate of the dice is not the only delay factor involved in producing the signal at the edge of the package. The package itself introduces a delay component fundamentally associated with the distance over which the signal has to travel through the package to reach the outside world. Two factors affect the magnitude of this delay component. One factor is the size of the chip-carrier and the length of the bonding wires connecting the dice pads to the package I/O conductors; the second factor is the dielectric constant of the material used to fabricate the package. The size of the chip-carrier should be as small as possible because, in general, this minimizes the length of bonding wires and the overall distance a signal has to travel from the pad to the package I/O. The length of the package, however, should be consistent with the I/O density capability of the printed circuit board where the chip-carrier is mounted. In other words, the I/O spacing of the package should not be more dense than the interconnect spacing allowed by the board. The small package size allows also for shorter interconnection lengths that a signal has to travel between communicating chips, thus allowing for shorter intercomponent delays and therefore higher system performance.

The speed of a signal on a line (as discussed in Chapter 2) is inversely proportional to the dielectric constant of the material into which the line is deposited. Therefore, the propagation delay through the chip carrier can be minimized if the chip carrier is made from low dielectric constant material. For example, a 1 cm × 1 cm ceramic package would add about 50 psec delay to the input or output signals to the chip, but the same size carrier made from a material with a lower dielectric constant such as SiO_2 would add only 35 psec to the I/O signals.[1]

5.1.2 Package Loading

While the on-chip signal lines are unterminated and, in general, voltage and timing differences that exist along a signal line can be ignored, this approach does not work for board level design. Here, signal lines must be treated as distributed transmission lines where consideration must be given to discontinuities and propagation delays. In a transmission line interconnection network, the chip package appears as an open-ended transmission line load (or a stub) to the lines carrying high-speed I/O signals.[2] To minimize ringing or voltage standing waves (VSW), the loading of the transmission line attributed to the package must remain minimal. There are three techniques commonly used to minimize the package loading of the line: (a) shortening of the stub length, which in this case comes in the form of wire bonds and package leads—typically the propagation delay of the stub should be less than or equal to one third of the rise or fall time of the signal driving the line; (b) increasing the impedance of the stub to a point consistent with crosstalk requirements; (c) adding a resistor in series with the stub, which effectively increases the stub impedance and can also provide impedance matching between the stub and the transmission line.[1] A rigorous analysis for the justification of these techniques is based on principles of impedance transformation and matching in transmission line systems and is beyond the scope of this book. The interested reader can consult on this subject with Reference 2 as well as with any of the many texts on modern electromagnetic theory.

5.1.3 Crosstalk

In addition to issues concerning the timing and integrity of individual signals, special consideration should be given to signal coupling and crosstalk, which can cause logic errors. Two main coupling problems require consideration—namely, signal line to signal line coupling and common mode coupling. Signal line coupling can occur when densely packed signal lines run parallel to one another, while common mode coupling occurs when signal lines share common ground or power supply connections. Common mode coupling is most critical where signal lines are connected from the pads to the package or from the package to the printed circuit board. The coupling effect becomes more pronounced as the number of signal lines sharing a ground connection increases, because the coupling noise induced by the simultaneous switching of the circuits driving these lines is additive. Since crosstalk can result in logic errors, shielding of high-speed signal lines within the package is essential. To reduce the effect of common mode coupling, high-speed signal lines can be shielded by alternating ground wires between these signal lines, and using multiple power supply connections. The alternating wireband approach, however, implies that the number of interconnections required to implement the same functional circuit increases, thus increasing the number of I/O pads on the chip and pin-outs on the package. However, a higher number of package pin-outs implies a larger size package and, hence, a slower package. Therefore, a trade-off must be made in deciding the level of crosstalk that can be tolerated against the amount of delay that results when using a larger size package.

5.1.4 Power Supply Decoupling

Decoupling of power supply lines from the logic circuitry is also a major consideration, because a noise pulse is generated in the power supply circuit when high-speed switching of large amounts of current passes through to the series impedance of the power supply and common wire-bound inductance. This power supply noise can be reduced by reducing the power supply source impedance, which can be accomplished by using very low series inductance decoupling capacitors, or capacitive layers fabricated as part of the package, normally called power planes. For slower logic technologies the decoupling capacitors can be placed on the printed circuit board in close proximity to the chip. For GaAs circuits, however, it is necessary to incorporate at least part of the decoupling function within the chip-carrier and within a short wire bonding distance from the GaAs dice.[3] Here again, it is critically important to provide low impedance connections to the actual circuit on the dice.

5.1.5 Temperature Effects

The thermal characteristics of the package are important because temperature variations will influence the operation of the circuit in switching speed or logic levels. A typical situation arises when interconnected chips in different parts of a system must operate at different temperatures with the consequential danger of mismatch between logic levels. In most cases it is not so much a question of one circuit not being able to drive another operating at different temperature, but of a much-reduced system noise immunity.

Heat removal from the package is facilitated when the package material exhibits inherently high thermal conductivity and some form of heat sink is used. The semiconductor junctions in the circuit have a maximum operating temperature, and it may be necessary to calculate the actual temperature to determine the margin of safety for reliable operation.

The junction temperature is given by

$$T_j = \Theta_{ja} P_d + T_a \quad (5.1)$$

where P_d is the power dissipated, T_a is the ambient temperature and Θ_{ja} is the thermal resistance between junction and ambient. This factor can be expressed as:

$$\Theta_{ja} = \Theta_{jc} + \Theta_{cs} + \Theta_{sa} \quad (5.2)$$

where Θ_{jc} is the thermal resistance from junction to package, Θ_{cs} is the thermal resistance from package to heat sink, and Θ_{sa} is the thermal resistance between the heat sink and ambient. The value of these variables depends on the material and the geometry of the package and the heat sink configuration used. Given these coefficient values for a specific package, the junction temperature can be calculated for a circuit and compared to the value specified by the process for reliable operation.

5.2 CHIP-PACKAGE DESIGNS

The design of chip-packages for GaAs circuits is considered critical in exploiting the potential of GaAs technologies, because the circuit performance can be compromised by packaging inadequacies. It is generally recognized that innovative package designs, beyond those which are commercially available in the late 80s, will be needed to accommodate digital circuits with the high pin counts that are routinely required in a wide range of applications.

High-performance packages for GaAs circuits have been developed primarily by a small number of systems and component houses to package their own GaAs components. This section examines some of the representative package designs and their characteristics.

5.2.1 Multilayer Ceramic Packages

A multilayer ceramic package using four layers of metalization is introduced by GigaBit Logic in Reference 1. A cross section of this package is shown in Figure 5.1. As shown, the bottom metal layer provides the pads for connecting the package to a printed circuit board. This layer also provides for metalized vias, which are used to increase the thermal conductivity of the package, thus enhancing the transfer of heat generated by the circuit to the environment. The next layer is the ground plane, which shields the signal lines and provides for an attachment pad for the GaAs dice. The next layer provides for the wire bonding connection between the circuit pads and the package and the input and output signal lines. High-impedance strip lines are used for the input signals while 50 ohm strip lines are used for the output signals. The I/O propagation delay attributed to the package with this line structure is 50 psec, while the maximum crosstalk for a

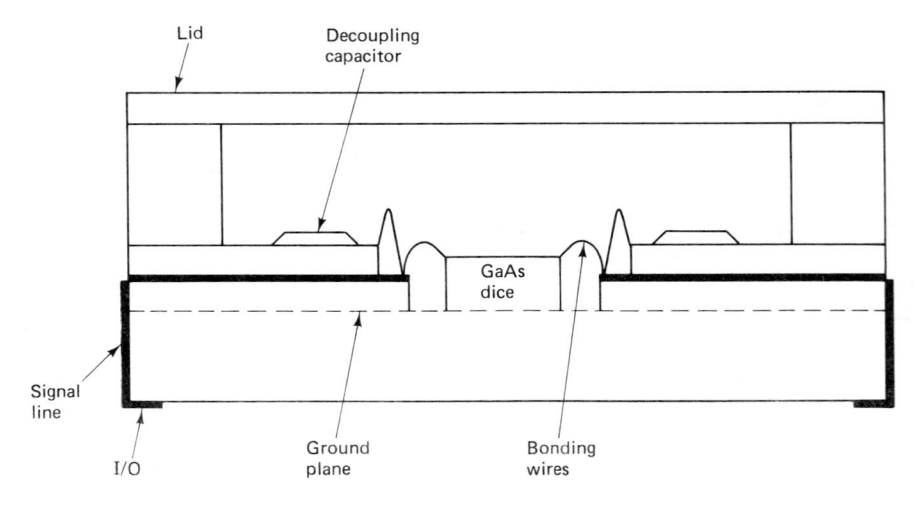

Figure 5.1 Cross section of a multilayer ceramic package.

1 V signal with 100 psec rise (fall) time is 60 mV. Another ground plane is provided by the fourth metal layer, which is used for mounting low inductance bypass capacitors for common mode decoupling. These capacitors have 100 pF capacitance and a series inductance of 2.5 mH. The overall size of this package is 1×1 cm^2. It has 36 pins and can accommodate up to 3 GHz bandwidth signals. The package is leadless—the most commonly used chip carrier for GaAs circuits. A leadless package offers slightly better speed performance than a package with leads (i.e., pin grid arrays, flatpacks), because it allows a signal to propagate over a constant-impedance transmission line right up to the package rather than propagating across an impedance discontinuity imposed by the lead. A photograph of this multilayer ceramic chip-carrier is shown in Figure 5.2.

A family of leadless multilayer ceramic packages with higher pin counts was developed by TriQuint Semiconductor.[3] This family includes a 44-pin and a 132-pin package. The pads on the 132-pin package are placed on 0.635 mm centers, while the spacing of pads in the 44-pin package is 1.27 mm. Both packages include power planes, which provide equivalent bypass capacitors of 350 pF for power supply decoupling. Provisions allow for two power supplies and decoupling is provided for each of them. The provision for two power supplies makes this package suitable for depletion-mode circuits. The signal transmission is carried out through a modified microstrip structure which, for the 132-pin package, offers a 100 psec propagation delay for the longest connection from the contact pad to the circuit bonding pad. Of the 132 pins, 64 are used for high-speed signals, 4 for power supply, and the remaining pins are used for ground con-

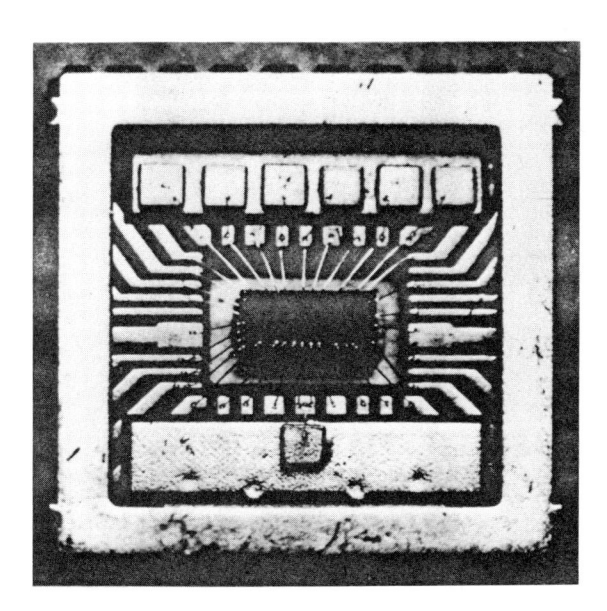

Figure 5.2 Photograph of an assembled 36-pin, multilayer ceramic chip carrier.

nections. The use of a large number of ground connections is not surprising in this case, because alternating ground wires shield the high-speed signal lines and thus reduce the coupled noise. The useful bandwidth for this family of packages was determined through measurements to be up to 2 GHz.[3] The size of the 132-pin package is 2.41×2.41 cm^2.

Another effort to develop low and high pin-count packages was undertaken by the Mayo Foundation in 1986. This package design effort focused on two categories of packages: a low pin-count (up to 64-signal I/O) package operating in the GHz range, and a high pin-count (up to 240-signal I/O) package operating in the 25–200 MHz range. The packages designed during this effort were an 88-pin package (64 I/O, 24 power and ground pins) and a 216-pin package (192 I/O, 24 power and ground pins). Figure 5.3 shows the 88-pin package. Each corner of the cavity is provided with two power contacts and a ground contact. The ground contact is made wide enough so that a dual parallel plate capacitor can be used adequately to decouple the power lines. This power-ground system at the corners of the package provides the required power feeds for the core of the chip. The pad drivers receive power feeds from eight dedicated contacts (two on each side of the package), while four termination plan contacts are used for on-chip ter-

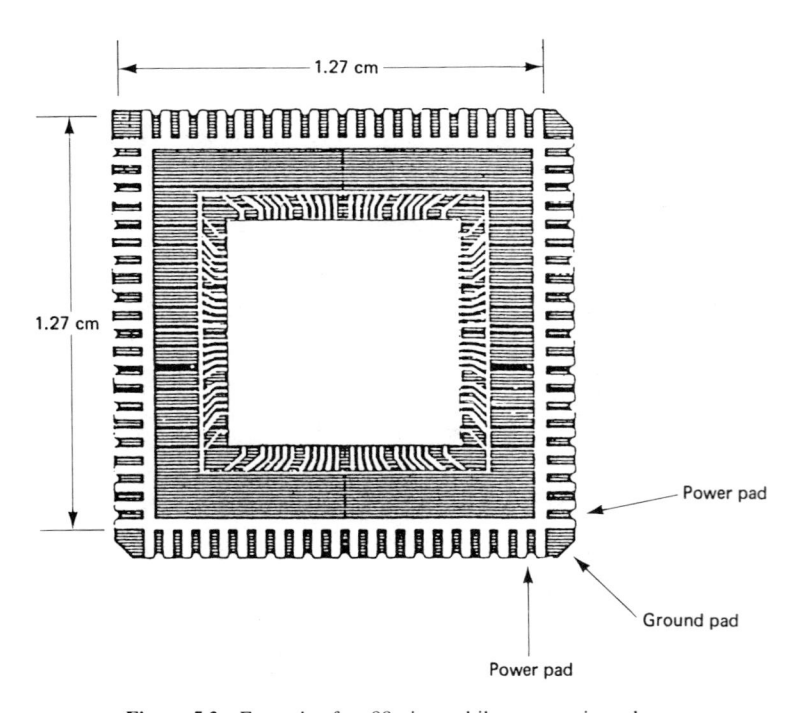

Figure 5.3 Example of an 88-pin, multilayer ceramic package.

Sec. 5.2 Chip-Package Designs

mination resistors. The chip carrier is designed with a metallic base, which can be soldered to the circuit board, providing a low resistance thermal path for heat removal from the chip. The signal transmission lines are designed to be 50 ohm microstrip or striplines and the signal line bondwire lengths range from 0.317 mm to 0.482 mm. These signal line lengths are achieved by placing the perimeter pins at a 0.50 mm pitch. Although the line impedance is controlled, the bond finger will behave as a stub when the line drives multiple logic gates.

The interesting part of this effort was the fabrication of the 88-pin package using four different materials, namely cofired alumina, cofired beryllia, copper/polyimide, and thin film beryllia. The package performance varied with the different materials, as expected, because of the differences in their dielectric constants. Measurement results obtained by the Mayo Foundation and published in Reference 4 are shown in Table 5.1.

If the 0.50 mm pitch is maintained for the 216 I/O package, the package size will be 3.04 × 3.04 cm. To decrease the chip carrier size, a very fine pitch of 0.25 mm was proposed, resulting in a package size of 1.65 × 1.65 cm^2 and bond finger lengths of 3.17 mm to 4.44 mm. The design of this package is very similar to the design of the 88-pin package described above.

TABLE 5.1 Characteristics of an 88-Pin Package Fabricated Using Four Different Materials

COMMON CHARACTERISTICS	
Package Size :	1.27 × 1.27 cm^2
Cavity Size :	0.635 × 0.635 cm^2
I/O Pad Pitch :	0.50 mm
Bond Finger Length :	3.17 to 4.82 mm
I/O Pins :	64
Power/Ground Pins :	24

SPECIFIC CHARACTERISTICS				
	Cofired Alumina	Cofired Beryllia	Copper/ Polyimide	Thin Film Beryllia
Bandwidth	3.0 GHz	4.5 GHz	8.0 GHz	Not available
Dielectric constant	9.5	6.5	3.5	6.5
Transmission line	Microstrip	Stripline	Stripline	Microstrip
Signal line width	0.10 mm	0.10 mm	0.10 mm	0.07 mm
Signal to ground plane spacing	0.15 mm	0.20 mm	0.12 mm	0.07 mm
Thermal resistance	0.3° C/W	0.25° C/W	0.3° C/W	0.25° C/W
Special features	Hermetic seal	Hermetic seal	–	Internal Resistors
Vendor	Interamics	Brush/Wellman	Augat/Microtec	Gen Microwave

5.2.2 Silicon-Based Package

A silicon package has been proposed in References 5 and 6 for different circuit technologies; a novel implementation of this idea for GaAs circuits has been developed by GigaBit Logic as reported in Reference 1. The package is constructed by first attaching the silicon chip carrier to the outer package and then mounting the GaAs die on the silicon chip. Conductive epoxy is used for both attachments. The compliance of the epoxy eliminates possible stress problems due to the mismatch of the thermal coefficients of expansion of GaAs and silicon. Wire bonds are used to connect the die pads to the silicon chip and then silicon chip with the loaded or leadless chip-carrier. Gold connections are used from the GaAs die to the silicon chip and aluminum connections are used from the silicon chip to the package pins. Bypass capacitors and terminating resistors are included on the package. The bypass capacitors are implemented as reverse biased p-n junctions while the terminating resistors are implemented using diffusion. The transmission line environment in the package is coplanar microstrip. Signal losses in the silicon package are attributed mainly to the dielectric loss in the Si and the resistive and skin losses in the aluminum lines. The dielectric loss depends on the resistivity of the Si substrate, and it can be made negligible by using a high-resistivity Si substrate (i.e., over 100 ohm-cm).[1] The loss due to the aluminum resistance and skin effect can be reduced by increasing the cross section of the metal lines in the package interconnection.

The characteristics of the silicon package developed by GigaBit Logic were found to be better than the characteristics of an equal size multilayer ceramic package developed by the same company, described in Section 5.2.1. The 36-pin, 1×1 cm^2, silicon chip package has an I/O propagation delay of 35 psec, thermal resistance (chip-to-package) of 5° C/W, and maximum crosstalk, for a 1 V signal with 100 psec rise and fall times, of 40 mV. The bypass capacitor used in this case is 300 pF, which is larger than the capacitor used in the ceramic package.[1]

Figure 5.4 shows the configuration of the silicon chip package. Extensions of the silicon chip package design may include power supply voltage generation and regulation circuits that potentially can reduce the requirement for multiple power supply leads for GaAs circuits.

5.2.3 Hybrid Packages

Hybrid packages are multichip packages containing a number of dice with minimal chip-to-chip spacing while providing a controlled impedance transmission environment within, to, and from the package. These packages must incorporate advanced thermal management technologies to control the high heat load that results from the high chip density in the package. Figure 5.5 illustrates a hybrid package suitable for GaAs systems. The hybrid package is configured similarly to a single dice package. A cofired ceramic package is utilized to provide hermeticity, electrical connections to the outside world, and distribution of

Sec. 5.2   Chip-Package Designs   243

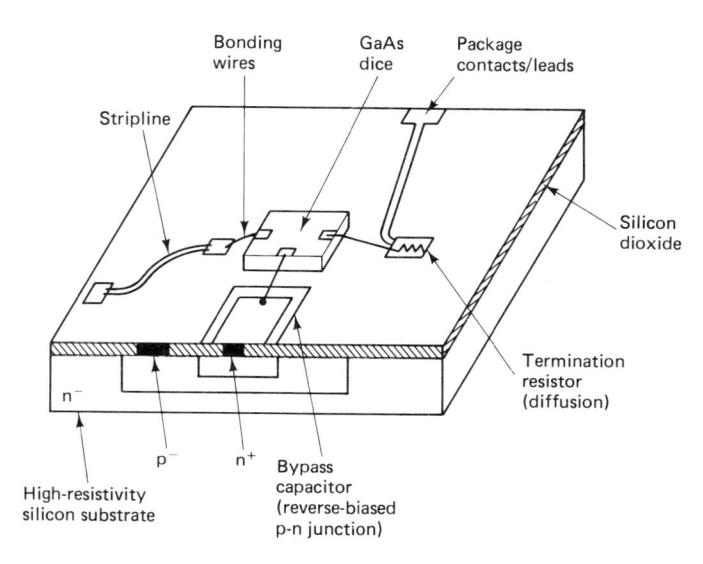

Figure 5.4 Configuration of the silicon-based package.

power and ground. The signal interconnections between the die in the package transmit through the interconnection plane—one of the critical design issues of the package, since it must provide a controlled impedance environment for high-speed signal routing. A combination of copper and polyimide can be used in this layer to take advantage of the low resistivity of copper, which will result in low loss, and the low dielectric constant of polyimide, which will allow high-density routing in a fine-line microstrip or stripline environment. Power and ground planes can also help minimize potentially hazardous voltage transients. Since several dice are housed in the package, the heat transfer becomes a major design issue. Increasing temperature will affect the performance of the circuits in the package, eventually limiting the viability of the packaging scheme. An approach for obtaining high heat transfer rates from the dice to the base of the package is to cut vias through the polyimide interconnection plane and bond the dice directly to the package base through the vias. Furthermore, the heat transfer works better if the package base is metallic instead of ceramic due to better thermal conductivity of the metallic material. The trade-off in this case is the reduction of the area available for signal routing in the interconnection plane.

The advantage of hybrid packages is the high chip density that can be achieved at the system level without compromising the performance of individual dice. Integral functions such as memory or processor-memory combinations can be housed in a hybrid package to achieve high circuit performance at the system level, due to the close dice proximity allowed by this packaging scheme and the absence of the large discontinuities associated with the chip-to-board interface. However, hybrid packages are viable only when the heat removal issues have

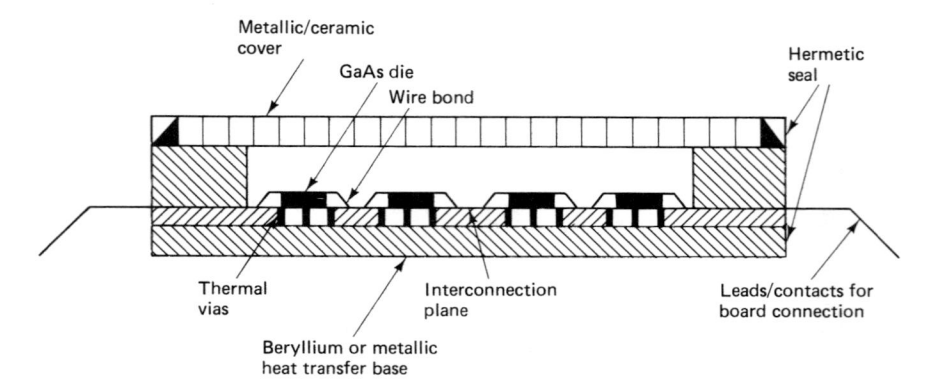

Figure 5.5 Cross section of a hybrid package.

been resolved, because high temperatures will cause reliability problems and performance degradation.

PROJECTS

5.1. Study and present to class a paper of your choice on GaAs packaging, with current results on this subject. Potential sources for your paper selection:

1. *Proceedings of GaAs ICs Symposium*
2. *IEEE Journal on Solid State Circuits*
3. *Electronic Packaging and Production*
4. *Semiconductor International*
5. *Electronic Design News (EDN)*
6. *Electronics*

REFERENCES

1. T. R. Gheewalla, "Packages for Ultra-High Speed GaAs ICs," *Proc. GaAs IC Symp.*, p. 184, 1984.
2. C. H. Durney, C. C. Johnson, *Introduction to Modern Electromagnetics* (Section 8.7), New York: McGraw Hill, 1969.
3. D. H. Smith et al., "New Approaches to Packaging for High Speed GaAs IC Applications," *Proc. GaAs ICs Symp.*, p. 151, 1985.
4. D. J. Schwab, B. K. Gilbert, "Development of High Lead Density Mini Chip Carriers for Gallium Arsenide Digital Integrated Circuits," *Proc. GaAs ICs Symp.*, p. 177, 1986.
5. D. J. Bodendorf et al., "Active Silicon Chip Carrier," *IBM Tech. Disc. Bulletin*, vol. 15, no. 2, p. 656, July 1972.
6. M. Ketchen et al., "A Josephson Technology System Level Experiment," *IEEE Electron Dev. Letters* EDL-2, no. 10, p. 260, October 1981.

6

HIGH-SPEED TESTING AND DESIGN FOR TESTABILITY

A crucial aspect in the development and effective utilization of GaAs digital integrated circuits is their testing. This is the ability to determine whether or not a circuit is functioning according to its design specifications and also the ability to isolate the cause of a malfunction to a specific section of the circuit. An integrated circuit design must be testable at all different stages of its lifetime:

1. prototype
2. production
3. in the field

There are different test requirements for these stages. In the prototype stage, tests are performed to validate the circuit design, correct possible design mistakes, and characterize the circuit in terms of static and dynamic performance. In the production stage, tests are performed to identify circuits that are inoperable, mainly due to fabrication defects. In the field, tests are performed at the board level to identify a circuit or a collection of circuits that causes the malfunction of the board. Failure of a circuit at this level is attributed mainly to environmental and operating conditions, as well as aging—assuming, of course, that the circuit was fault-free during production testing.

There are three facets in testing an integrated circuit at any stage of its lifetime:

1. test generation
2. test application
3. test evaluation

Test generation refers to the activity of developing the appropriate logic stimulus to exercise the function of the circuit under test. *Test application* refers to the actual application of the logic stimulus to the circuit under test and the capturing of the circuit's response. *Test evaluation* refers to the process of interpreting the response of the circuit under test and determining the presence and the location of assumed faults with some predetermined level of confidence.

Although extensive research has been performed on the testing of integrated circuits, many problems remain and the emerging GaAs technologies only compound the difficulty of providing technically feasible and cost-effective solutions. In fact, up to the time of writing this book (1988), little progress has been made in developing general purpose techniques and structures for testing GaAs digital ICs.

This chapter discusses the problems encountered in high-speed testing and presents potential design techniques that can enhance circuit testability. Although integrated circuit testing is a very broad topic, this chapter highlights all pertinent general issues and focuses on specific testing problems associated with the inherent performance characteristics of GaAs integrated circuits.

6.1 TESTING ISSUES

The testing of an integrated circuit has been treated as an activity separate from its design for a long time. With emerging LSI and VLSI technologies, however, the ability to generate test patterns automatically and evaluate these test patterns via fault simulation has waned drastically. As a result, semiconductor and system manufacturers are taking rigorous approaches to integrate design and test into a single activity that includes a collection of techniques known as *design for testability*. These techniques are aimed at providing design characteristics that make possible the testing of an integrated circuit in a cost-effective way. In addition to test generation and evaluation complexities, a major problem is test application. This is a recurring problem appearing each time testing of circuits, fabricated with a new highest performance technology, is attempted. This problem arises because circuits based on a new technology are available before the compatible test equipment, often using circuits of the same technology, can be manufactured. What greatly amplifies the problem for GaAs circuits, however, is the toggle rate and the rise and fall times that are characteristic of the circuit operation. These speed characteristics are comparable to signal propagation delays over very short distances, so the test equipment must have very short delays as well as short interconnection discontinuities. Such a tester will have to rely heavily on GaAs

Sec. 6.1   Testing Issues   247

circuits and a special architecture to provide the required features for testing emerging high-speed GaAs components. Until such a tester becomes available, alternative means must be provided for test application. Table 6.1 shows the requirements for a tester suitable for GaAs integrated circuits in contrast to what is offered by state-of-the-art testers.[1]

TABLE 6.1 Tester Requirements

Features	Tester Requirements for GaAs ICs	State-of-the-Art Testers
Data rate	0.5 GHZ–2 GHz	50 MHz–100 MHz
Signal rise and fall times	100 psec–150 psec	700 psec–1500 psec
Timing measurement resolution	10 psec–20 psec	300 psec–700 psec
Voltage measurement resolution	10 mV	1 mV–10 mV
Number of high-speed I/O pins	64–128	128–256

This section discusses in some detail the issues and problems related to the three facets of IC testing: test generation, test application, and test evaluation.

6.1.1 Test Generation

Test generation for an integrated circuit can be performed only after a fault model is assumed. A fault model represents the logic manifestation of design errors, fabrication defects, and operational failures in the circuit behavior. Although extensive work has been performed in fault modeling for Si circuits, there is insufficient data in this area for GaAs circuits. Since the circuit structures in the two technologies are different, failure mechanisms could differ between Si and GaAs devices. This indicates that further research is required to establish fault models for GaAs circuits and to determine how closely they represent the effect of possible failures on the circuit behavior. In the meantime, the fault model assumed for test generation is the classical and widely accepted stuck-at (s-a) type fault model. According to this model, a failure mechanism causes a signal at a circuit node to become fixed at a constant value, irrespective of the value of the logic signals that determine the value of the signal at that node during fault-free operation. The stuck-at model is used as stuck-at-one (s-a-1) or stuck-at-zero (s-a-0) to indicate that a node carries a fixed value of logic 1 or 0, respectively.

The tests generated based on the s-a model are aimed at both fault detection and fault isolation. There are several techniques available for generating tests for s-a faults. The complexity and cost of test generation, however, depend on the complexity of the circuit under test. For a combinational network of k inputs, a test set of 2^k, k-bit words is required for detecting all detectable faults. In case the circuit under test is sequential and in addition to its k inputs includes l storage elements, then the minimum number of test vectors in the generated test set must be $2^{(k+l)}$ for detecting all detectable faults. It is evident that even for small values of k and l the number of vectors in a test set can be staggering. In fact, it is known that in general the generation of a test set that detects a predefined number of faults in a circuit is an NP-complete problem.[2] A problem P is said to belong to class NP (nondeterministic polynomial) if any instance of P can be solved in time bounded by a polynomial of the problem size by a nondeterministic algorithm. A problem P_1 is defined as NP-complete if (a) it belongs to NP and if (b) every problem P in NP is polynomial transformable to P_1.

In the case of GaAs circuits where the scale of integration is small compared to Si circuits, test generation can still be manageable as long as the circuit design takes testing into consideration. As noted above, generated test sets can detect only *detectable* faults. Faults can be undetectable if, in a circuit that realizes a function f, there is a fault where, when present, the circuit function $fa = f$, where fa is the circuit function in presence of fault a. Such a circuit is said to be redundant with respect to fault a. Although the presence of a redundant fault will not affect the circuit behavior, it is important to recognize the existence of redundancy during design, because the presence of a redundant fault may cause an otherwise detectable fault to be undetectable.[3] However, no simple technique exists to determine redundancy, other than demonstrating that no tests exist for some faults. The ability to do so is, again, determined by the circuit complexity. It is important to note at this point that test generation, in addition to the fault model, also assumes that faults are permanent (i.e., they are present when the test is applied) and single (i.e., one fault present at a time).

Although at the prototype stage faults that occur during manufacturing generally affect more than one circuit area, it has been found in practice that single fault test sets suffice for the detection of multiple faults but not for the isolation of multiple faults.[3] Again, the reason for all these assumptions is management of the complexity of the problem. For example, if combinations of faults are considered for a network with n nodes where each node can be fault-free, s-a-1, or s-a-0, then all possible network state combinations will be $3^n - 1$. Assuming the network contains 100 nodes (i.e., a small circuit), we would have to consider $3^{100} - 1$ or approximately 5×10^{47} different combinations of faults—far too many faults to assume for any realistic analysis to be made.

Figure 6.1 illustrates the way the s-a model is used to simulate the effect of a fault on the circuit behavior. The faulty gate perceives the input s-a-1 as 1,

Sec. 6.1   Testing Issues   249

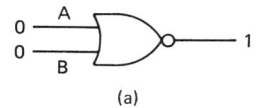

Figure 6.1 Test for an input s-a fault: (a) fault-free operation of the NOR gate and (b) faulty operation when A is s-a-1.

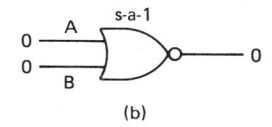

irrespective of the logic value placed on this input. Figure 6.2 illustrates an example of logic redundancy. As the truth table in Figure 6.2(b) indicates, the fault-free operation of the circuit is the same as its operation when A is s-a-1. Therefore, A-s-a-1 is undetectable.

Many algorithms have been proposed over the years for test pattern generation. Some of the proposed approaches were of more theoretical than practical importance and therefore they were not used for practical problems. A few approaches are of practical significance and they became the basis of tools used for automatic test pattern generation. The most widely used algorithm is the D-algorithm[4] which is a complete test generation algorithm and can generate a test for any logical fault that is detectable. An area where this algorithm has been demonstrated to be inefficient is test generation for circuits containing many XOR (exclusive-OR) gates. This deficiency of the D-algorithm is improved by a test generation algorithm called PODEM (Path-Oriented Decision Making).[5] Finally, a fanout-oriented test generation algorithm, called FAN, generates test vectors more efficiently than the D-algorithm or PODEM.[6] All these algorithms generate tests assuming specific faults, and the overall test generation process is deterministic. While the results obtained using these algorithms are satisfactory for combinational circuits, generating tests for sequential circuits is a much more difficult

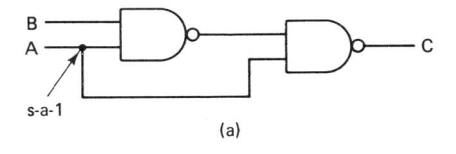

A	B	C	C(A-s-a-1)
0	0	1	1
0	1	1	1
1	0	0	0
1	1	1	1

Figure 6.2 (a) Circuit with logic redundancy; (b) fault-free operation and operation in the presence of a fault on the redundant node.

task. This is due to the difficulty of controlling and observing the states the sequential logic can assume. However, the controllability and observability of sequential circuits can be raised to the level of combinational circuits by using design for testability techniques, to be discussed later in this chapter.

In addition to algorithmic test pattern generation, test patterns are also generated randomly. Random test patterns can be generated using a random number generation routine and a specified probability distribution. Most frequently, uniformly distributed random tests are used. Several circuit techniques can also be used to generate tests of this nature. These techniques are classified as built-in test techniques and will be examined in greater detail. The test patterns generated with these techniques are pseudorandom, and their effectiveness depends on the circuit logic structure. The effectiveness of random patterns has been a research subject for some years now, and the results obtained are applicable to combinational circuits. However, they largely depend on the fault assumptions made.

6.1.2 Test Application

Test application is the process of providing the generated test patterns to the circuit under test, capturing its response, and comparing it to the expected response of a "good" circuit. A generic setup for testing logic circuits is shown in Figure 6.3. The test patterns, generated using either deterministic or random techniques, are applied by the test controller, via the test interface circuits, to the circuit under test (CUT); and the response of the CUT is captured and made available in a display or hard-copy form. The test controller, along with the interface circuits and the test adapter, constitutes the tester or test equipment. The main problems of testing high-speed circuits concern test applications. This is primarily due to three basic problems associated with testing a new, high performance technology:[7]

1. uncertainty in test parameter validity
2. tester performance limitations
3. unreliability in providing tests directly through the interface between the CUT and the tester

The validity of functional verification testing and performance characterization of a circuit depends on the conformity of the signal characteristics and test conditions used for the test to the actual parameters of the circuit when it is in "normal operation." For high-speed circuits, in addition to basic parameters such as logic swing, polarity, and noise margin, dynamic parameters such as switching speed, rise time, and fall time, are important for the normal operation and should be reliably verified during testing. If it is not possible to measure and calibrate the testing signals used, there is no assurance that they actually conform to the

Sec. 6.1 Testing Issues

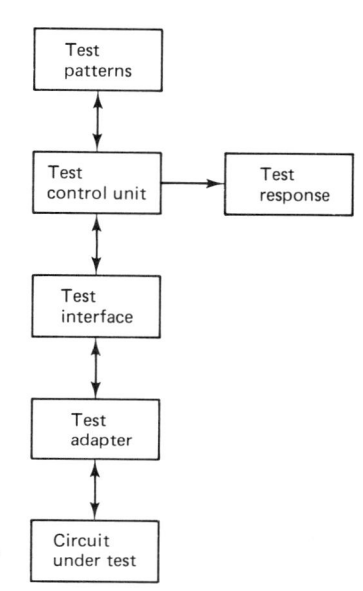

Figure 6.3 Generic setup for testing logic circuits.

characteristics required by normal operation. A significant deviation of the test parameters from those present in normal operation will render the test invalid.

Closely related to test-parameter conformity is the problem of tester performance limitations. This problem is well known in logic testing of integrated circuits. New circuits being developed are normally intended to operate at higher speeds than the existing ones. With a tester of lower performance than the CUT, the test parameters provided by the tester will deviate from those required for the normal operation, and the testing will become insufficient. This problem is particularly acute for GaAs ICs. With switching speeds in the 1–3 GHz regime and slew rates between 120 psec and 2 nsec, GaAs ICs outperform by far Si-based testers—as Table 6.1 indicates. Testing high-speed circuits at low speed is not always possible. As discussed in Chapter 2, dynamic GaAs circuits require a minimum clock frequency for correct operation. When there is no minimum frequency requirement, low-speed testing may verify functional correctness, although it does not provide enough insight into the performance of the circuit required by its design specification. Even at low speed, however, the rise and fall times specified in the circuit design should not be exceeded by the tester, because this could result in double clocking and cause the CUT to fail.

The problem of unreliability in providing tests directly through the interface has many facets. A crucial one is the difficulty of continuous synchronization of two systems (i.e., the CUT and the tester) with independent timing disciplines. For circuits developed with a new technology, it is usually desirable to be able to vary their cycle timing so that tolerances as well as performance limits can be determined. Therefore, it is preferable to test the CUT using its normal operation timing rather than the one arbitrarily determined and limited by the tester. The idea of synchronizing the tester, which in general is a sizable data-processing system, with the CUT timing is in most cases impractical. Therefore, the CUT and the tester will be two systems with independent cycle timing that may not be continuously synchronized. This implies that during testing the direct feeding of dynamic signals to CUT through the interface should be avoided. For reliable transfer of dynamic test signals between the tester and the CUT, utilization of buffering on the receiving side of the interface may be necessary to achieve synchronization.

From the above discussion, one can see that the test engineer's dilemma of trying to "use yesterday's hardware to test tomorrow's circuits" is greatly amplified with the development of GaAs circuits. Although no tester has been available by 1988 that can meet the requirements for testing GaAs ICs, the technology for generating and capturing high-speed signals accurately will become available through utilization of GaAs components. Additional requirements for a high-speed tester for GaAs ICs arise from the need for extremely short discontinuities in the propagation of dynamic test signals over transmission lines.[1] Test adapters such as wafer probe-cards, sockets, and load-boards on which switches and termination resistors are mounted must maintain a constant impedance at frequencies that can be as high as 5 GHz. This implies that the discontinuities posed by connectors, pins, and probes should maintain an inductance below 2 to 3 nH, which limits the length of these fixtures to about 2 to 3 mm. Therefore, the future GaAs IC tester will have to use advanced materials and innovative interface circuit techniques in addition to GaAs components for accommodating the requirements of both the generation and sensing of dynamic test signals, as well as the propagation of these signals from the tester to the CUT and vice versa.

6.1.3 Test Evaluation

Test evaluation or test verification is the process of proving that a set of tests is effective in meeting the test objectives. In most cases, test effectiveness is measured by the fault coverage (or test coverage) of a test, which is defined as the fraction of faults that can be detected or located within the circuit under test. The fault model used to estimate fault coverage is the same as the fault model used to generate the test to be evaluated.

To date, formal proof for test effectiveness has been impossible in practice. The best alternative yielding a quantitative measure of test effectiveness has been fault simulation. Fault simulation is also employed for analyzing the operation of a circuit under various fault conditions to detect circuit behavior not considered by the designer. For example, faults can create race conditions that normally do not exist in the fault-free circuit or inhibit the initialization (i.e., reset, preset) of a circuit.

Let C be an arbitrary circuit and L a set of faults, $f = 1, 2, \ldots, n_f$. The fault-free condition for the circuit in this case is denoted with $f = 0$, while $C°$ is the fault-free circuit and C^f denotes the circuit under fault condition f where $f \in \{0, 1, 2, \ldots L_f\}$. Given a test sequence $X = x(1)x(2)\ldots x(z)$, then the circuit response to $x(k)$, $k = 1,\ldots,z$, on the outputs $y1, y2,\ldots ys$ of the circuit under fault condition f, will have a steady state value of $y1^f(k)$, $y2^f(k), \ldots, ys^f(k)$, where $yj^f(k) \in \{0,1\}$. The input test sequence X is said to detect fault f if and only if for some k and j, $yj^o(k) \neq yj^f(k)$ for $yj^o(k)$, $yj^f(k) \in \{0,1\}$. This principle is used by all fault simulators in performing their function. The test sequence and fault conditions are specified by the user. The faults assumed in most cases are based on the stuck-at fault model.

Although the above principle is used by all fault simulation methods, the way in which the circuit is described, the input is applied, the faults are injected, and the circuit response is captured, vary. Three typical fault simulation methods include parallel, deductive, and concurrent fault simulation.[3] The performance of these methods depends strongly on the size of the circuit to be simulated (number of gates) as well as its structure (combinational, sequential) and the number of faults considered. All fault simulation methods, however, are very time consuming and costly, especially for large circuits.

Test evaluation for GaAs ICs is manageable for the SSI and MSI scales of integration with increasing costs as the scale of integration increases. It is very important to perform fault simulation to observe operating conditions that were not anticipated during normal design—especially race and timing skews that can cause major problems in the function of high-speed circuits. Also, the cost of developing and producing GaAs circuits should be a motivating factor for ensuring high fault coverage for the produced test vectors that subsequently ensure high confidence in the testing of prototype circuits.

6.2 TESTING APPROACHES

Until automated, reconfigurable, high-speed testers become available, the testing of GaAs circuits will be conducted using combinations of dc testing and ad-hoc techniques to perform dynamic testing. Dc testing includes parametric and functional testing of the circuits at low speeds. Dc testing can be performed using test equipment that is available for Si circuits. This type of testing is very important and should be the first step in the testing strategy, because it can be used as a screen to reduce the number of circuits to be tested by a high-speed tester. The three most common failure modes that can be detected by dc testing are:[1]

1. fabrication and/or design defects such as opens and shorts
2. poor noise margins, failing to meet the design specifications
3. poor transconductance, yielding inadequate switching currents

Although dc testing is useful for circuit screening or early rejection, no general theory can be used to predict the circuit high-speed performance by measuring the circuit's dc parameters. At-speed testing or close to operating-speed testing is essential to establish whether the circuit will perform its function as specified.

High-speed testing is performed with ad-hoc test system configurations made of radio frequency signal generators, sampling oscilloscopes, and available GaAs components. This type of testing set up is often assembled for testing a specific component; reconfiguration of the testing system may be required for testing another component with different I/O characteristics. Needless to say, the limitations of such systems are many. Signal synchronization (i.e., clock and data) and I/O limitations, as well as the difficulty in reconfiguring for testing of different components, are some of the major difficulties of this testing approach. However, this approach is very valuable in testing simple GaAs components that subsequently can be used to build automated test equipment, suitable for testing more complex GaAs circuits.

6.2.1 An Example of Testing a GaAs Chip

A custom-made setup based on the techniques discussed above was used for functional verification testing of the error detection and data reconfiguration chip whose design was presented in Chapter 4. The major issue to be resolved before any testing setup can be configured relates to testing fixtures—specifically, to the test adapter or test evaluation board that the chip is placed on during testing. The evaluation board provides the needed interface environment between the chip and the test set-up, as well as the means for removing the heat generated by the chip during testing. The evaluation board used in this case is manufactured by TriQuint Semiconductor, and its two sides are shown in Figure 6.4 This evaluation board provides a 50 ohm environment from the chip carrier to the coaxial connectors (bottom side). An integral heat sink (top side) is used for efficient cooling without requiring the use of heat sink thermal components or epoxy attachment. Input lines are terminated near the chip carrier socket with 50 ohm resistors (top side), while locations are provided for bypass and decoupling capacitors for the power lines. The capacitor bypass locations are used only when necessary since some chips provide integral decoupling capacitors for high-frequency components. The leadless chip carrier is placed into the socket and its assembly with the heat sink and connection to the I/O board lines are facilitated by a set of elastomeric connectors. Figure 6.5 illustrates the assembly of the socket with the chip carrier and the heat sink.[8]

The test setup used for functional verification testing of the chip is shown in Figure 6.6. The RF generator is used to provide the clock to both the circuit under test and the circuit used to provide the input data. The clock in this case is a high-frequency (≈ 1 GHz) sine wave. A 4-bit synchronous counter (available as a standard component from TriQuint Semiconductor) is used to provide the input data to the circuit under test. The purpose of adding a delay line to the clock circuit is to be able to adjust the phase between the clock and the input data. The output is captured with a sampling scope that is triggered with the most significant bit of the counter, thus providing an indirect way of synchronizing the sampling of the output with the clock. All cables used for component interconnections are 50 ohm coaxial cables.

Sec. 6.2 Testing Approaches   255

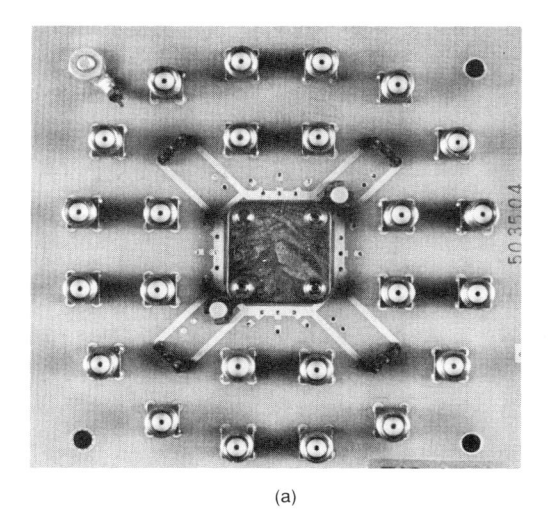

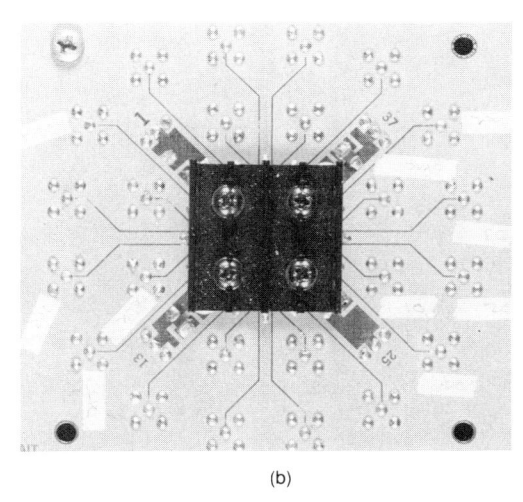

Figure 6.4 Evaluation board: (a) top side and (b) bottom side.

Using this testing arrangement, the function of the chip can be verified for a range of clock frequencies and the speed performance can be demonstrated. In this case, the highest frequency for correct operation was found to be 1.002 MHz, which is in agreement with the design specification of 1 GHz (see Chapter 4). Although this ad-hoc test arrangement can be used adequately to verify the function of the chip and assess its speed performance, certain observations can be made about its shortcomings. The only way input data can be provided at GHz rates is with a GaAs component (the 4-bit counter in this case) that can provide the desired patterns. In this example such a component is available, but for many chip applications such a component may not exist. Since the clock signal is a sine wave, if a chip requires that two nonoverlapping clock phases be externally

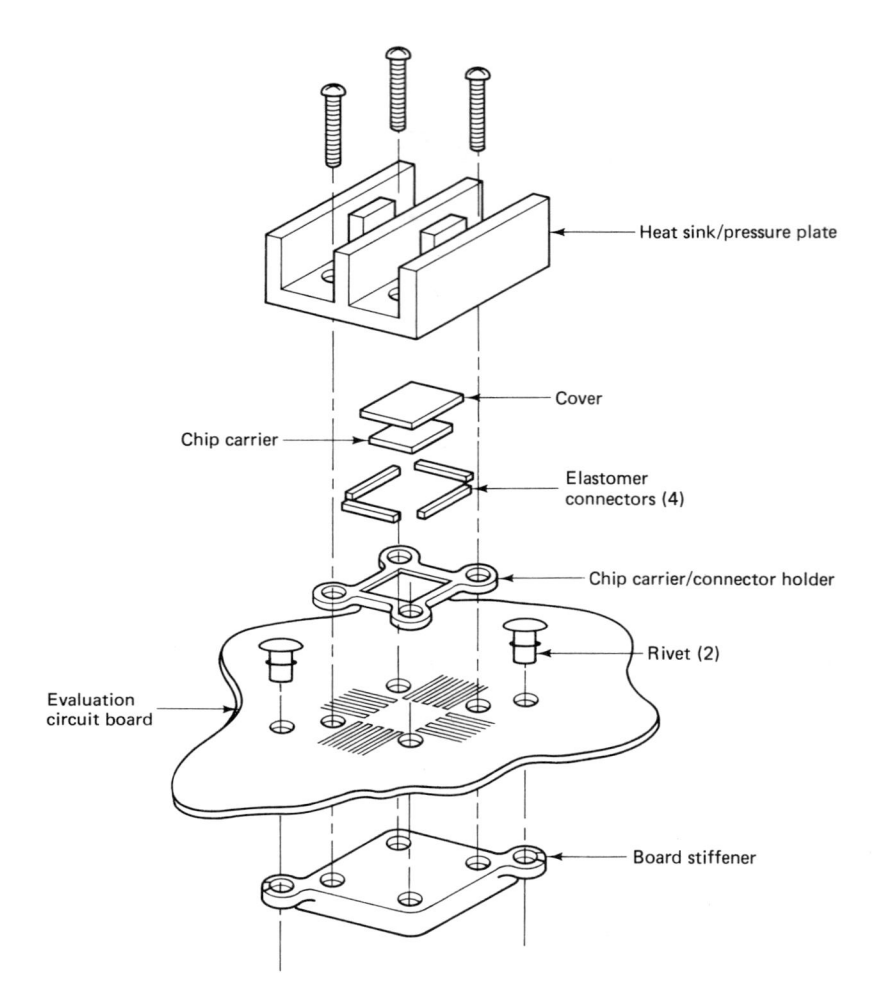

Figure 6.5 Assembly of the chip carrier with the socket and the heat sink on the evaluation circuit board.

provided, this clocking scheme will be of little use. The sampling scope allows only two outputs to be observed at a time. Although this limitation works in the chip example used here, it can cause severe limitation in general. Moreover, the test output data can only be observed. It cannot be stored and automatically analyzed against expected data or printed for a large number of input patterns and manually analyzed. For a chip with a large number of states this type of functional verification testing is unacceptable even for modest fault coverage requirements. Therefore, until testers in the GHz range become available, ad-hoc high-speed testing arrangements will be used only to assess speed performance for a

Sec. 6.3 Design for Testability   257

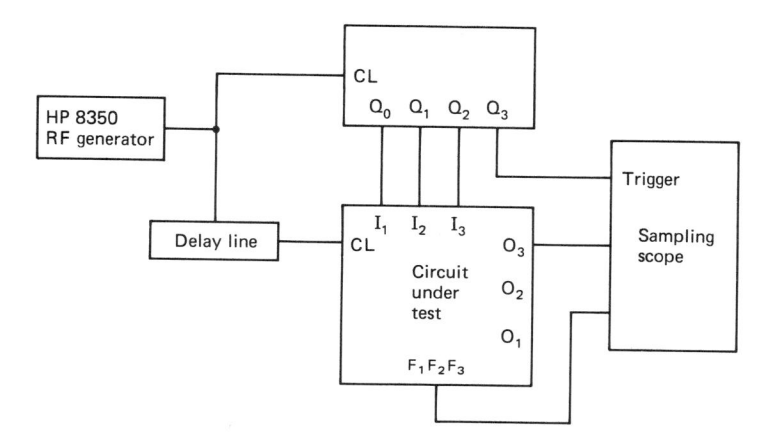

Figure 6.6 Testing setup for the error detection and data reconfiguration chip.

limited number of input vectors, while dc testing is used for initial circuit screening, and low-speed testing is used for detailed logic function verification.

Given the multifaceted problem of high-speed testing of GaAs chips using external testing equipment, an alternative may be the incorporation of testability features into chip design that will facilitate its at-speed testing with moderate demands on external test equipment. A general approach in this direction is design for testability (DFT) and built-in test (BIT), which are discussed next.

6.3 DESIGN FOR TESTABILITY

Design for testability was motivated by the difficulty of generating and applying tests to LSI and VLSI circuits, manufactured using Si technologies, and the cost associated with the testing process. It is interesting to point out that most of the practical design for testability approaches are developed by system manufacturers, not by semiconductor manufacturers. This is largely due to the fact that products were discarded because there was no adequate way to test them in production quantities, or systems were failing in the field because ICs considered fully functional were in fact defective, but inadequate testing was unable to determine this during system integration. The cost in both cases was unacceptable.

Design for testability ranges from general guidelines to well-structured design rules and special circuitry. In any case, the objective of a design for testability technique is to reduce the complexity of test generation and test verification. Each technique is associated with an implementation cost and return on investment in terms of fault coverage obtained and reduction in testing costs of a component. In the case of GaAs ICs, the reasons and motivations for using design for testability relate primarily to test application. Although the complexity of test generation and test verification for GaAs ICs will increase as the levels of circuit integration increase, the at-speed testing is the primary problem where design for testability can offer some solutions. Therefore, the emphasis in this section will be placed on techniques that potentially can be used on-chip to facilitate at-speed testing with test stimulus generated on-board, thus limiting the requirements for external test equipment. These techniques are classified as built-in test techniques and are presented in detail later in this section. The basic concepts of design for testability guidelines and structured design for testability techniques are also discussed.

6.3.1 Practical Guidelines for Enhancing Circuit Testability

The guidelines presented in this section indicate good engineering practice in circuit design as far as the testability of the circuit is concerned. Utilization of these guidelines during design is aimed at improving the test pattern generation, test application, and fault isolation processes involved in testing an integrated circuit.

Avoid logical redundancy. A circuit node is logically redundant if all output values of the circuit are independent of the binary value of the node for all input combinations or state sequences.[3]

Logic redundancy is often intentionally incorporated in circuits in order to achieve fault tolerance or to mask possible static hazard conditions, or it is present unintentionally due to nonoptimized design. The problem with a logically redundant node is that, by definition, it is not possible to make a primary output value dependent on the value of the redundant node. This implies that certain faults on the node cannot be detected; this may cause two problems. First, a fault condition may reintroduce the hazard condition that the circuit was designed to eliminate. Second, a fault on a redundant node can prevent detection of a second fault on a nonredundant node. The first case is illustrated in the example of Figure 6.7.

In this circuit, gate U_3 is included to eliminate the possibility of a static hazard when switching from the $d_1 \, d_3$ term to the $\overline{d}_1 \, d_2$ term (i.e., U_3 provides the "bridging" term). From a logical (i.e., Boolean) point of view, however, the output of U_3 is redundant. If an s-a-1 fault occurs on this line, this fault is undetected because it is not possible to propagate its effect through U_4. This is so because the conditions to set this line to 0 are inconsistent with those required to set the outputs of U_1 and U_2 to 1, thus making impossible the establishment of a sensitive path through U_4. The implication of this fact is that the circuit will continue to operate correctly but now a static-1 hazard (i.e., negative glitch) is possible on U_4 output while d_1 changes from 1 to 0 with d_2 and d_3 held at 1. The width of this pulse is determined by the difference in the signal propagation paths and may be wide enough to preset the flip-flop. In a configuration with redundant nodes, therefore, it is important to make these nodes directly observable, so certain fault conditions can be covered during testing.

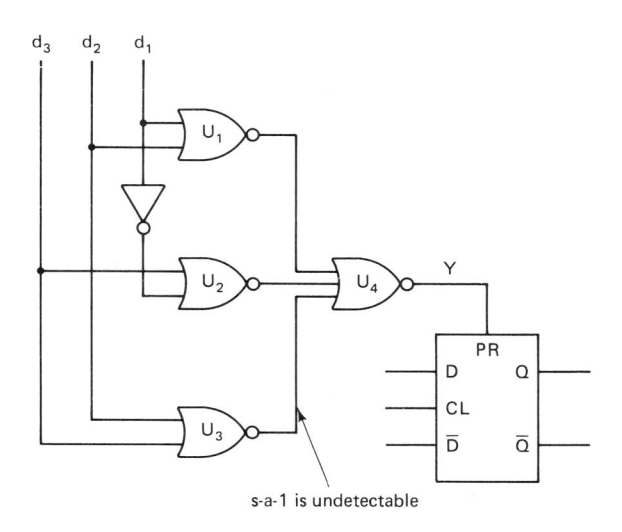

Sec. 6.3 Design for Testability

s-a-1 is undetectable

$$Y = d_1 d_3 + \bar{d}_1 d_2 + d_2 d_3 = d_1 d_3 + \bar{d}_1 d_2$$

Figure 6.7 Undetectable fault reintroducing a hazard condition.

The second case is illustrated in the example of Figure 6.8.

In this case the fault s-a-1 is undetectable at Y because of the incompatible requirements d_1, d_2, and d_3 to sensitize the path from input to output. The fault b s-a-0 is detected at Y only by the test input $d_1 d_2 d_3 = 110$. Therefore, if s-a-1 is present, the fault b s-a-0 cannot be detected.

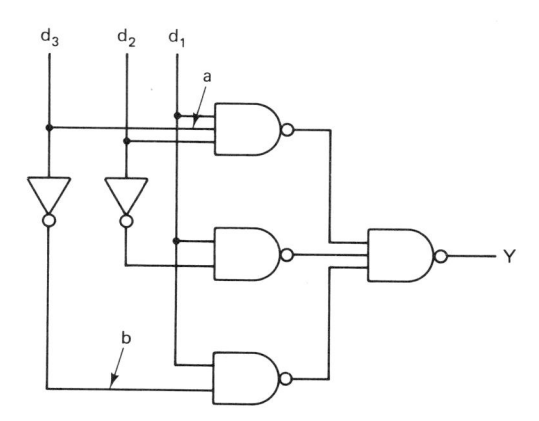

Figure 6.8 Fault detection masking of a normally detectable fault.

Avoid asynchronous logic. Asynchronous logic uses memory devices (i.e., flip-flops) and global feedback, but the state-transitions are determined by the sequence of transitions on the primary inputs. There is no system clock to indicate circuit transition to the next state.

The advantage of using asynchronous logic is speed of operation, because the speed at which state-transitions occur depends only on the propagation delay of gates and interconnect. In this respect, however, design of asynchronous logic is more difficult than synchronous logic (i.e., clocked logic) and must be carried out with special consideration to the possibility of races. Test pattern generation for such circuits may prove to be very difficult. This is particularly true if the outcome of a race condition is not deterministic. The possibility of nondeterministic behavior can cause further problems during fault simulation. As a result, from the testing point of view, synchronous logic is preferable to asynchronous logic, even to the point of using synchronous counters instead of asynchronous ripple counters or asynchronously-coupled synchronous counters.

Avoid using stored-state circuits that cannot be initialized by applying a predetermined stimulus to the input pins of the chip. Initialization is a critical precursor to any practical test program and simulation run. Ideally it should be possible to set every stored-state circuit in the chip into a known state. A master RESET common to all stored-state devices is usually preferable because it performs the initialization at minimal I/O expense.

Problems of initialization can occur if the state of one circuit depends on the state of another, as in an asynchronous ripple counter. If independent control of each state of the counter is not provided, initialization can be achieved only by clocking the counter until a particular state is established and identified by the tester. This can be a long process depending on the size of the counter. Initialization of sequential circuits can significantly reduce the requirements for test vectors and thus test time.

Avoid uncontrollable feedback paths. Global feedback paths complicate both test pattern generation and fault diagnosis because they minimize the controllability of sequential circuits. Feedback paths can be broken and controlled by a variety of techniques, some of which are shown in Figure 6.9. The tester signals can be primary inputs, or they can be directly controllable from primary inputs to which the tester has access.

Avoid minimizing controllability and observability features. The ability to generate and apply tests to an integrated circuit depends on the ease of controlling and observing the values of the internal nodes of the circuit. In practice, controllability and observability of all nodes of a circuit are not possible due to I/O limitations and active area penalties. However, there are some key control points in a logic circuit whose controllability and observability should be emphasized during the design. Examples of key control points are:

Sec. 6.3   Design for Testability

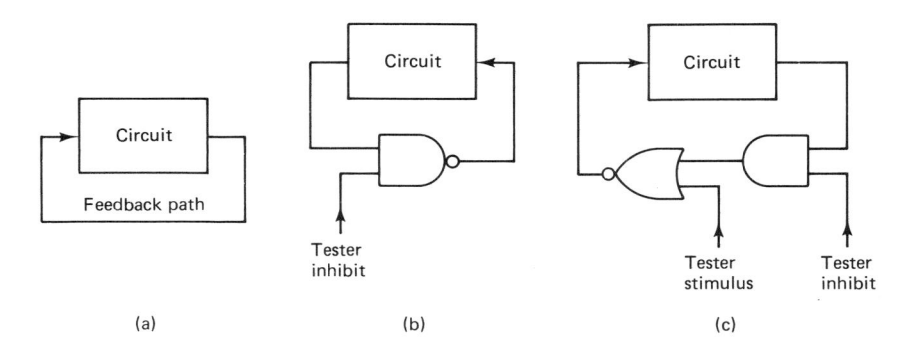

Figure 6.9 Techniques for controlling feedback paths.

1. CLOCK and SET/RESET inputs to devices with memory such as flip-flops, counters, and shift registers
2. DATA SELECT inputs to multiplexers and demultiplexers
3. TRISTATE CONTROL lines of circuits with tristate outputs
4. READ/WRITE/ENABLE inputs to memory circuits
5. CONTROL, ADDRESS, and DATA BUS inputs to any bus oriented design
6. ENABLE/HOLD inputs to a processor

Examples of key observation points are:

1. any not directly accessible control lines, such as those listed above
2. outputs from memory devices such as flip-flops, counters, and registers
3. outputs of data-funneling devices such as parity generators, priority encoders, and multiplexers
4. logically redundant nodes
5. the trunk section of nodes of high fanout
6. global feedback paths

Allow direct control of the clock circuitry.   For chips containing an on-board clock generator, it is important to be able to clock the circuit with clocks generated off chip by the tester. In case the clock generator does not perform as was intended, testing of the circuit can become excessively complicated and occasionally impossible. Having access to the output of the clock generator the clock can be replaced by externally generated clocks and the prototype circuit can be debugged. A technique for controlled internal clocks during test is shown in Figure 6.10.

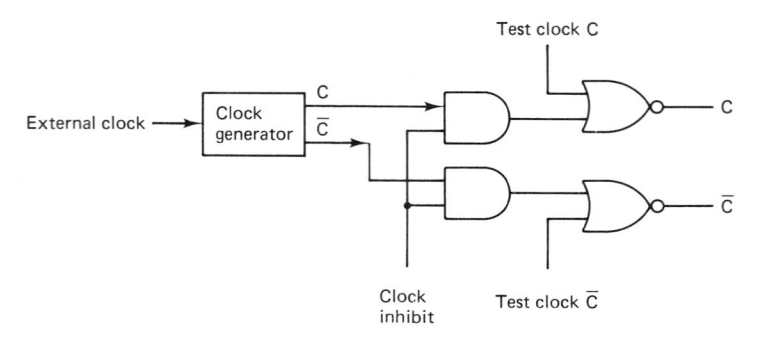

Figure 6.10 Technique for controlling internal clocks.

Avoid logic configurations with diagnostic ambiguity. Wired-OR and wired-AND junctions as well as high fanout lines create ambiguity problems in fault isolation testing. If possible, they should be avoided or the circuit should be modified to ease fault diagnosis.

6.3.2 Structured Design for Testability Techniques

Structured design for testability techniques were introduced in an effort to ensure testability and producibility of digital systems utilizing LSI and VLSI components. Highly structured design methodologies for these components are the only viable approach to producing testable circuits and systems at reasonable cost. These methodologies are based on the concept that if the values of all the memory devices in a circuit can be controlled to any specific value, and if they can be observed with a straightforward operation, then the test generation and possibly the fault isolation for a sequential circuit can be reduced to that of performing test generation and fault isolation for a combinational circuit. External control signals are used to switch the memory elements from their normal mode of operation to the test mode of operation, where they are directly controllable and observable from the circuit I/O.

For GaAs circuits, the test generation is relatively easy to handle for small- and medium-scale integration complexities, while it becomes of great concern for large-scale integration. At any scale of complexity, however, the test pattern application and test response capture are major issues and the structured design for testability techniques offer potential solutions. Furthermore, GaAs components designed to use structured design for testability methodologies will facilitate, or even make possible, the testing of system modules containing GaAs components. This is one of the major issues to be addressed in the early development of a new technology if the insertion of the technology into system design is to be viable. Although our purpose is to verify a GaAs integrated circuit functionally, provisions also have to be taken so that a board containing such a circuit can be

Sec. 6.3   Design for Testability   263

tested. That is to say, the circuit designer should keep in mind that the ultimate destination of an integrated circuit is a system which can be producible only if it is testable.

The structured design for testability techniques discussed in this section were developed mainly by systems manufacturers and have been used extensively in the design of Si circuits. All these techniques can be adapted by GaAs circuit designers with equally good results in circuit testability enhancements. The trade-offs in selecting a technique, however, may differ and should always reflect the chip and system design requirements.

The base circuit structures used in the design for testability techniques, as they were originally proposed, are well suited for implementation in Si technologies but not in GaAs technologies. The circuit structures used to illustrate the design for testability techniques in this section, while maintaining the functional principles of the originally proposed circuits, have been modified so they are suitable for implementation in GaAs. For illustration purposes the circuits presented are most suitable for implementation in BFL and DCFL.

Level-sensitive scan design (LSSD). LSSD was developed by IBM and is its discipline for structural design for testability. The need for LSSD grew out of the requirement for race-free testing and high test coverage of large sequential logic networks with few controllable inputs and observable outputs.

"Level Sensitive" refers to constraints on circuit excitation logic depth and the handling of clocked circuitry. "Scan" refers to the ability to shift into or out of any state of the network. A key element in LSSD is the "Shift Register Latch" (SRL) shown in Figure 6.11. Such a circuit is immune to most anomalies in the ac characteristics of the clock, requiring only that it remain in the "sample" state at least long enough to stablize the feedback loop, before being returned to the "hold" state.[9] This circuit is a simple modification of the flip-flop design shown in Figure 3.1. The clocking discipline is the same as in Figure 3.1, and the function of the circuit during normal operation also remains the same. The only difference is the addition of the testing latch (LT) section which allows testing patterns to enter into the latch through the input port, T, under the control of a separate clock, TCL. The overhead required for implementing this addition is minimal. In case of BFL implementation, for example, the overhead is two dual gate FETs. During normal circuit operation TCL is kept low and the LT section has no effect on the operation of the flip-flop. While in the testing mode, however, test patterns can be shifted into L1 and observed at the outputs of L2 by using clocks TCL and $\overline{CL}$, and by keeping CL low under external control. All SRLs are interconnected into a long shift register that scans test patterns in and out of the chip during testing. Figure 6.12 illustrates the on-chip interconnection of the SRLs. The same concept can be used at the board level where test patterns can be scanned in and out of all chips on the board. Figure 6.13 illustrates the interconnection of SRLs in this case. A chip design based on the LSSD approach has all storage elements implemented as SRLs and connected, as shown in Figure 6.14. Each of the

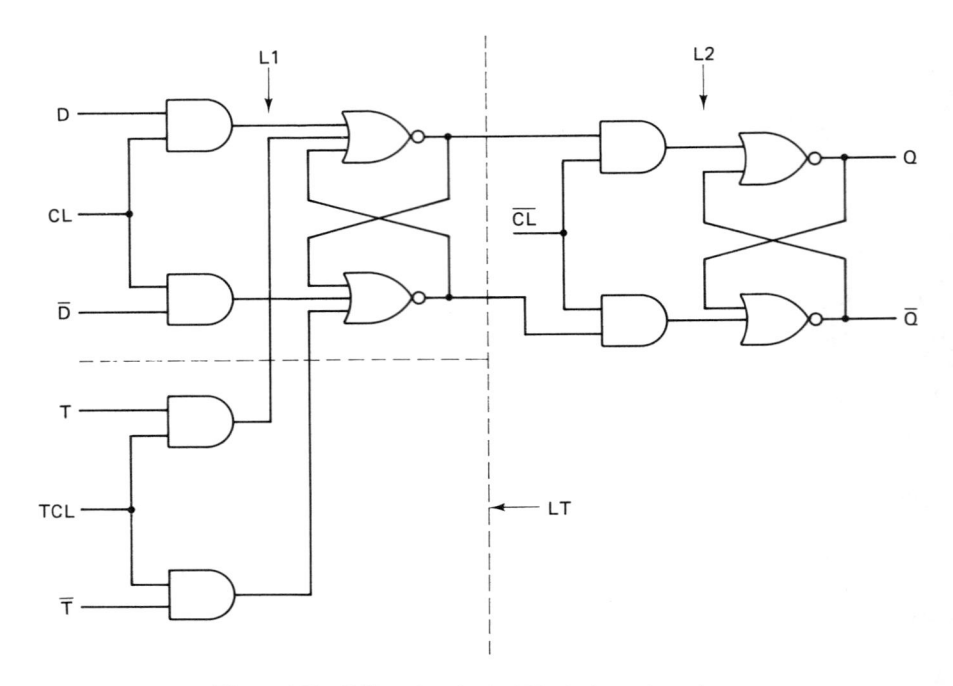

Figure 6.11 Shift register latch (SRL) logic configuration.

master-slave flip-flops is connected in series and clocked by two nonoverlapping clocks CL and $\overline{\text{CL}}$, during normal operation. In the test mode, all SRLs are chained to form a shift register chain. Test patterns are applied to the combination network by scanning them into the SRLs using the T and $\overline{\text{T}}$ inputs under the control of TCL, and by transferring them to the input of the network under the control of $\overline{\text{CL6}}$. At this time, CL is disabled using external control. The response of the combinational network is captured by the L1 section of the SRLs, under the control of CL which is now enabled and is shifted out serially through the scan out port.

The scan capability of LSSD reduces the sequential test-generation problem to a combinational one and makes possible the logic partitioning of the circuit, thus significantly simplifying the testing problem. Furthermore, test patterns are applied serially and loading the patterns can be done with a "slow" clock, TCL, while CL and $\overline{\text{CL}}$ are kept at the value specified in the design requirements. This scheme allows testing of most of the chip at high speed, thus solving some of the test application problems associated with at-speed testing of GaAs circuits. Note that during application of the test patterns through the SRLs the primary inputs to the combinational logic do not have to change necessarily at the same rate as the clocked inputs.

Sec. 6.3　Design for Testability　　　　　　　　　　　　　　　　　　　　　　　265

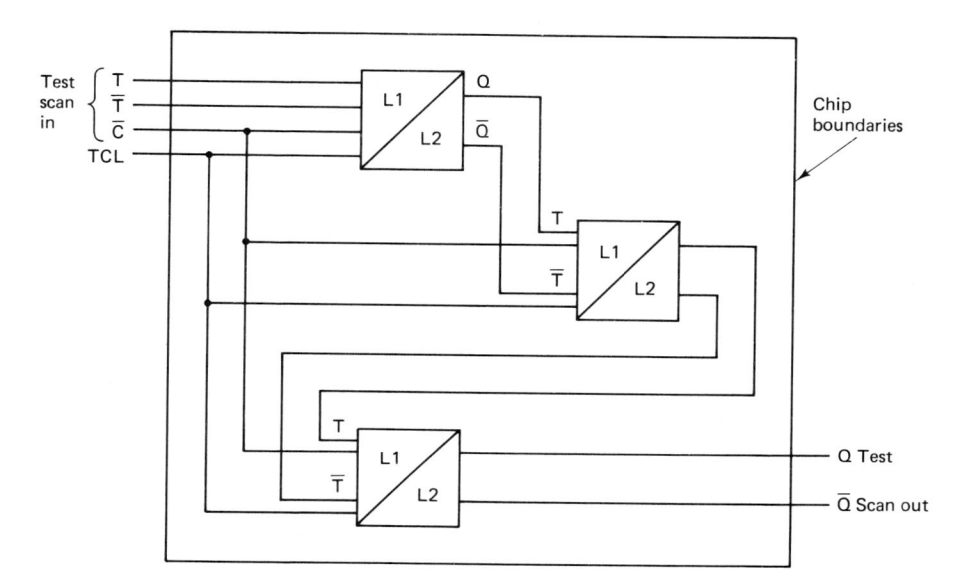

Figure 6.12　Interconnection of SRLs.

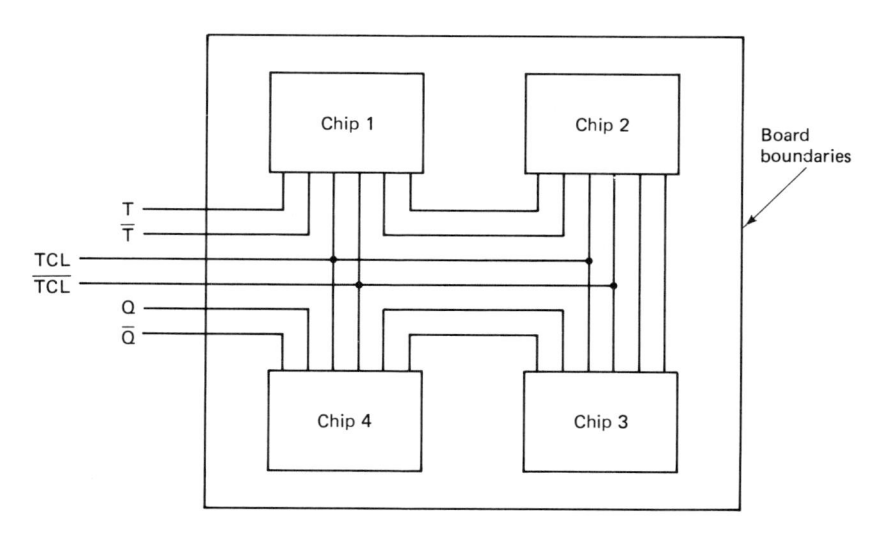

Figure 6.13　Interconnections of SRLs at the board level.

There are some negative aspects associated with the LSSD approach. First, there is hardware overhead (which can be kept to a minimum). Up to five additional I/O pins are required for a chip to provide test data and control the SRLs. External asynchronous input signals are not allowed to change more than once per clock cycle. External control of the chip clock is required. Finally, not all circuits are practical to implement using the LSSD methodology (i.e., RAM, ROM, PLAs).

The designer has to follow certain rules to implement LSSD circuits for which automatic test pattern generation may be available.[9] Seven rules that apply to a general circuit environment are given below:

R1: All storage elements are implemented as clocked latches. Data stored in these latches cannot be changed by any input when the clocks are "off."

R2: All storage elements are controlled by two or more nonoverlapping clocks such that:

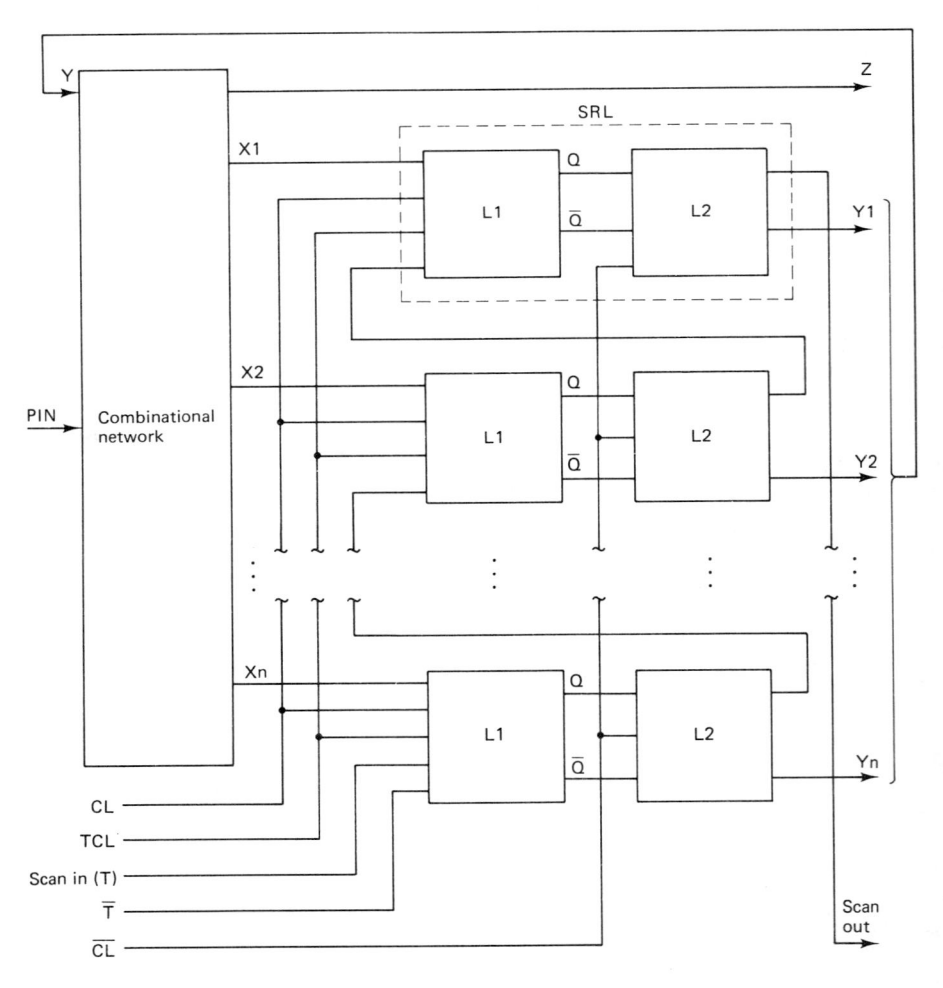

Figure 6.14 General LSSD configuration.

Sec. 6.3 Design for Testability

    a. the output of a latch A may be connected to the input of another latch B, if and only if the clock that sets the data into latch B does not clock latch A.
    b. latch A may gate a clock that drives another latch B, if and only if this clock does not clock A.

R3: All circuit latches are implemented as part of an SRL. All circuit SRLs are interconnected to one or more shift register chains, each of which has its inputs, outputs, and clocks as primary I/O of the integrated circuit.

R4: Any SRL output can be used in the circuit but outputs from both L1 and L2 of an SRL must not be used as inputs to the same combinational logic function.

R5: A set of clock primary inputs, from which the clock inputs to SRLs can be controlled, must be provided. No clock can be ANDed with another clock independently of its logic value (i.e., true or complement).

R6: Clock primary inputs may only feed the clock input to the latches or primary outputs but they should not feed the data inputs to latches either directly or through combinational logic.

R7: All SRLs must be able to shift in parallel. This implies that each SRL has its own scan-in primary input (from which it receives data) and its own scan-out primary output (from which it feeds data to other registers).

These LSSD rules apply to all functions normally implemented in standard logic circuits, including clock generation networks, which can be disguised as clock gating networks. However, these rules exclude certain types of networks such as memory and circuits with feedback loops not broken by SRLs or other means that make them controllable.

    Scan path. The scan path technique was originally introduced by Nippon Electric Company (NEC) as described in Reference 10. This technique has the same objectives as the LSSD approach and very similar implementation. Also, the Scan Path approach was the first practical implementation of shift registers for testing, which was included in a total system (NEC, FTL-700 system).

    The scan path approach uses as a memory element a master-slave flip-flop with scan path and a single system clock. The logic configuration of this device is shown in Figure 6.15. As can be seen, this circuit is very similar to the SRL used in LSSD, the only difference being the single system clock. During normal operation TCL remains low, thus disabling the test input from affecting the values of the input data and also leaving an affected operation of the slave flip-flop. During normal operation, the function of the flip-flop is controlled by the clock CL. During CL high, input data is sampled while it is shifted to the slave when CL goes low. During this transition, however, both the input of the master and the input of the slave are sensitive to input data and a race condition can occur if a feedback path exists from the output of the slave to the input of the master. The

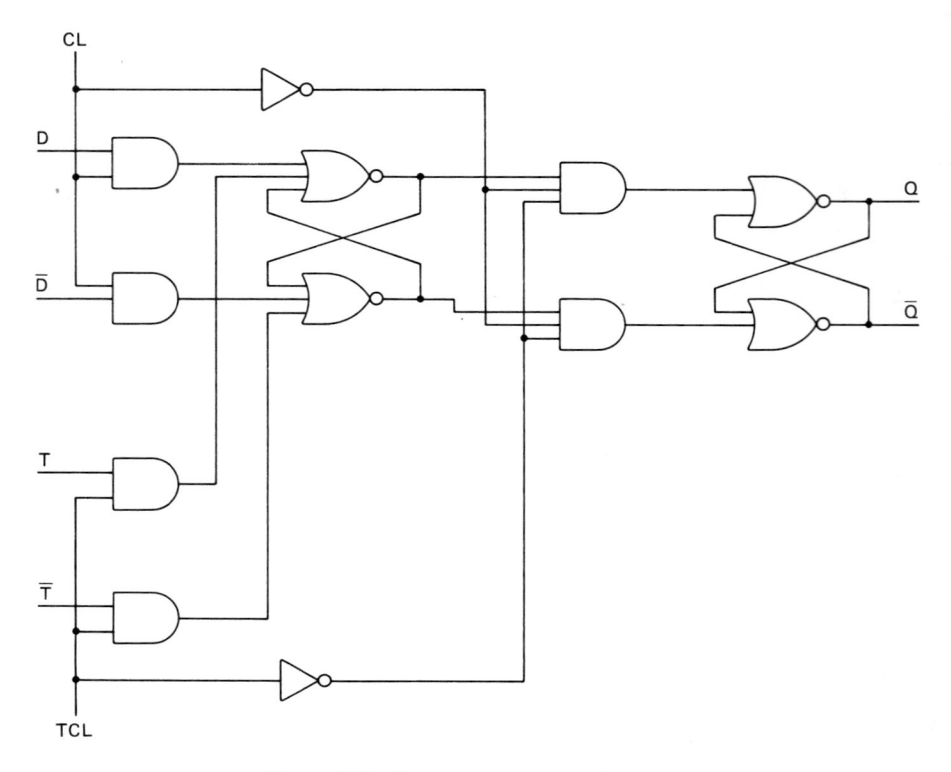

Figure 6.15 The scan path memory element.

reason for this condition, of course, is the utilization of a single clock for the circuit inverted internally. The effect of the race condition can be controlled by controlling the delay of the inverter that provides the clock to the slave. A similar situation exists for TCL which is used during testing. In that mode, CL remains low and test patterns can be scanned in and out of the flip-flop using the T, $\overline{T}$ input ports and the TCL. The operation of the memory devices during testing and the application of test patterns are very similar to that of LSSD. The basic difference between the two approaches is in the clocking scheme used to control their memory elements.

Scan/set logic. The scan/set design for testability technique, originally proposed by Sperry-Univac in Reference 11, is similar to LSSD and scan path in that it is aimed at increasing controllability and observability of sequential circuits. The basic idea of this technique is to have shift registers outside the data path providing test patterns and sensing test results at circuit nodes on the data path. Figure 6.16 shows an example of the scan/set logic. With this scheme the values of n nodes in the circuit can be loaded into the n-bit shift register with a single clock, and then they can be serially scanned out through the scan-output port and observed. This shifting occurs after test patterns have been applied at the primary inputs, and the scanned data are monitored, along with the primary output response, by the tester. The originally proposed scan/set logic technique used a 64-bit shift register to monitor 64 internal circuit nodes. The shift register is also used to provide a set capability from the external tester. A test pattern can be serially shifted into the shift register through its scan-input port and then gated into the memory elements of the functional logic. These two capabilities allow both the control and observation of internal nodes by an external tester.

Not all internal nodes, however, need to be scanned and/or set. Only the storage devices that are not controllable and observable through the primary inputs need to be given this capability. The scan/set logic provides this flexibility, since the shift register is not part of the functional logic and its size can be determined by the testing needs of the circuit. Another advantage of this technique is the ability to perform the scan/set operation during normal operation. This is possible only because the register is not part of the functional logic and this enables the user to have a "snapshot" of the internal nodes monitored by the register during normal operation, without affecting the performance of the circuit. The trade-off for this flexibility, however, is hardware overhead. Since the shift register is used only for testing, the circuit overhead can become significant if the number of nodes that need to be scanned and/or set becomes large.

Random-access scan. The random-access scan technique was originally introduced by Fujitsu, as detailed by Ando in Reference 12. This technique differs from LSSD and scan path in that instead of using shift registers it provides a direct-access path to each internal storage element. This direct-access path is created using an addressing scheme very similar to that of a random-access

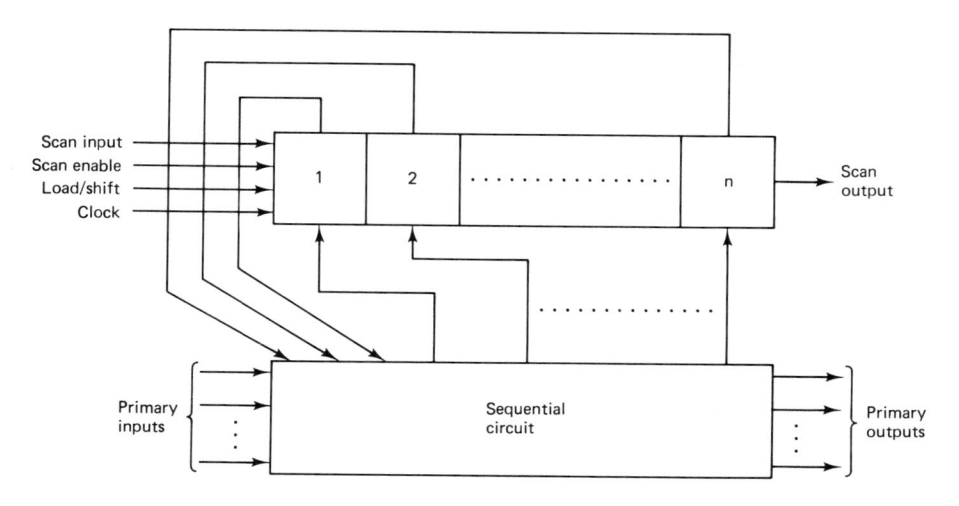

Figure 6.16 Scan/set logic configuration.

memory, thus allowing each latch to be selected uniquely and to be scanned in/out.

The basic storage element used by this approach is the polarity hold addressable latch shown in Figure 6.17. During normal operation, the latch uses the D and $\overline{\text{D}}$ inputs for sampling data under the control of the CL clock. In test mode, test patterns can be scanned in through the T and $\overline{\text{T}}$ ports, under the control of the TCL clock, but only when the latch is addressed by the X and Y addresses. The data can also be scanned out and observed through the scan-out port only when the latch is addressed. The same approach can be used with another type of latch where instead of scanning, data in the latch can be set or reset when it is addressed.

The configuration of sequential circuits designed using the random-access scan approach is illustrated in Figure 6.18. A pair of address decoders and additional test inputs are required as well as the addressable storage elements. These elements are accessed like a memory cell in an array by exercising the X and Y addresses. A single input for test data scan-in T/$\overline{\text{T}}$ is broadcast to all addressable elements but affects only the latch that is being addressed. The same configuration is used for the scan-out outputs of the addressable elements.

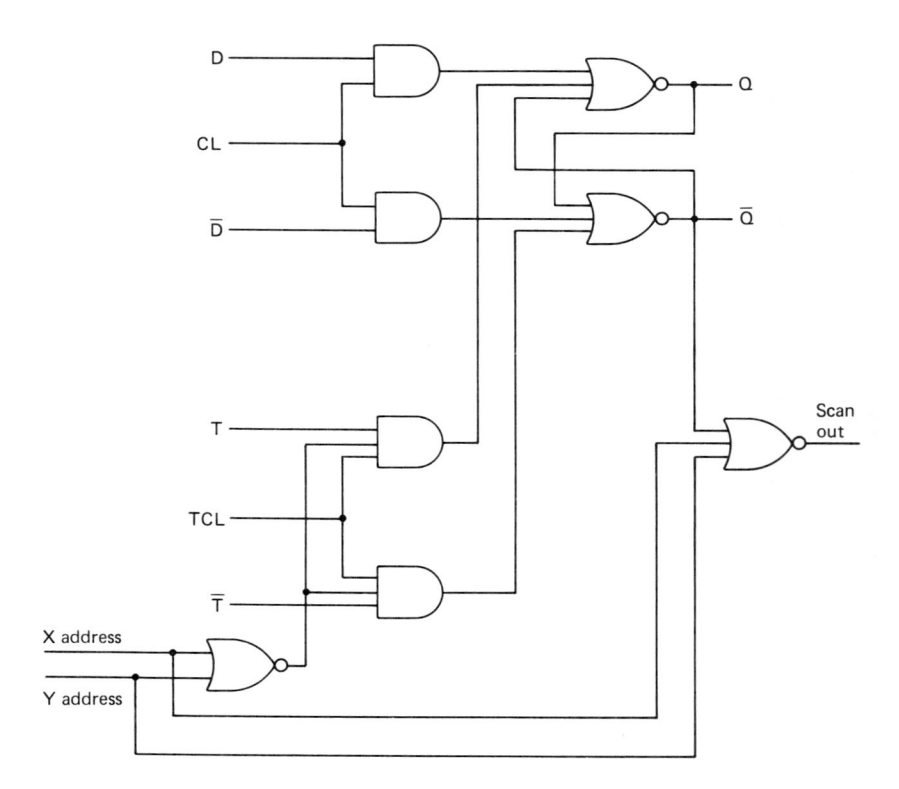

Figure 6.17 Polarity-hold-type addressable latch.

Sec. 6.3 Design for Testability

The advantage of the random-access scan is that each scan-out operation can be performed without disturbing the state of all latches. Also in many cases the hardware overhead can be minimal since, unlike scan path and LSSD, only a single latch is required for points to be observed and controlled. The trade-off in this case is the number of I/O ports needed for the address and the involvement of the external tester while in testing mode. In LSSD and scan path the external tester provides the test clock and a stream of test data. In addition to this information, in the case of random-access scan, the tester has also to provide the address.

Incomplete scan. Most of the scan design techniques presented base their implementation on the assumption that a circuit can be decomposed into a combinational part and into the scan path. Although this assumption applies to a large number of synchronous circuits, it cannot be applied to all circuits. For example, for some circuits this decomposition is not possible because of some asynchronous logic or prohibitive hardware overheads. In this case, the designer may need to consider an incomplete scan design where only some of the flip-flops in the circuit can be connected together to form the scan path.

A scan design is defined as *complete* or *incomplete* depending on whether all or a fraction of the circuit flip-flops are used to form the scan path. In case of

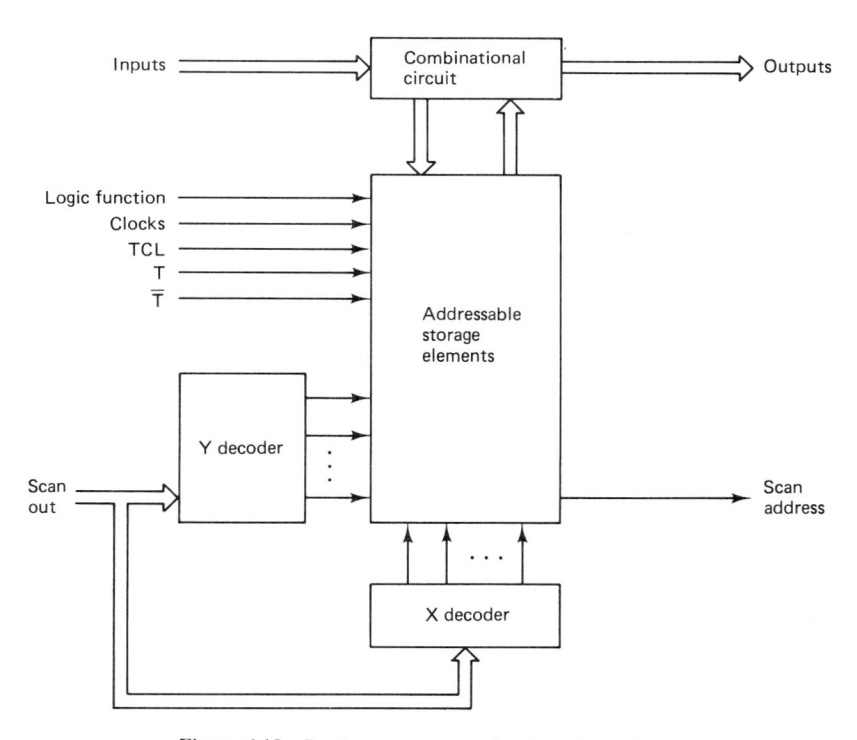

Figure 6.18 Random-access scan circuit configuration.

a complete scan design the test generation problem for a sequential circuit reduces to the test generation for a combinational circuit. In case of an incomplete scan design, however, after the scan path is removed the remaining circuit is still sequential. In this case the scan path should be used to reduce the sequential depth of the circuit outside the scan path. An approach to this objective was proposed in Reference 13.

Some recommendations for implementing the incomplete scan design approach are as follows:

1. The scan path should be used as a source of information to the circuit outside the scan path and not the other way around.
2. All flip-flops that are easily controllable and observable from primary I/O need not be connected to the scan path.
3. All flip-flops that are, directly or through combinational logic, controllable from primary inputs need not be connected to the scan path.
4. Flip-flops that are not easily controllable and observable must be connected to the scan path.
5. If asynchronous flip-flops, which are controllable, cannot be connected into the scan path, they must be isolated from the influence of the scan path.

Figure 6.19 illustrates different incomplete scan design cases that comply with, or work against, the above recommendations.

6.4 BUILT-IN SELF-TEST (BIST)

The structured design for testability techniques discussed in Section 6.3 are concerned with testing performed using externally provided test vectors. These test vectors have to be supplied to the circuit under test by test equipment and, as discussed earlier, there are several problems in interfacing test equipment to the circuit under test. Although the structured design for testability techniques improve dramatically the circuit controllability and observability, they do not affect the requirements for applying test vectors at the circuit primary inputs. These requirements may render impossible at-speed testing of GaAs integrated circuits until GHz throughput testers become available.

One way to resolve the test application problem is to use built-in/self-testing techniques. Self-testing (ST) is used in a broad sense here to refer to various testing approaches in which test patterns are applied internally to the circuit under test without the use of external test equipment. The term built-in-test (BIT) is used to indicate that the circuits which generate the test patterns and apply them to the circuit under test are actually part of this circuit. This section is concerned only with off-line, built-in/self-testing approaches. When these approaches are

Sec. 6.4 Built-In Self-Test (BIST)

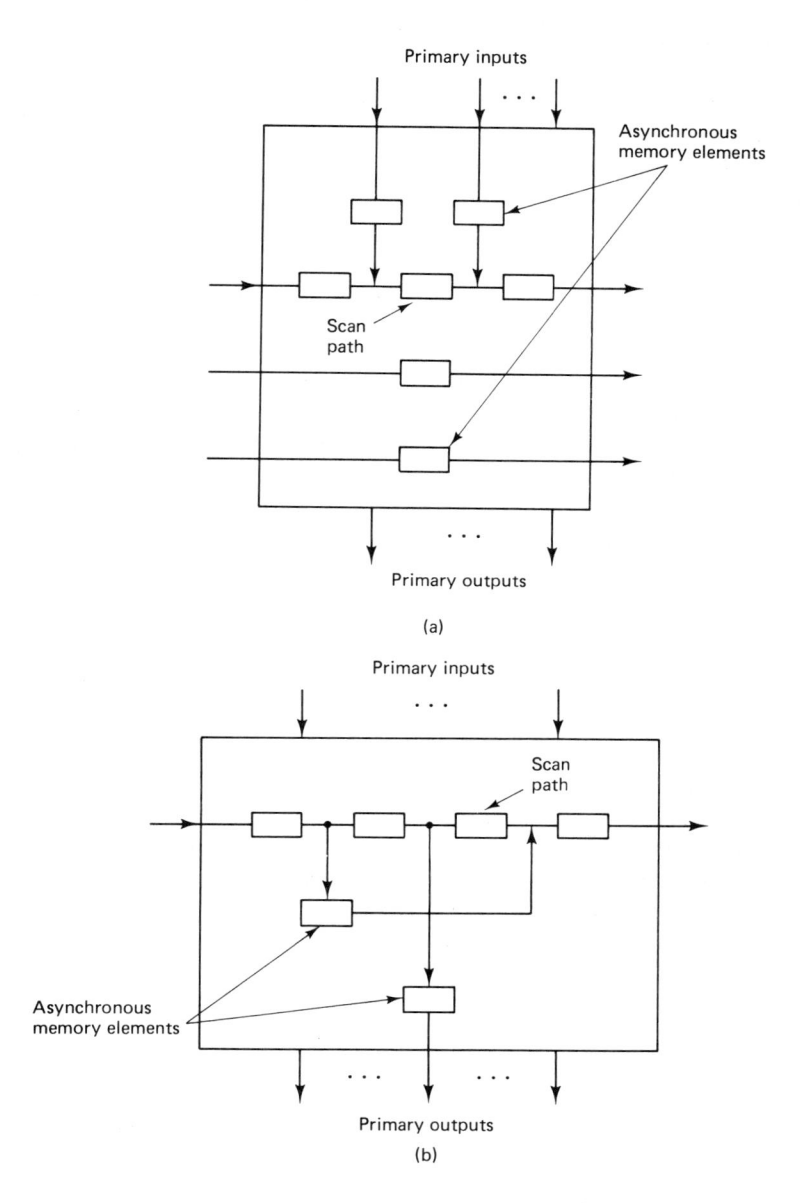

Figure 6.19 Incomplete scan design configurations: (a) optimal and (b) nonoptimal.

used to perform tests, the circuit cannot perform its intended function with actual data. This is in contrast to well-known concurrent, or on-line, approaches based on error-detecting, error-correcting, and even self-checking codes.

6.4.1 Signature Analysis

During testing of a circuit designed using one of the approaches discussed in Section 6.3, all circuit responses to test inputs are compared with correct reference values to determine the presence of a fault. Since a correct reference value has to be stored for every test response, the amount of data that needs to be stored may be large and the test time may be extensive if the comparison of the reference values to test responses is manual. In addition, in the case of GaAs circuits the recording of test responses at high speed is a problem, and using a sampling scope to perform this function is a tedious and often inaccurate process. A BIT approach, called compact testing, can avoid the difficulty of analysis of test responses and the storage of a large number of reference values. In compact testing, the entire input test pattern set produces a compressed circuit test response that is used for comparison with a single or a few reference values. The circuit's compressed response is often called a "signature," which lends the name to the signature analysis technique.

The signature analysis technique can be a very attractive BIT technique for GaAs circuits, for three reasons. One is that test patterns can be generated on-chip at the speed the chip is required to operate. The second reason is that the test response also can be captured and compressed on-chip, thus reducing the whole test output to a single or a few signatures for the entire test. The test response is captured and compressed at the operating speed of the signature. Finally, the complexity of GaAs circuits allows for relatively easy test evaluation through fault simulation.

Test patterns for signature analysis testing can be generated using a linear feedback shift register (LFSR). The design of LFSRs and their pattern generation properties were discussed in Section 3.4. The LFSR presented in that section and shown in Figure 3.34 is called an external XOR type due to the placement of the XOR gates on the feedback. A variation of this design is called internal XOR type and is shown in Figure 6.20. Both types of LFSRs can be used for data

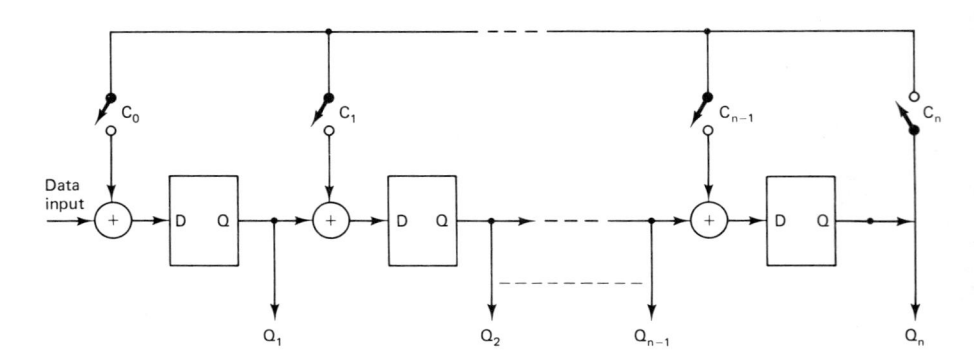

Figure 6.20 Internal XOR–type LFSR.

Sec. 6.4 Built-In Self-Test (BIST)

compression. The data compression technique using LFSRs is based on the cyclic redundancy check (CRC) codes, which have been discussed extensively in algebraic coding theory.[14]

The serial-data input stream into the LFSR in Figure 6.20 corresponds to a polynomial that can be written as:

$$d(x) = D_n x^k + D_{n-1} x^{k-1} + \ldots + D_1 x + D_0$$

while the characteristic polynomial of the LFSR is defined by the feedback constants as:

$$f(x) = C_n x^n + C_{n-1} x^{n-1} + \ldots + C_1 x + C_0$$

The input data enters into the LFSR, one bit at a time, with D_n coming first and after all flip-flops have been reset to 0. The LFSR divides any polynomial $d(x)$ by its characteristic polynomial $f(x)$. After a total of K shifts, the quotient of this division has been received, in a bit-serial fashion, from Q_n while the remainder of the division is stored in the LFSR flip flops. This remainder is called the signature of $d(x)$. Therefore, the input data stream has been compressed to another bit stream, the dimension of which equals the length of the LFSR. This operation can be described by the expression:

$$d(x) = f(x) q(x) + s(x)$$

where $q(x)$ is the quotient of the division and $s(x)$ is the signature. By analyzing the signature one can determine the presence of a fault in a circuit. The general configuration for signature analysis testing is shown in Figure 6.21. An LFSR is used to generate input test patterns for the circuit under test. The outputs of the circuit drive another LFSR which is used to compress the test output data and produce the test signature. The obtained signature is compared with the signature of a "good" circuit and a disagreement indicates the presence of a fault in the circuit under test. The signature of a "good" circuit is determined through simulation and it is known before actual testing is performed. The LFSRs, which produce the test patterns and the signature respectively, can be integrated in the same

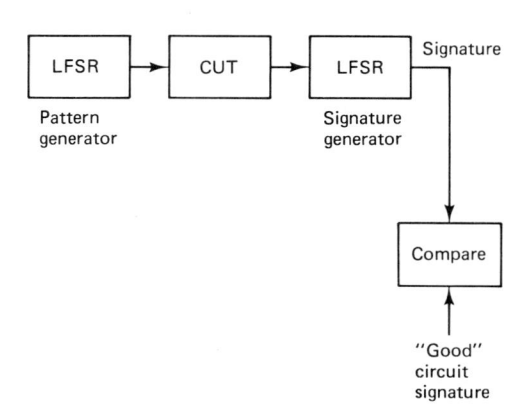

Figure 6.21 General configuration of signature analysis testing.

chip with the circuit they are testing, while the signature comparison can be performed externally to the chip. There are some distinct advantages to using this approach. The most important one is that testing can be performed at speed, without external test setup requirements for providing patterns and capturing test responses.

The LFSR circuits presented so far have a single input and therefore, compression of a circuit response with parallel outputs would require the addition of a parallel-to-serial converter as an interface between the LFSR and the circuit under test. This is not an effective approach, however, from an implementation point of view. An effective alternative in this case is a multiple input LFSR which is often called a parallel signature analyzer. The operation of this type of LFSR is the same as the operation of the single input LFSR discussed above. Figure 6.22 shows two types of parallel signature analyzers.

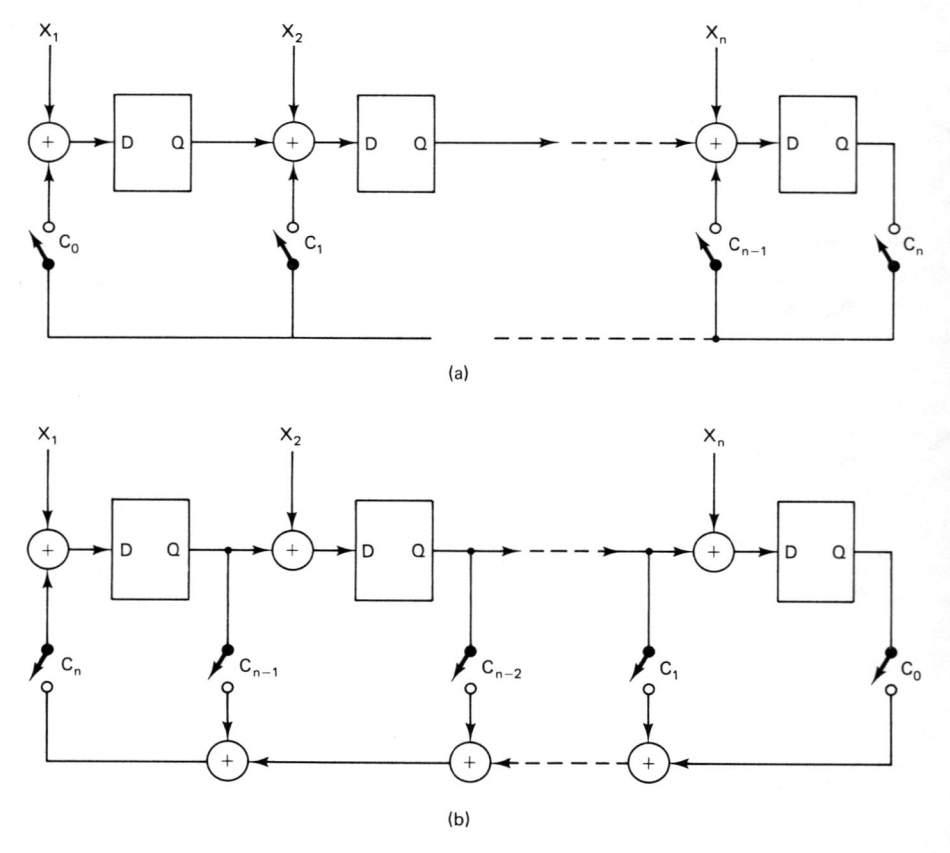

Figure 6.22 Parallel signature analyzers; (a) internal XOR type and (b) external XOR type.

Since error detection with the signature analysis technique is performed using pseudorandom test patterns, one issue to be resolved before employing this technique for testing a chip is its effectiveness in meeting the fault coverage testing requirements for the chip. If an error is present in the circuit under test, then the erroneous input data stream to the signature analyzer can be represented by:

$$de(x) = d(x) + e(x)$$

where $d(x)$ is the error-free input polynomial and $e(x)$ is the corresponding error polynomial. An error is not detected by the signature analysis technique if $d(x)$ and $de(x)$ have the same signature. As was stated above, the LFSR operation can be described by the operation:

$$d(x) = f(x)q(x) + s(x)$$

while in presence of an error that produces the same signature it becomes

$$de(x) = f(x)qe(x) + s(x)$$

and since $de(x) = d(x) + e(x)$, we have

$$e(x) = f(x)[de(x) - d(x)]$$

Therefore, for an error polynomial $e(x)$, $d(x)$ and $de(x)$ have the same signature if and only if $e(x)$ is a multiple of $f(x)$. The probability of occurrence of this event was presented in Reference 15 as a measure of the effectiveness of signature analysis. Here the theorems that evaluate the error detection capability of LFSRs are presented without proof. The interested reader should consult References 15 and 16 for the proofs of these theorems:

Theorem 6.1. Assuming that all possible error patterns produced by a circuit under test are equally likely, and the input pattern stream to an r-bit LFSR has a length of k bits, then the probability that the signature generator will not detect an error is:

$$\frac{2^{k-r} - 1}{2^k - 1}$$

For long sequences, the probability of not detecting an error approaches 2^{-r} and hence the error coverage approaches $1 - 2^{-r}$, which can be very high. For example, for a 16-bit LFSR, the result of this theorem indicates that there is certainty of 99.998 percent that if an error is present it can be detected.

Theorem 6.2. An LFSR with a characteristic polynomial with two or more nonzero coefficients detects all single bit errors. The result of this theorem can be used to design the feedback circuit of the LFSR, which defines the LFSR's characteristic polynomial.

Theorem 6.3. Assuming that all possible error patterns produced by a circuit under test are equally likely, and the response sequence of an r-bit parallel signature analyzer consists of L vectors of m bits each, then the probability of failing to detect an error in the response sequence by the parallel signature analyzer is:

$$\frac{2^{mL-r} - 1}{2^{mL} - 1}$$

Although the above theorems give an indication of the effectiveness of signature analysis for fault detection in combinational logic, their validity is questionable in a general sense because the error assumptions made are not generally applicable to all circuits. Therefore, the designer should use these results as a guide but should employ fault simulation to determine the effectiveness of the signature analysis configuration chosen as a BIT technique for a chip.

6.4.2 Built-In Logic Block Observer

One BIT approach that combines some of the features of the scan design techniques and signature analysis is the built-in logic block observer (BILBO) which was first introduced in Reference 17. A BILBO integrates scan design with signature analysis by utilizing an LFSR as a scanable register, pattern generator, and signature analyzer. An 8-bit BILBO module, with its different modes of operation, is illustrated in Figure 6.23. The control inputs CN1 and CN2 are used to select one of the four functional modes of the BILBO module:

Mode 1: CN1 = 0, CN2 = 1. In this mode the register is reset.

Mode 2: CN1 = 1, CN2 = 1. In this mode the BILBO module operates as a parallel-in, parallel-out register where the inputs are loaded through the input ports, X1,...,X8, and they are available through the output ports, Q1,...Q8.

Mode 3: CN1 = 0, CN2 = 0. In this mode the BILBO module operates as a linear shift register, where data is loaded serially through the input port S_{in} and is unloaded through the output port S_{out}.

Mode 4: CN1 = 1, CN2 = 0. In this mode the BILBO module operates as a parallel signature analyzer, where the parallel input patterns are loaded through the input ports, X1,...,X8. In this mode of operation, if the state of the inputs X1,...,X8 is kept constant, then the BILBO module becomes a pattern generator capable of generating a maximum length sequence (i.e., with the feedback shown in Figure 6.23(d)) of $2^8 - 1$ patterns.

A circuit configuration using the BILBO BIT approach is shown in Figure 6.24. The operating principle of the BILBO modules is the same. First the BILBOs are initialized, using their CN1, CN2 control signals, to their operating mode.

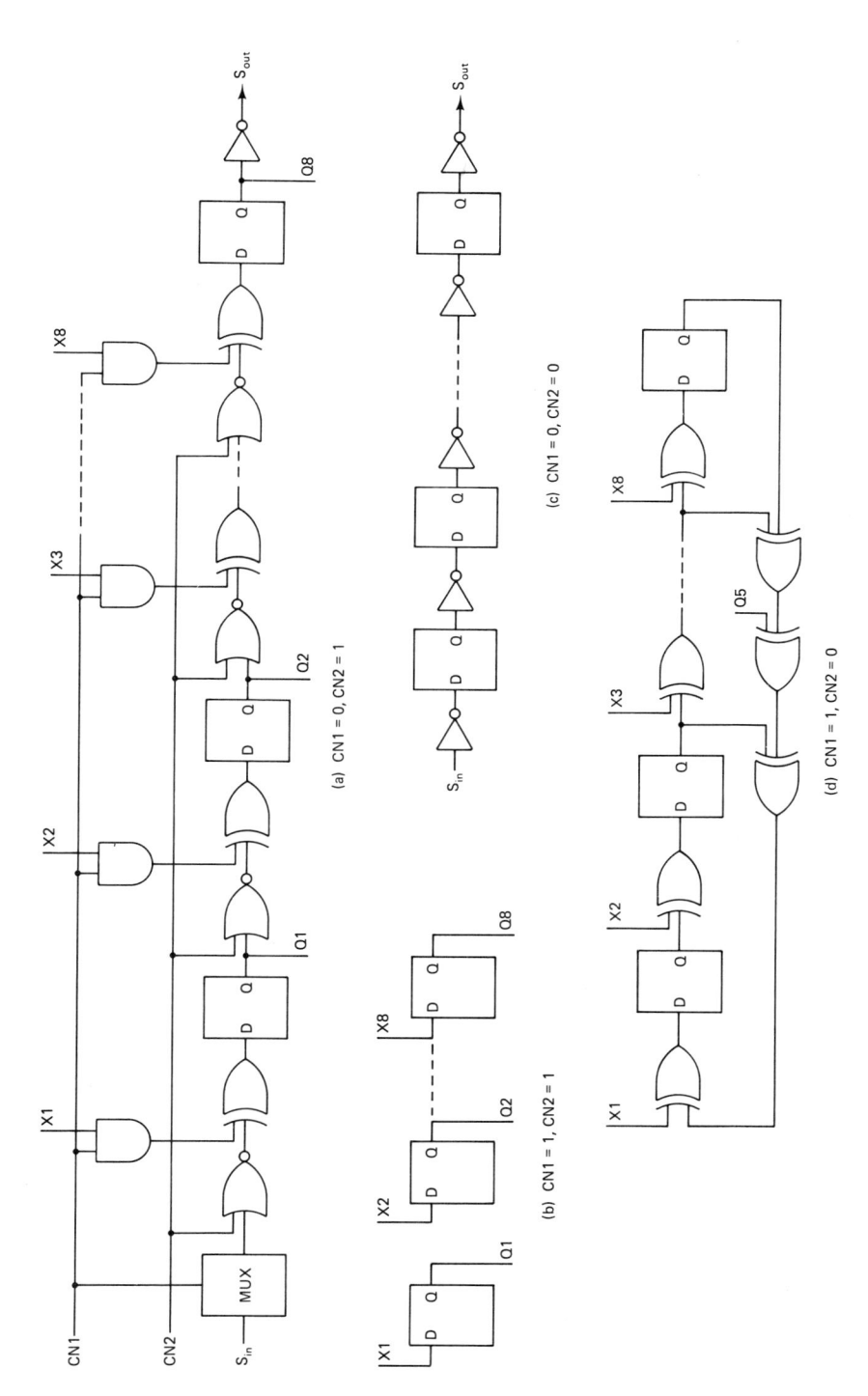

Figure 6.23 Eight-bit BILBO configuration; (a) logic diagram; (b) parallel-in, parallel-out mode; (c) linear shift register mode; and (d) parallel input LFSR mode.

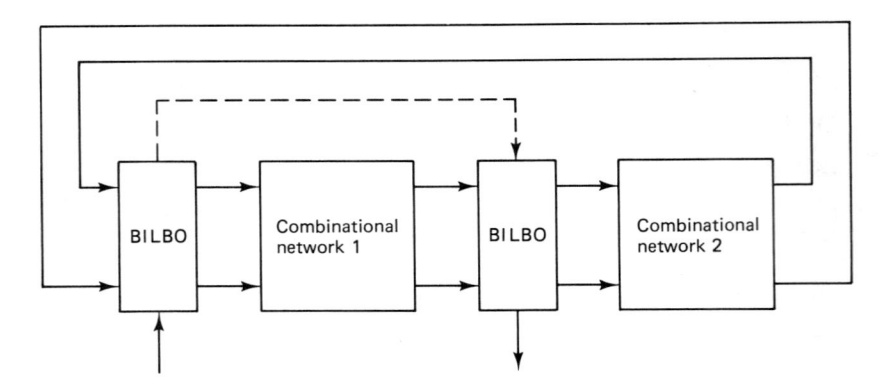

Figure 6.24 A circuit configuration using the BILBO BIT approach.

The BILBO at the input of the circuit under test assumes the pattern generation mode, while the BILBO receiving the output of the circuit under test operates as signature generator. When the signature has been obtained, the signature generator BILBO assumes its linear register mode and shifts out the obtained signature to be compared with a "good" signature. The roles of BILBOs are reversed when testing of the circuit-2 begins. Connecting the BILBOs together in a scan path minimizes the I/O overhead required for shifting and the signatures of subcircuits.

The versatility of the BILBO module offers many advantages in using this approach. As in signature analysis, test patterns are generated on-chip and the test response is a signature, so there is test data reduction through data compression. Testing is performed at speed and simultaneous testing of modules is possible. The BILBO module can be used as a parallel-in, parallel-out register, or as a linear shift register (Fig. 6.23(c)), during normal operation thus minimizing the hardware overhead required just for testing purposes. On the other hand, implementation of the BILBO module is rather complicated, due to the XOR gates in between the flip-flop modules. These gates pose a significantly higher overhead compared to the simple LFSRs used in signature analysis. A BILBO module has also higher overhead than LSSD and scan path registers.

When the BILBO approach is used as a BIT technique, the designer should perform fault simulation to determine the effectiveness of the test patterns, generated by the module, to detect faults in the circuit tested with these patterns.

6.4.3 Built-In Self-Test at the Board Level

The BIST techniques discussed so far address the problem of functional verification testing of GaAs chips. However, even if individual chips are functionally verified at full speed using these techniques, the same problem arises when GaAs chips are incorporated into a system. In fact, functional verification testing of circuit boards containing GaAs components may be even more difficult

Sec. 6.4 Built-In Self-Test (BIST)

than testing individual chips, especially when fault isolation to a small group of chips is required. This difficulty arises from the fact that a circuit board is inherently more complex than a single chip and will, in general, require a large number of I/O signals to be controlled and observed during testing. This poses a severe problem for testing setups for GaAs logic, as was discussed in Section 6.1. A solution to external testing of circuit boards would be the employment at the board level of some of the BIT techniques presented for chip level functional verification testing. There are two cases which one can encounter in circuit board testing. One case is when the chips on the board already have testability features built-in or the chips are self testable. Here, the board design should allow for the creation of a scan path based on the testability features of the chips. Figure 6.25 illustrates this approach. Every chip has its inputs and outputs latched. The I/O latches can be scanable registers used with some of the DFT approaches, or they can be BILBO registers. In either case all registers are serially connected as shown in Figure 6.25, thus forming a scan path. This scan path is externally controlled and is activated only for testing purposes. Before any testing of the chips begins, the scan path is tested first to determine its correct operation. This is accomplished by serially shifting data through the scan-in input and observing the outcome at the scan-out output. This requires only a single high-speed signal in addition to the clock and sampling of a single output signal. Upon determining that the scan path is fault-free, each chip can be tested independently. The approach to testing, of course, depends on the testing features of the chip. If the chip uses some of the DFT scan techniques, then patterns are shifted serially through the scan-in port, the results are captured at the output latch of the chip, and, through the scan path, the results are observed at the scan-out output. If a BIT technique is used, the test patterns are generated on the chip as is the final signature which is transferred to the scan-out port where it can be observed. Some very important observations can be made for this design approach. First, if external patterns are required, they have to be provided through a single port and a single port has to be monitored to receive the test response. Testing can be performed at speed when BIT is incorporated into chips, and almost at speed when the chips are using a DFT approach. In the latter case, a GaAs register can be used to convert a relatively slow stream of parallel patterns to a fast stream of serial input as expected by the scan-in input. In this case, the parallel-in serial-out register should be added to the board. Since the chip inputs and outputs are latched, when the chips are in testing mode, there is a natural braking of possible feedback paths from chip to chip on the board. This is an important feature that contributes to increasing the test resolution. If a feedback path cannot be broken or controlled, then the detection of a fault cannot be isolated to a single chip. In this case, the chips in the feedback loop are said to belong to an ambiguity group (AG). In general, an *ambiguity group* is defined as the smallest number of chips where fault isolation can be achieved. This approach allows not only the testing of the chips on the board but also the testing of interconnections between chips. This is achieved by setting the output scan registers in the normal operation mode

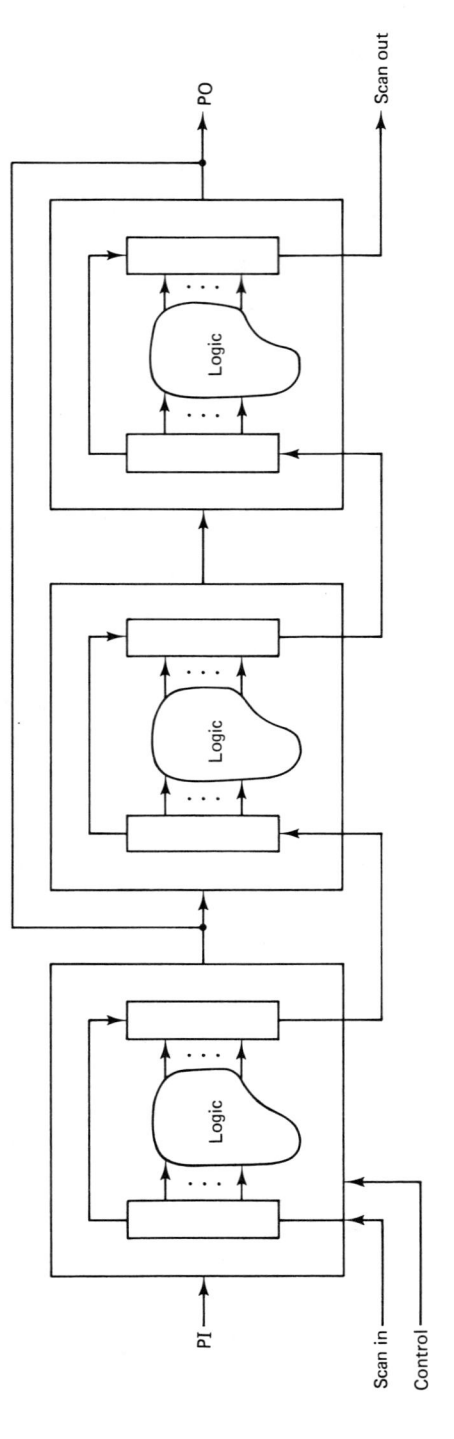

Figure 6.25 Scan path for board level testing of chips with DFT/BIT features.

and by transferring data to the input of the next chip followed by a serial scan and observation of the result at the scan-out output. This testing procedure checks the function and performance of the output drivers of the transmitting chip, the board interconnection lines, and the input drivers of the receiving chip. This parallel transfer of data between chips can be performed at the clock speed to be used during normal operation of the board.

The board design case discussed above will be ideal for design verification testing, but it is practiced only in very controlled design environments and for systems with high testability requirements. In most board design cases, the system designer uses off-the-shelf chips that may not have any testability features. In this case, testability features have to be added to the board design to make possible functional verification testing for given fault detection and isolation requirements. A general purpose module which can be used as a BIT module for board level design is introduced in Reference 18. This module can be integrated on a single chip and it can be used in different board design configurations to facilitate fault detection and isolation. This module, called the testing switch, provides pseudorandom test patterns and generates signatures during testing, while it acts as a pass-through component during normal operation. One of the major advantages of this module is the capability of programming the feedback of the test pattern generator. This capability allows the user to predetermine the length of the test pattern sequence as well as the order of the patterns in the sequence. This feature can be extremely important, especially for testing sequential logic.

The design of the testing switch is standardized for 16-bit data paths and it can be implemented, for example, using an E/D gate array. Such an implementation is generic enough to accommodate most GaAs chip configurations on a circuit board. A fundamental reason for using 16-bits is to maximize the probability of fault detection with the generated pseudorandom patterns, approximately equal to $1 - 2^{-r}$, where r is the number of stages in the pattern generator (see Section 6.4.1). Figure 6.26 shows a functional description of the testing switch module. A two-phase nonoverlapping clocking methodology is assumed in its design. All the flip-flops used have a common **reset** except for those explicitly indicated.

ctrl1 and **ctrl0** are the bits that determine the data exchange paths of the exchange network. Either the normal data inputs (**din1–din16**) or the test pattern generator outputs (**a1–a16**) can be passed to the normal data outputs (**dout1–dout16**). Similarly, either of the inputs can be passed to the **d1–d16** lines as input to the signature analyzer.

The test pattern generator operation is controlled by the signals **fbcoefin, genctrlin, gendatain,** and **genshiften**. Figure 6.27 shows an external exclusive-OR implementation of the programmable feedback LFSR test pattern generator while Figure 6.28 depicts an internal exclusive-OR implementation. The programmable feedback feature of the test pattern generator offers considerable flexibility in controlling the pseudorandom nature of the generated test vectors, as well as in sequencing of the desired test vectors. **genctrlin** is low during the shifting of the feedback coefficients (binary 0 or 1) and initialization of the LFSR state.

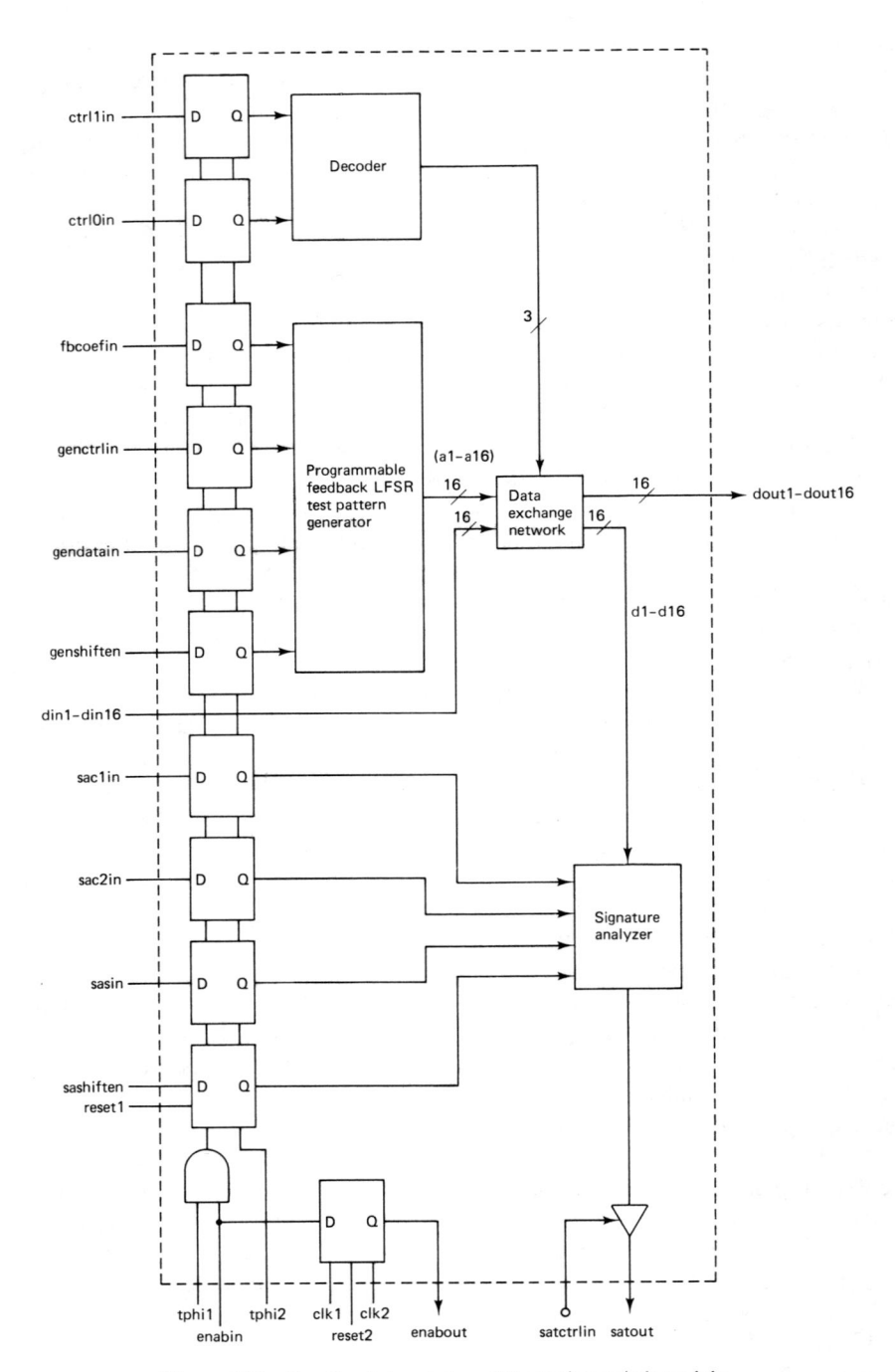

Figure 6.26 Functional description of the testing switch module.

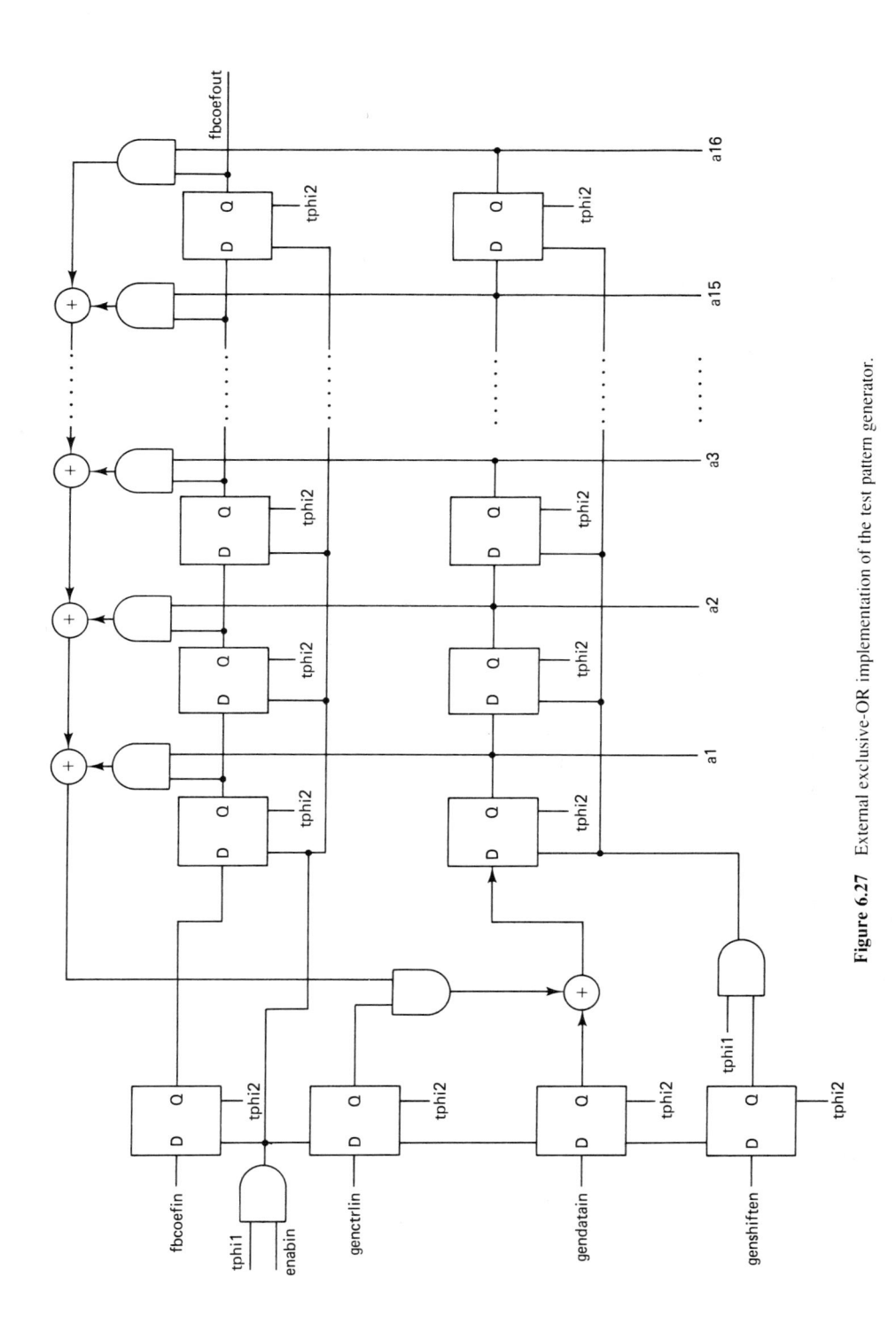

Figure 6.27 External exclusive-OR implementation of the test pattern generator.

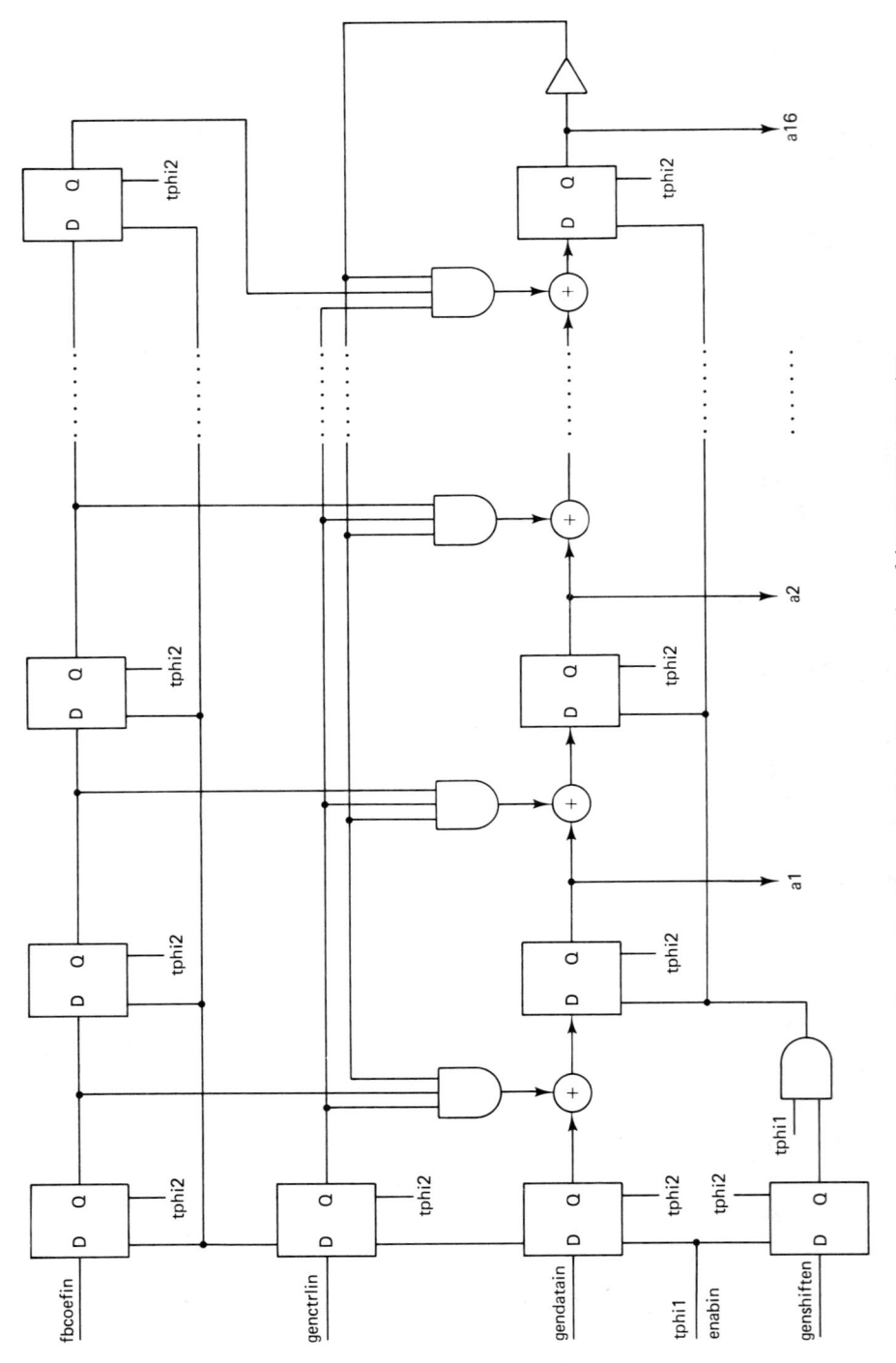

Figure 6.28 Internal exclusive-OR implementation of the test pattern generator.

After initialization, **genctrlin** becomes high, **enabin** goes low, thus latching the feedback coefficients and the **genctrlin** signal. After **enabin** goes low, pseudo-random test patterns become available every clock cycle at **a1–a16** outputs provided that the latched value of **genshiften** is high.

The signature analyzer operation is controlled by the signals **sac1in, sac2in, sasin,** and **sashiften**. Figure 6.29 shows an implementation of the signature analyzer. **c1I** = 1, **c2I** = 1 corresponds to the parallel data compression mode, whereas **c1I** = 0, **c2I** = 1 corresponds to the serial shift mode; **reset1** and **preset1** are chosen to be different from the master **reset** (or **preset**) of the other flip-flops in order to control the signature analyzer operation properly when used in conjunction with several other signature analyzers in the system. They may be used exclusively to enable and disable shifting of all the signature analyzers in the system at precisely the same times. In such a case, **reset1** and **preset1** are common to all the switches in the system. Signature analyzer contents can be shifted out through **tout** under the control of the **tctrl** signal.

Figure 6.30 shows a typical scheme for the control of the data exchange network. As shown in the figure, whenever the test pattern generator outputs are passed on to the normal data outputs, the normal data inputs are passed on to the signature analyzer for compression. The **enabin** to **enabout** delay can be an integral multiple of the **tphi1/tphi2** clock period by a suitable choice of the clocks **clk1** and **clk2**. Such an integral multiple delay may be required, depending on the scheme chosen for setting the various switches in the system, as will be illustrated.

Note that the implementation of the testing switch module presented here is only one of several possible ones and the specific implementation appropriate for a given situation may additionally be guided by other constraints. For instance, in order to reduce the number of pins of the testing switch chip, all the control signals could be serially shifted into a serial-in, parallel-out register and be transferred to the control flip-flops. Note, however, that these control signals are used to set up the module to the appropriate configuration and they do not change (except the clock) during at-speed testing, thus making possible the utilization of relatively slow test equipment for providing these control signals.

Board level usage of the testing switch module. The switch module is designed for use on boards to assist in fault detection and isolation to a given number of chips forming an ambiguity group (AG). The AGs are assumed to have reset or preset capability so that AG outputs can assume logic "0" or "1," which may be necessary to test for pin level faults. Furthermore, the AGs are assumed to be comprised of chips using synchronous logic.

Figure 6.32 illustrates an example of incorporation of the testing switches into a board containing the interconnected AGs shown in Figure 6.31. Note that typical interconnection features such as feedback and fanout between AGs are included in Figure 6.31. Even though not explicitly considered, it is assumed that control and data inputs to an AG are passed through different testing switches so that deterministic control over the AG functions may be possible.

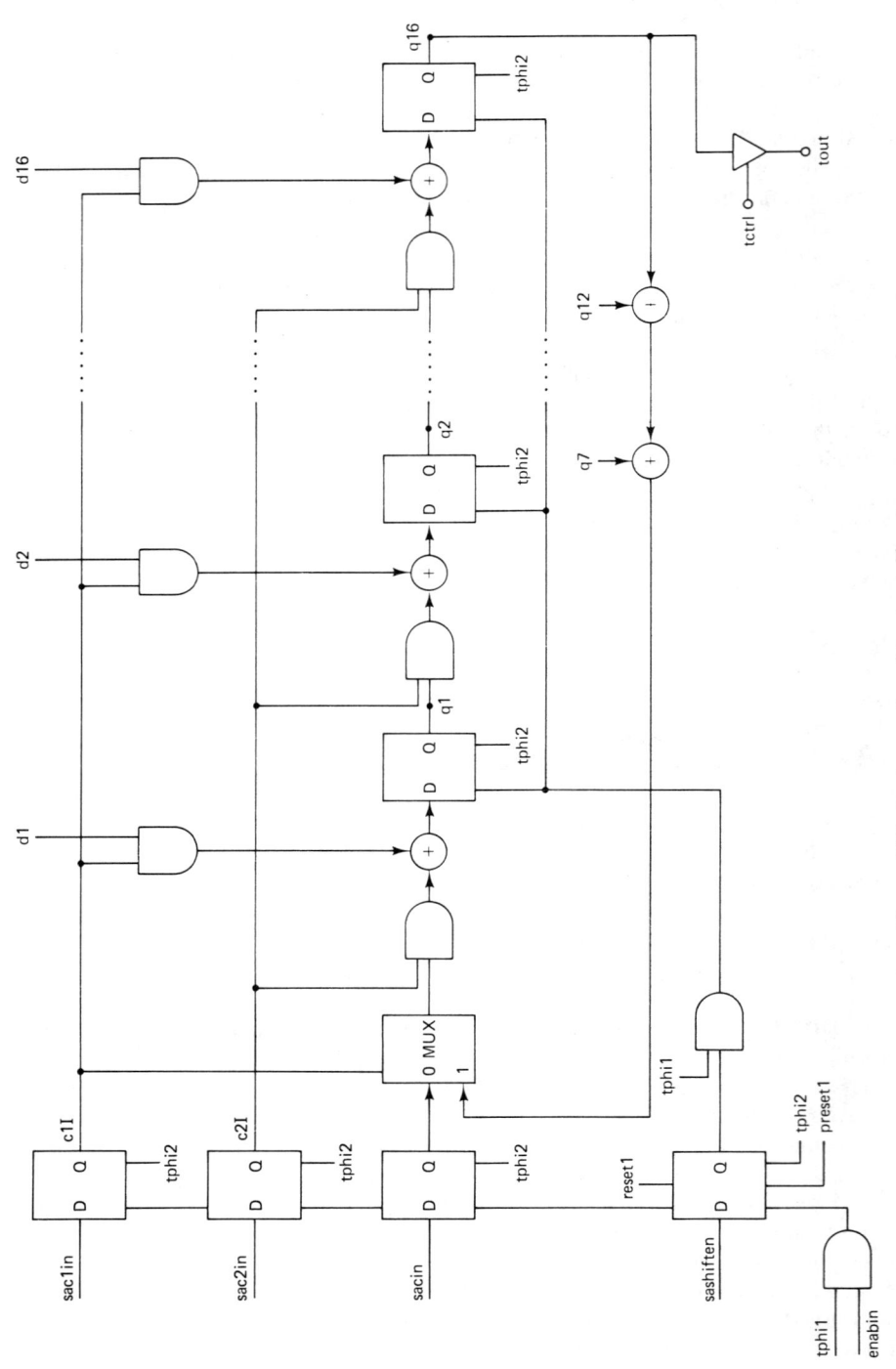

Figure 6.29 Implementation of the signature analyzer.

Sec. 6.4 Built-In Self-Test (BIST)

Note: a1, a2,..., a16 inputs come from the test pattern generator;
d1, d2,..., d16 outputs are connected to the signature analyzer.

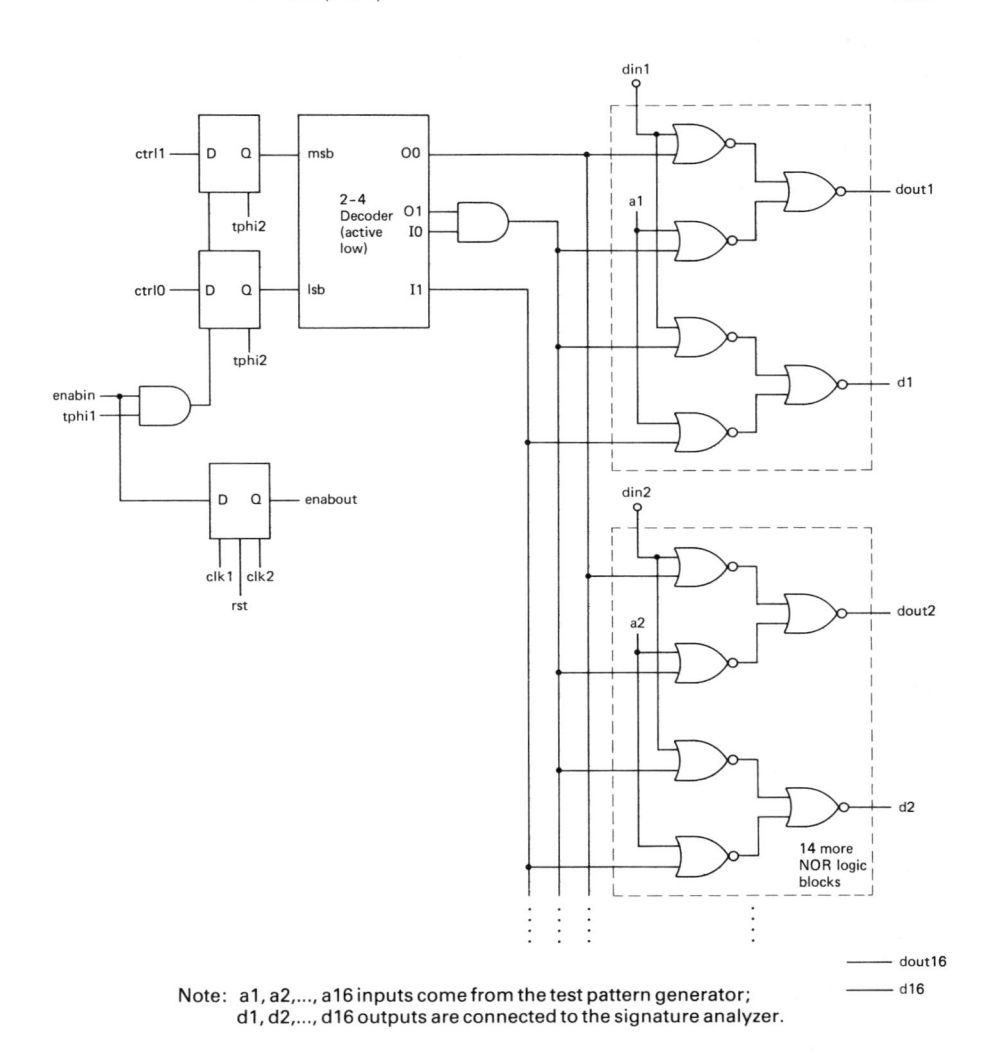

Figure 6.30 Control of the data exchange network.

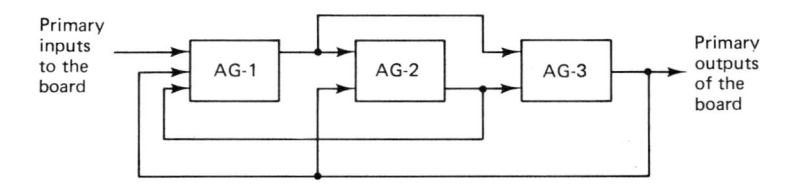

Figure 6.31 Typical interconnection of the ambiguity groups (AGs).

Fault isolation and fault detection testing. Upon activating the board for fault isolation, the following two preliminary tests are performed:

1. Each switch is selected through its **enabin** input, and the test pattern generator to signature analyzer path is enabled. After a predetermined number of clock cycles, the signature is shifted out through its **tout** node and checked against the expected signature. This test verifies the integrity of the test pattern generator, part of the data exchange network, and signature analyzer of each switch, as well as its **tout** and **tctrl** pins.

2. Reset and/or preset of all AGs is performed, and the normal data inputs-to-signature analyzer path is enabled in each switch. After one clock cycle (i.e., signature analyzer in each switch simply acts as a parallel-in, serial-out register), the signature analyzer contents are shifted out and checked to verify that the AG outputs are indeed reset (or preset). This test assists in isolating faults to an AG-switch(es) combination.

After the above two preliminary tests, the AGs are tested simultaneously, if possible, or sequentially using the pseudorandom test patterns generated by the test pattern generators in the testing switch modules. Control inputs to AGs can be held at the desired fixed logic values by using the **disable-shift** capability of the test pattern generators of the control input switches.

Simultaneous testing of AGs may not always be possible because of the requirement that certain AG outputs must be kept at predetermined states while testing other AGs. This is in case feedback paths between AGs exist, as in the example of Figure 6.31.

In this example, all AGs must be checked sequentially, i.e., one after another. It is assumed that the order of testing of AGs is predetermined based on appropriate knowledge of the board. However, if we assume that the feedback from AG-3 to AG-2 is absent, then it can be verified that AG-1 and AG-3 can be simultaneously tested. Simultaneous testing of AGs can always be accomplished independent of the feedback and fanout nature of the interconnections if each feedback and fanout path is passed through a separate switch—as illustrated in Figure 6.33, for the example under consideration. In this case, the constraint of reset/preset of AGs during testing of other AGs can be relaxed. It is apparent,

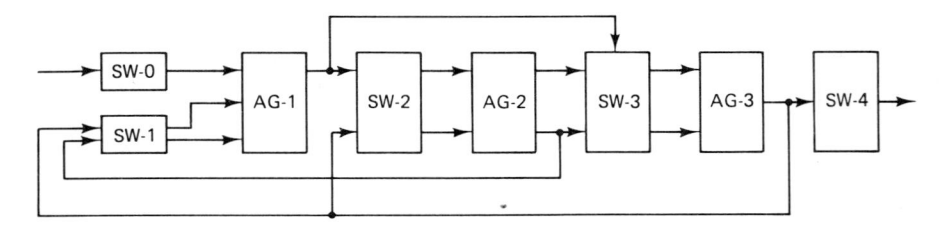

Figure 6.32 Example insertion of the switch modules into the design of Figure 6.30.

Sec. 6.4 Built-In Self-Test (BIST)

however, that simultaneous testing of AGs is achieved at the cost of an increased number of testing switches on the board.

A fault detection or go/no-go test of a board without fault isolation to AGs can also be performed using the testing switch approach. In Figure 6.32, go/no-go testing may be performed by using the test pattern generator in SW-0 to apply inputs to AG-1 and using the signature analyzer on SW-4 to compress the outputs. The remaining testing switches SW-1, SW-2, and SW-3 are set to the normal mode of operation (i.e., the normal data in–normal data out paths are enabled in these switches).

Proper setting of the testing switches on the board plays an important role in fault isolation and go/no-go testing. This setting of the testing switches can be performed in a predetermined order as assumed in the original description of the switch module. One approach for predetermined order setting is shown in Figure 6.33. Since the serial setting of the testing switch can take several clock cycles (16 **tphi1/tphi2** clock cycles in the chosen implementation), the delay of **enabin** to **enabout** may have to equal several **tphi1/tphi2** clock cycles. With this requirement in perspective, the **clk1** and **clk2** ports of the **enabin-enabout** flip-flop are separated from the **tphi1** and **tphi2** ports. A less-flexible way to accomplish the same objective would be to realize directly the required delay by adding more flip-flops to the original **enabin-enabout** flip-flop and clocking them with the same **tphi1/tphi2** clocks. Based on the proposed scheme, **clk1/clk2** can be derived from **tphi1/tphi2** using a divide-by-n frequency divider circuit, where n is the number of stages of the LFSRs in the testing switch module.

Regardless of the technique chosen for functional verification testing of a board, evaluation of the test vectors used (i.e., deterministic or pseudorandom) should be performed through fault simulation. The DFT/BIT techniques presented enhance the testability of a board; however, they do not guarantee the quality of the test vectors. The quality of the test vectors should be determined by the designer and can be used as a criterion for selecting appropriate DFT/BIT techniques for a given design.

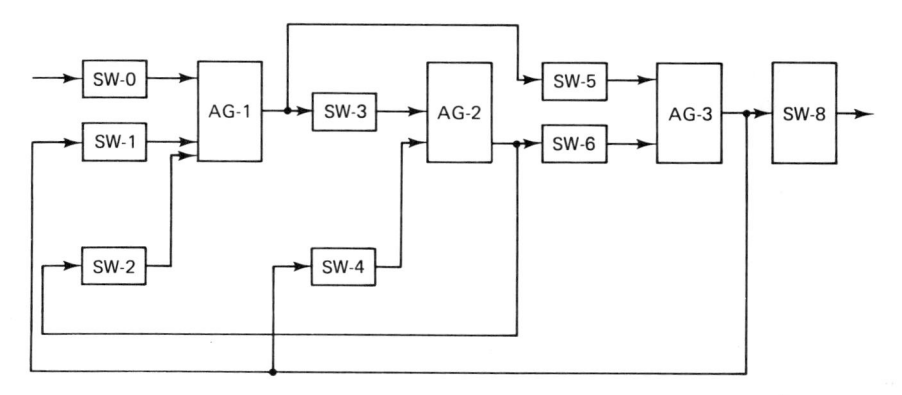

Figure 6.33 Example insertion of the switch modules into the design of Figure 6.30.

PROJECTS

6.1. Derive the test vectors for the standard cells designed in Project 4.1. Assuming the single stuck-at fault model, provide fault simulation results that demonstrate your test vectors achieve 100 percent fault coverage. If you cannot obtain 100 percent fault coverage, explain why. Perform the fault simulation at the transistor level for the simple gate-cells and at the gate level for the flip-flop cells.

6.2. Redesign the flip-flops in Project 4.1 so they are scanable. Select your favorite design for testability technique, and provide a discussion for your choice. Remember that minimizing area overhead and number of I/O is crucial for the implementation of GaAs circuits.

6.3. Incorporate built-in testing in the comparator chip described in Project 4.2. The built-in test circuit should be activated using an "enable" signal (single bit) externally provided and the test result should be indicated through a single bit (flag). Provide gate-level fault simulation results that demonstrate the effectiveness of your BIT technique. What is the required time for completing the self-test function? What is the overhead for implementing your BIT technique (gate and I/O overhead)? Is the performance of the circuit affected by incorporation of BIT? If it is, explain why. If the testing requirement for this chip is 98 percent fault coverage (for single stuck-at faults), does your BIT scheme meet this requirement? If not, propose design revisions (in the functional circuit, the BIT circuit, or both) that will make it possible to attain this level of fault coverage.

6.4. The test requirements and design for test constraints for the median filter chip described in Project 4.4 are as follows:

1. 99 percent fault coverage (single stuck-at faults)
2. At-speed testing of the core logic (i.e., all logic but I/O)
3. No more than 4 I/O, for test purposes only, are allowed
4. No more than 10 percent gate overhead, for test purposes only, can be afforded

Design a test plan for the chip including test vectors and a tester environment that will allow you to test the chip with these test vectors. You are free to use any design for testability or built-in test technique that you determine appropriate for meeting your test requirements. In any case, however, you are required to demonstrate through fault simulation that your technique meets the test requirements. You are also required to demonstrate that the implementation of your technique is within the design constraints listed above.

6.5. In a systems environment, it is highly desirable to be able to test the interconnection between chips on a board. Design this capability into the median filter chip so that when two of these chips are interconnected, the system designer can determine that the output of one chip is correctly received by the input of the other.

REFERENCES

1. T. R. Gheewala, "Requirements and Interim Solutions for High-Speed Testing of GaAs ICs," *Proc. GaAs IC Symp.*, p. 143, 1985.
2. O. H. Ibarra, S. K. Sahni, "Polynomially Complete Fault Detection Problems," *IEEE Trans. on Computers*, C-24, p. 242, March 1975.

3. M. A. Breuer, A. D. Friedman, *Diagnosis and Reliable Design of Digital Systems*, Potomac, Md.: Computer Science Press, Inc., 1976.
4. J. P. Roth, "Diagnosis of Automata Failures: A Calculus and Method," *IBM J. of R/D*, p. 278, October 1966.
5. P. Goel, "An Implicit Enumeration Algorithm to Generate Tests for Combinational Logic Circuits," *IEEE Trans. on Computers*, C-30, p. 215, March 1981.
6. H. Fujiwara, T. Shimono, "On the Acceleration of Test Generation Algorithms," *IEEE Trans. on Computers*, C-32, p. 1137, December 1983.
7. F. F. Tsui, "In-Situ Testability Design (ISTD): A New Approach for Testing High-Speed LSI/VLSI Logic," *Proc. of the IEEE*, vol. 70, no. 1, p. 59, January 1982.
8. TriQuint Semiconductor, *Operating Manual: Quick-Chip Evaluation Board*, Beaverton, Oreg.: TriQuint 1984.
9. E. B. Eichelberger, T. W. Williams, "A Logic Design Structure for LSI Testing," *Proc. 14th Design Automation Conf.*, p. 462, June 1977.
10. S. Funatsu, N. Wakatsuki, T. Arima, "Test Generation Systems in Japan," *Proc. 12th Design Automation Symp.*, p. 114, June 1975.
11. J. H. Stewart, "Future Testing of Large LSI Circuit Cards," *Digest of Papers, 1977 Semiconductor Test Symp.*, p. 6, October 1977.
12. H. Ando, "Testing VLSI with Random Access Scan," *Digest of Papers, Compcon 80*, p. 50, February 1980.
13. E. Trischler, "Incomplete Scan Path with an Automatic Test-Generator Methodology," *Proc. International Test Conf.*, p. 153, October 1980.
14. S. Lin, D. J. Costello, *Error Control Coding: Fundamentals and Applications*, Englewood Cliffs, N.J.: Prentice-Hall, 1983.
15. J. E. Smith, "Measure of the Effectiveness of Fault Signature Analysis," *IEEE Trans. on Computers*, vol, C-29, p. 510, June 1980.
16. D. K. Bhavsar, R. W. Heckelman, "Self-Testing by Polynomial Divison," *Proc. International Test Conf.*, p. 208, October 1981.
17. B. Koenemann, J. Mucha, G. Zwiehoff, "Built-In Logic Block Observation Techniques," *Proc. International Test Conf.*, p. 37, October 1979.
18. N. Kanopoulos et al., "A New Implementation of Signature Analysis for Board Level Testing," *Proc. International Test Conf.*, p. 730, September 1987.

7

GaAs INSERTION INTO SYSTEM DESIGN

The benefits of using a new semiconductor technology for circuit fabrication are determined by the impact this technology has on system performance and cost. A new technology becomes viable when it can be introduced into system design to increase the system's performance and/or reduce its cost. System reliability, producibility, and maintainability are also affected by the insertion of a new technology into system design.

The availability of high-speed digital GaAs circuits offers attractive possibilities for system performance improvement and even new system applications not possible with high-performance Si-components. However, the role of GaAs and Si circuits will be different in system design, and system performance will gain the most from selective insertion of GaAs components into system designs containing Si-components. The inherent speed advantages of GaAs circuits will benefit the system performance only if major architectural improvements take place along with the GaAs insertion into system design. This is because the chip-to-chip interconnection delays may remain constant, thus limiting the system performance despite improvement on chip gate delays.

A number of design issues must be resolved during GaAs circuit insertion into system design. These issues relate to system architecture, functional partitioning, timing, and functional verification testing. Issues at a lower level of design hierarchy include intercomponent communication, power distribution, heat removal, component interfaces, and system packaging.

This chapter presents the system application areas that will be most highly impacted by GaAs insertion and discusses architectural approaches and system design issues most relevant in using GaAs components at the system level.

7.1 SYSTEM APPLICATIONS TO BE STRONGLY IMPACTED BY GaAs INSERTION

As discussed in Chapter 4, circuit architectures take maximum advantage of the GaAs properties when they retain data on a chip for completing a function that can be computed repeatedly with minimal external communication requirements. This same feature is the main characteristic of system architectures to be most highly impacted by GaAs insertion. Therefore, we will be looking at system applications where operations are performed using well structured algorithms that involve computations requiring minimum branching and interrupts and data structures that can be handled with simple addressing schemes. Also of interest are general purpose computing applications where certain subsystems have the above mentioned characteristics. In any case, however, the system designer should think in terms of technology insertion, not necessarily in terms of total system implementation in GaAs. This will allow trade-offs to be made and advantage to be taken of both the large scale of integration available through Si-circuits, and the high speed available through GaAs circuits while both are selected for low power.

It is recognized that technology insertion cannot be separated from development of system architectures that exploit the characteristics of the new technology.[1] In turn, system architectures are closely related to algorithms used to perform the required computations by the system application. The development of algorithms best suited for environments where operations are performed by high clock rates with minimum decision branching during execution is therefore essential for effective utilization of GaAs components into system design. The development of algorithms that exploit the low gate speed of GaAs components rather than only the high parallelism possible with VLSI Si-components is gaining interest and should be further pursued to increase the benefits from GaAs insertion into system design. Although the development of such algorithms has been a field of research in the late 80s, there are certain system applications that naturally lend themselves to architectural features best suited for implementation in GaAs. This section presents the computational requirements for these applications and possible system design areas where GaAs insertion is most suitable. The list of application areas presented here is by no means complete. As with any new technology, applications are created as the technology evolves. However, the applications discussed here are good representatives of application requirements and features best suited for GaAs implementation.

7.1.1 High-Throughput Signal/Image Processing

A Digital Signal Processor (DSP) is a discrete parameter system specialized to operate upon and react to externally originated streams of data, generally, on a real-time basis. The external sources generating the data streams are usually sensors such as radar, sonar, optical and infra-red receivers, medical electronic probes, seismic arrays, etc. The processing throughput requirements depend on the characterization of the signals and the operations that need to be performed to get the information of interest into a desired form of representation. In general, there are two application-dependent aspects of signal characteristics:

1. the basic signal parameters such as frequency content, dynamic range, and signal-to-noise ratio, which affect both the sample rate and the sample quantization requirements
2. the signal modeling, which determines how the signal is interpreted to obtain information

Based on the signal models, the signal parameters, and the specific objectives of an application, a signal processing structure is formulated by defining the order in which the signal is to be manipulated to get the information of interest into a desired form of presentation. The actual signal manipulations, generally, are based on a relatively small set of basic signal processing operations such as convolution, correlation, or difference equation calculations, Discrete Fourier Transforms (DFT), and vector or matrix arithmetic operations. The signal processing application of interest specifies the appropriate combination of these operations to accomplish the required processing and the performance requirements.

The signal characteristics, modeling, and sequence of processing operations determine the processing throughput requirements. However, the selection of algorithms for performing required signal processing operations and the selection of semiconductor technology for implementing signal processing components determine successful implementation of systems for meeting throughput requirements. Applications requiring high throughput signal processing use high bandwidth signals and include primarily radar signal processing, electronic countermeasures, surveillance, and image processing. In all high throughput processing applications, real-time signal processing requires the availability of very fast data acquisition and storage components as well as fast arithmetic components. Although increased throughput can be obtained through extensive parallelism, not all signal processing algorithms can be computed in parallel. Also certain operations such as data acquisition and storage of the incoming samples are sequential in nature. These areas, therefore, benefit the most from the inherent speed of GaAs components. GaAs A/D converters allow high-frequency data acquisition required in radar systems, and GaAs memory can be used to store the incoming samples.

This capability allows direct processing of the incoming signal without requiring down conversion in frequency as is the current practice in radar signal processing. Real-time processing of such signals requires performance levels in the range of billions of arithmetic operations per second, well beyond the capability of existing Si-based digital signal processors. Introducing GaAs arithmetic circuits into signal processor architectures can make possible the use of digital processing techniques for applications of high interest which, however, have throughput and memory requirements well beyond what is achievable with Si-based signal processors. One such application that combines signal and image processing operations is the synthetic aperture radar (SAR), which can achieve high spatial resolution in the direction of motion of the radar platform. This type of radar is used aboard satellites and space shuttles to image the earth's surface. Operating at a frequency of 1.3 GHz (designated as L-Band), this type of radar has been used to produce images of the earth's surface with 25 m × 25 m resolution over a continuous strip of 100 km. To achieve this resolution each pixel in the SAR image is generated by coherently processing a large number of radar returns, which requires extremely high throughput and fast memory capabilities for real-time processing. Since these requirements are well beyond the capabilities of current signal processing systems, SAR data have been processed using optical techniques. However, these techniques do not have the accuracy and adaptability features, inherent in digital systems, which are essential for high-accuracy SAR imaging. The development of GaAs digital circuits can lead to developing digital techniques and special purpose systems for processing SAR data. The key, however, for successful utilization of GaAs components in high throughput signal/image processing systems, is the development of architectures that exploit the inherent speed of the GaAs components and avoid the drawback of large numbers of interconnections.

High-throughput signal processing systems of the 1980s are dominated by small- and medium-scale integration ECL components in the critical path of data acquisition, storage and arithmetic operations. Substitution of ECL components with GaAs ones will increase system performance while decreasing size and power requirements. This substitution will start at the front-end data acquisition and scratchpad memory sections and will proceed to the main memory and arithmetic sections. An optimum signal processor configuration, however, may result from an architecture that uses the advantages of both VLSI Si components and MSI/LSI GaAs components.

7.1.2 Biomedical Research

The use of digital signal processing techniques is becoming widespread in medical applications such as the analysis of EEG and ECG signals, and computer-aided tomography which is the creation of two- and three-dimensional images of the interior of the body from x-ray projections and ultrasonics. The image-reconstruction operation, used by computer tomography (CT) scanners, is performed by computationally intensive algorithms. The reconstruction of the gray-scale value of a pixel in a CT image may require $10^3 - 10^4$ arithmetic operations. For image resolution of 512 × 512 pixels, the minimum computational capabilities incorporated into CT scanners are of the order of 10^7 arithmetic instructions per second. Even at this performance level, however, the physician may have to wait for processed results for one to two hours after completion of the scan sequence, when multiple cross-sections are required for a given examination.

7.1.3 Communications Applications

Digital signal processing techniques influence almost all areas of the general field of communications. Digital techniques have been applied to the problems of signal modulation, multiplexing, encoding, and data rate compression. In telecommunications, digital techniques are used for tone detection, echo cancellation, and digital switching networks. GaAs insertion into communication systems will address, as with previous applications, high bandwidth operations and real time operations. In fact, the communications field is one of the main applications that drive the development of standard off-the-shelf GaAs components.

One area with high bandwidth processing requirements is the error detection and correction coding (EDAC) of messages that is used on signals transmitted between satellites and earth stations or between two satellites to ensure uncorrupt data transmission. Data corruption due to interference and noise can be minimized using larger transmission antennas and/or a higher power signal. However, both these options increase the weight and cost of a satellite, which makes the EDAC approach a more attractive one since the cost trade-offs are minimal. With satellite communication bandwidths approaching 10^9 bits/sec, both encode and decode EDAC circuits must operate at rates of hundreds of megabits/sec exceeding the capabilities of silicon-based EDAC circuits that operate at a few megabit/sec rates. GaAs insertion, in this case, is expected to provide higher quality transmissions. A related application is signal encryption and decryption for military communications and surveillance. In this case, the signal is encoded before transmission and decoded at reception, using special coding schemes that intend to safeguard the information against interception by unauthorized stations.

One area of communications where GaAs insertion started in the late 80s is fiber optic communications systems operating at gigabit/sec rates. GaAs circuits find application both at the transmitter and at the receiver ends of the system, offering the required speed performance at moderate power dissipation not available by the fastest (100 K, ECL) silicon circuits. At the transmitter end, GaAs components are used as time division multiplexers, laser diode drivers, and bitstream encoders, while the receiving end includes time division demultiplexers, transition detectors, and decoders. In all these circuits, the main requirement is the highest speed of operation that will still accommodate the system bandwidth. Since the complexity of these circuits is relatively small, more than one of these circuits can be integrated on a single chip, thus avoiding synchronization problems and possible performance degradation due to external interconnect attributed delays.

7.1.4 Scientific Computers

Scientific computers are used to solve computationally intensive problems requiring a large number of computations with high precision. The field of scientific computing in the 80s is dominated, in terms of performance and cost, by the so-called supercomputers such as the Cray 1 and Cray 2. These machines are implemented using medium-scale integrated ECL circuits for memory and arithmetic operations. One metric often used to access the performance of a scientific computer is the number of floating point operations per second that can be performed by the arithmetic logic unit (ALU). This metric is strongly dependent on the cycle time of the critical path of the ALU. An estimate of the length of the ALU critical path can be computed by using the formula:

$$t_c = n \cdot (\text{gate delay}) + m \cdot (\text{chip crossing delay}) + (\text{latch time}) + (\text{clock skew})$$

where n is the number of gates on a chip in the critical path, and m is the number of chips in the critical path. GaAs insertion will affect the t_c because of the short gate delays achievable with GaAs. However, the computer architecture should allow for small m for effective increase of machine performance. The value of m strongly depends on the circuit integration level of the chips in the critical path and utilization of GaAs DCFL circuits may result in a reduction of m compared to high power ECL circuits used in supercomputers.

GaAs insertion is also expected to increase the performance of large mainframe computers. In this case the metric used to assess the machine's performance is number of instructions per second that can be executed by the central processing unit (CPU). Executing an instruction is a function of the CPU cycle which strongly depends on the access time of the cache memory of the computer. A simple formula for computing this cycle is:

$$t_c = n \cdot (\text{gate delay}) + m \cdot (\text{chip crossing delay}) + (\text{memory access})$$
$$+ 1 \cdot (\text{card crossing delay}) + (\text{latch time}) + (\text{clock skew})$$

Based on the above simplistic formula, the machine cycle time can be reduced by incorporating GaAs circuits with short gate delays and GaAs memory circuits with small access time. Maintaining a small m and one card crossing (as the above formula assumes) is essential in realizing any overall performance benefits from utilizing fast GaAs circuits.

Based on our brief discussion so far, one can argue that GaAs insertion into computer design will increase performance by increasing the rate at which the ALU performs arithmetic operations, and/or by decreasing the access time of cache memory. Several design trade-offs have to be made, however, to realize an overall increase in machine performance. These trade-offs may include rethinking of the processor architecture as well as the memory design. A number of trade-offs the designer has to make are presented in greater detail in the next section.

7.2 INSERTION ISSUES AND TRADE-OFFS

In general, there is not a unique computational algorithm, nor a unique system architecture, for solving a particular problem expressed in terms of a particular computational theory. The system designer selects algorithms and develops architectures so that the solution of the problem is achieved within certain performance constraints, and within a certain budget. In most cases, the architecture of modern digital systems is influenced more by function and application than by implementation based on selection of specific semiconductor technology. In the case of GaAs insertion into system design, the consideration of implementation issues is equally important to function and application, because the performance of GaAs systems can greatly suffer from technological problems that arise if the system architecture does not facilitate the design features that can result in efficient implementation in GaAs. Most of the technological problems that affect the architectural design of GaAs based systems are related to intercomponent communications. The issues to be addressed at the architectural level that can affect data communications relate to algorithm selection, data structures, memory configuration, and processor organization.

This section discusses some of the issues and trade-offs involved in incorporating GaAs components into system design. These issues are related primarily to the difficulty of propagating fast signal transitions from chip-to-chip and board-to-board, and also to the difficulty of functional verification testing of high-speed systems. Although they are not included here, other decisive factors that must be considered when implementing GaAs based systems are reliability and cost.

7.2.1 Design Partitioning

Design partitioning is one of the first, and one of the most important, stages in the design of a digital system. During this design stage, the system designer makes decisions concerning resource allocations for performing the system functions according to system specifications. Ideally, resource allocations should consider hardware, software, and testing resources by taking into account the technology of design implementation. In Si-based systems, system partitioning is rarely concerned with the characteristics of the technology to be used for system implementation. When GaAs technology is used, however, system partitioning becomes a critical issue because of the unique characteristics of GaAs components and the potential effect of these characteristics on overall system performance. An ideal approach to system design, where resource allocation takes place early in the design phase, is shown in Figure 7.1.

Different trade-offs may be required for resource allocation when a system, or parts of a system, is implemented with GaAs components. Especially when deciding what functions are to be implemented in hardware or software, the system designer is bounded by the low scale of integration of GaAs chips and

Sec. 7.2　Insertion Issues and Trade-offs　　　　　　　　　　　　　　301

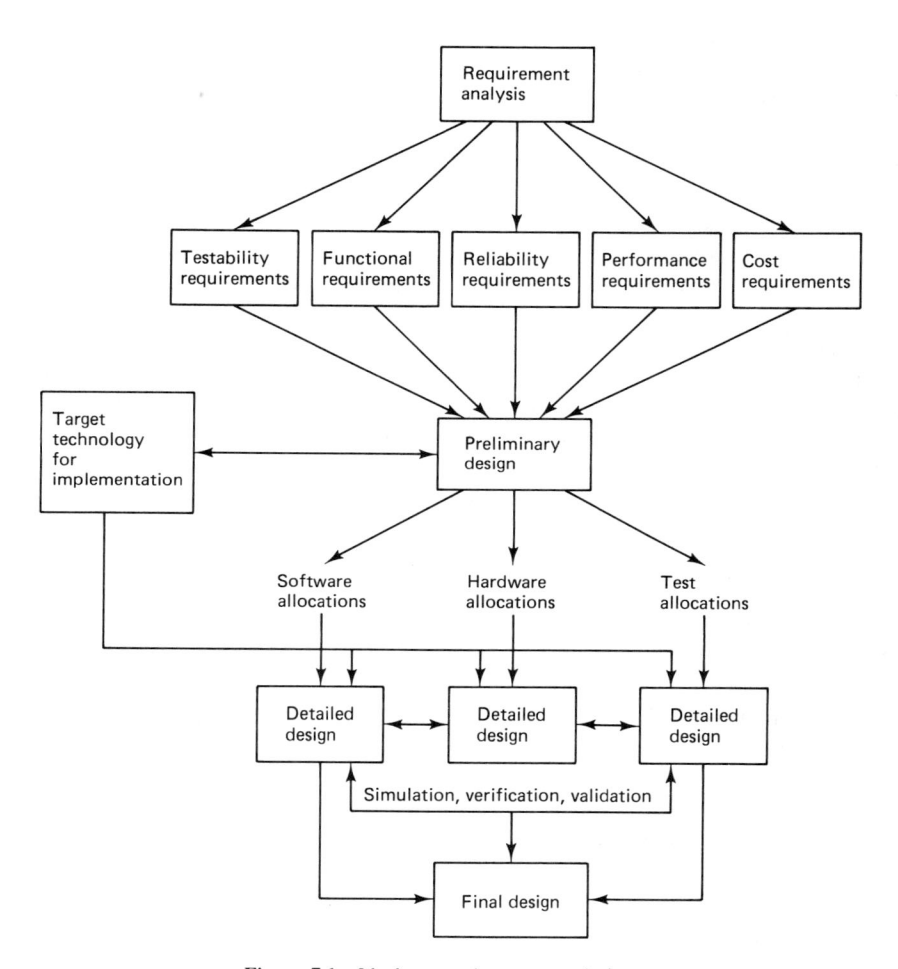

Figure 7.1 Ideal approach to system design.

performance degradation when crossing chip boundaries. On the other hand, the high on-chip speed can be utilized to perform functions efficiently, using a large number of repeated simple instructions without any performance penalty compared to Si-based structures. This suggests that some traditional hardware functions may need to migrate into software to take full advantage of the performance of GaAs components in system design. This suggestion is in agreement with efforts to design and implement GaAs processor chips and computer systems. All efforts in this area are concentrated in architectures with a reduced instruction set which favors simplified instructions with few format options or addressing modes.[2,3] This architecture requires selection of instructions based on justifications concerning the cost of implementing an instruction over the frequency of use of this instruction in application programs. This results in a simple instruction set that can be executed optionally by a system with low hardware requirements, and an especially simplified control structure that can alleviate the off-chip data communications burden. A major decision to be made during system partitioning is the selection of the instruction set and the approach for its implementation and validation of its performance. An important factor in achieving efficient implementation of a simplified instruction set is the ratio of the processor's instruction cycle time to the time required by its peripheral components (mainly memory) to support the instruction cycle. Minimization of the instruction cycle itself and also of the time required mainly by memory to support the instruction cycle can be achieved in two ways—by migrating some hardware functions to the system compiler and by optimizing compiler design based on the GaAs component characteristics. An interesting comparison of the compiler technology-GaAs technology relationship to the CAD–VLSI relationship is made in Reference 4. In that article, the authors contend that just as the full capabilities of VLSI could not be fully utilized without appropriate advances in CAD technology, so also GaAs processors will struggle to obtain their potential performance without assistance from a sophisticated compiler technology.

One final, but very important observation that should be made from Figure 7.1, is that unlike the detailed design of chips and systems, there are no (in the late 1980s) CAD tools available to address adequately system requirement analysis and preliminary design. These processes have been largely based on a case-by-case approach using ad-hoc techniques and field experience acquired by systems engineers. CAD tool development in these areas will enhance the overall system design process and will facilitate insertion of new technologies into system design.

7.2.2 Intercomponent Connection Problems

Figure 7.2 illustrates a typical placement of chip carriers on a multilayer Printed Circuit Board (PCB). During system operation, signals have to propagate

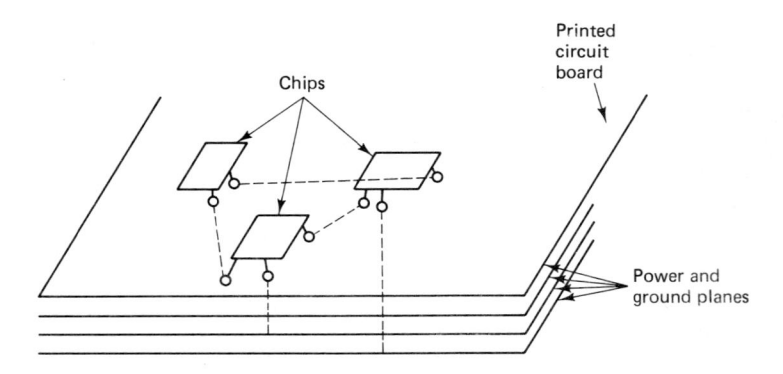

Figure 7.2 Chip interconnection on a board.

Sec. 7.2 Insertion Issues and Trade-offs

among chips to complete a required function. The signals may be delayed by both the integrated circuits and the board interconnection network, and the delays incurred affect the system performance. The major contribution to the average delay is due to the integrated circuits for lower-performance systems, while the interconnection delay will dominate high-performance systems employing GaAs components. The designer of a digital system has to pay special attention to potential interconnection related problems in order to meet typical design criteria which can be summarized as follows:

1. All signals in the system must meet the specified timing requirements.
2. The signal waveform must be within given tolerance.
3. Signal coupling between interconnection lines must be less than a given upper boundary so that coupled signals do not cause improper switching of the logic circuits.
4. Voltage transients on the power (and ground) distribution network induced by switching circuits must be limited to a small fraction of the power supply dc voltage(s).
5. External electromagnetic interference should not result in false switching of the logic circuits.

The question now becomes: What must the designer do to avoid violation of the above design criteria when using GaAs components in system design? Commercially available GaAs ICs generate output signals with rise and fall times in the vicinity of 150–200 psec and clock frequencies greater than 2 GHz. With signals of such speed, even very short interconnection lines (i.e., a few millimeters) will behave like transmission lines.[5] The two most practical transmission line structures that can be used for intercomponent communication in high-speed systems are the microstrip and the stripline. Figure 7.3 illustrates a cross section of a microstrip line with dielectric thickness h, strip width w, and strip thickness τ. The exact analysis of the microstrip line is beyond the scope of this book, so the characteristic impedance and the expression for the propagation delay are quoted from Reference 6:

$$Z_o = \frac{87}{(\varepsilon_r + 1.41)^{1/2}} \ln\left[\frac{5.98\,h}{0.8w + \tau}\right] \quad (7.1)$$

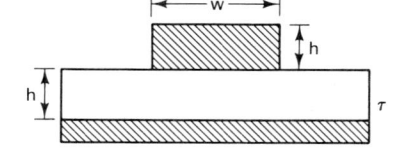

Figure 7.3 Microstrip transmission line structure.

$$t_{pd} = 1.017\,(0.475\,\varepsilon_r + 0.67)^{1/2} \quad \text{nsec/ft.}$$

where ε_r is the dielectric constant of the board. The stripline shown in Figure 7.4 is a line configuration closely related to a microstrip and most commonly used in multilayer boards. The characteristic impedance and the propagation delay of the stripline are given by Reference 6:

$$Z_o = 60\,\varepsilon_r^{-1/2} \ln\left[\frac{4h}{0.67 \cdot w\,[0.8 + (\tau/w)]}\right]$$

$$t_{pd} = 1.017 \cdot \varepsilon_r \quad \text{nsec/ft.}$$

(7.2)

The dielectric constant of the board depends on the board material used and the thickness of the conductors, and the distance between conducting planes is determined by the overall board thickness and number of planes. Therefore, the parameters ε_r, h, and τ are determined by board processing considerations while the conductor width w is the only parameter that can be affected by design to control the characteristic impedance of an interconnection network. Based on the expressions above, the interconnection traces should be as wide as required for achieving the normally used 50 ohm standard lines. For any given size board, the wider the traces, the fewer traces the board can accommodate. Therefore, the interconnection density of a PCB designed with GaAs components may be significantly lower than a similar board populated with lower speed components.

An architectural solution to the low density interconnections of high-speed boards is the minimization of the amount of intercomponent communication required for performing the function intended for the board. Although global clock and control signals may not be able to be eliminated, structures such as high-speed wide buses should be avoided.

Another interconnection characteristic that can affect system performance is the interconnection length. As Equations 7.1 and 7.2 indicate, the signal propagation delay is a function of the dielectric constant of the board material and the length of the interconnect. For any given board material, the designer should maintain the shortest possible interconnection length which becomes the dominant performance factor for high-speed systems. Another problem attributed to long interconnects is signal attenuation. The attenuation is usually negligible at low speeds (i.e., TTL and MOS) or over short traces (i.e., a few inches). However, when signal traces extend to as much as 12 inches, signal alteration at high frequency can cause rise time degradation and lower noise margins. The signal

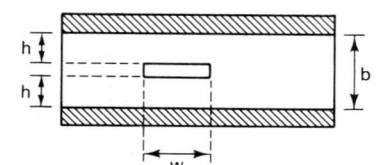

Figure 7.4 Stripline transmission line structure.

Sec. 7.2 Insertion Issues and Trade-offs

attenuation is affected by a number of board material and construction factors and is proportional to the trace length,[6] which the system designer can affect.

Ideally, all off-chip signals should be propagated over constant impedance transmission lines that are terminated in matching resistors. In practice, however, this rarely occurs and the designer should pay special attention to signal-line reflections and ringing caused by lack of impedance matching. This is important because reflections and ringing can degrade the system dynamic noise margin and cause inadvertent switching. The factors contributing to the rise of these irregularities are mainly the terminating resistors' inequality to the transmission line impedance, nonuniform transmission line impedances, inductive vias connecting different board layers, unterminated line sections (stubs), and nonideal resistors.[7] Figure 7.5 illustrates two cases of signal reflection and ringing. In 7.5(a) ringing is due to the length of the connection between the terminating resistor and the chip and the resistor size, while in 7.5(b) the rise time has been degraded due to a series inductance in the transmission lines introduced by the via that provides the connection to those lines that run on different board planes. A similar degradation to rise time can also be attributed to the shunt capacitance of load devices or connection pads. The maximum series inductance (L_s), shunt capacitances (C_p), and stub length (S_l), that yield acceptable discontinuities, can be calculated using the following expressions:

$$L_s \leq t_r \frac{Z_o}{3} \qquad (7.3)$$

$$C_p \leq \frac{t_r}{3Z_o} \qquad (7.4)$$

$$S_l \leq \frac{t_r}{3} \qquad (7.5)$$

where t_r is the signal rise time and Z_o is the transmission line characteristic impedance. Another important design consideration is the placement of the terminating resistor and its value, which depends on whether, in a fanout configuration, the fanout points are distributed along the line or all are lumped together at the end of it. Figure 7.6 illustrates these two cases. In 7.6(a), matching with a parallel resistance to ground at the input of the transmitting chip will add load to the output driver while a series resistance to Z_o would allow only half the driving output voltage swing on the line. The chips at the end of the line would receive the full voltage swing if that end were left as an approximate open circuit, while the chips along the line receive a second signal from the reflection. Therefore, in this case the termination is best implemented at the end of the transmission line as parallel output matching. The configuration in 7.6b can use series input matching while the output end of the line is virtually an open circuit. Parallel matching does not affect the speed of the output driver but the distributed load on the transmission line increases the signal propagation delay. The delay in the presence of the additional load is:

$$t'_{pd} = t_{pd}\left[1 + \frac{C_d}{C_o}\right]^{1/2} \qquad (7.6)$$

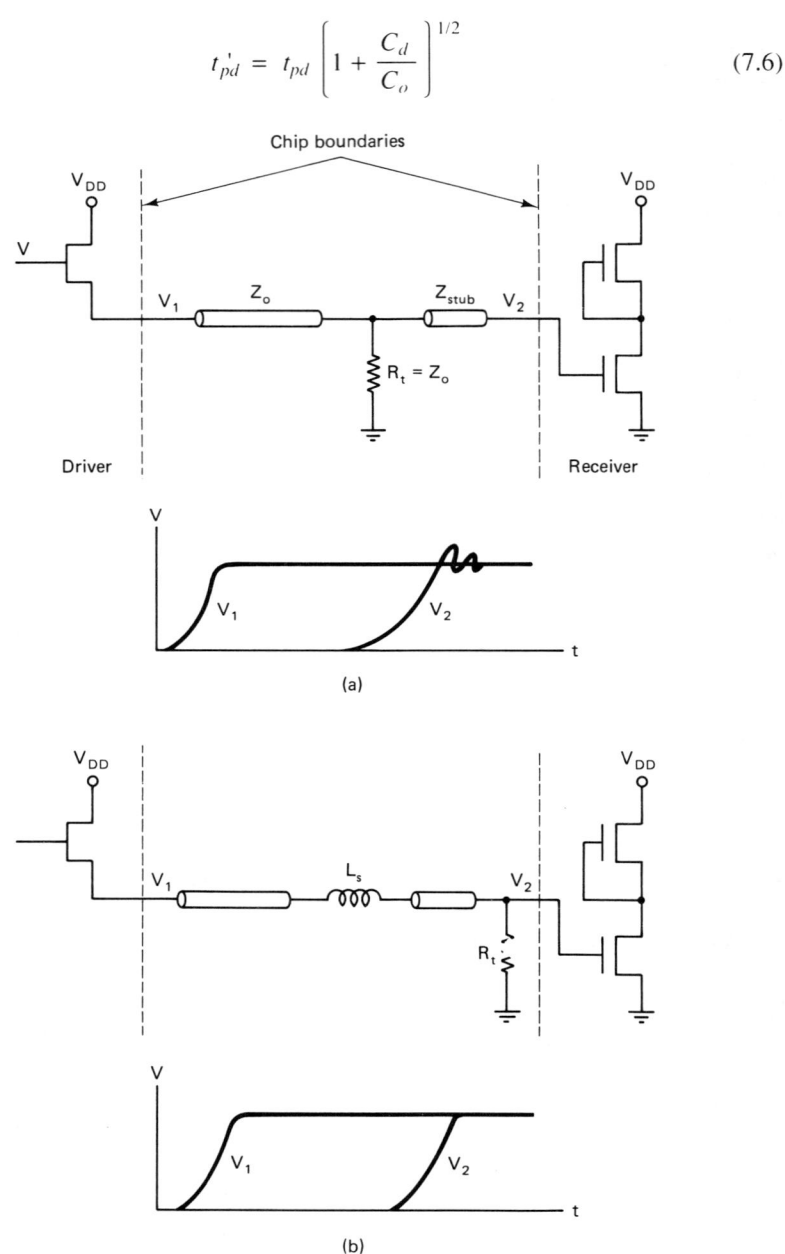

Figure 7.5 Reflections and ringing.

Sec. 7.2 Insertion Issues and Trade-offs

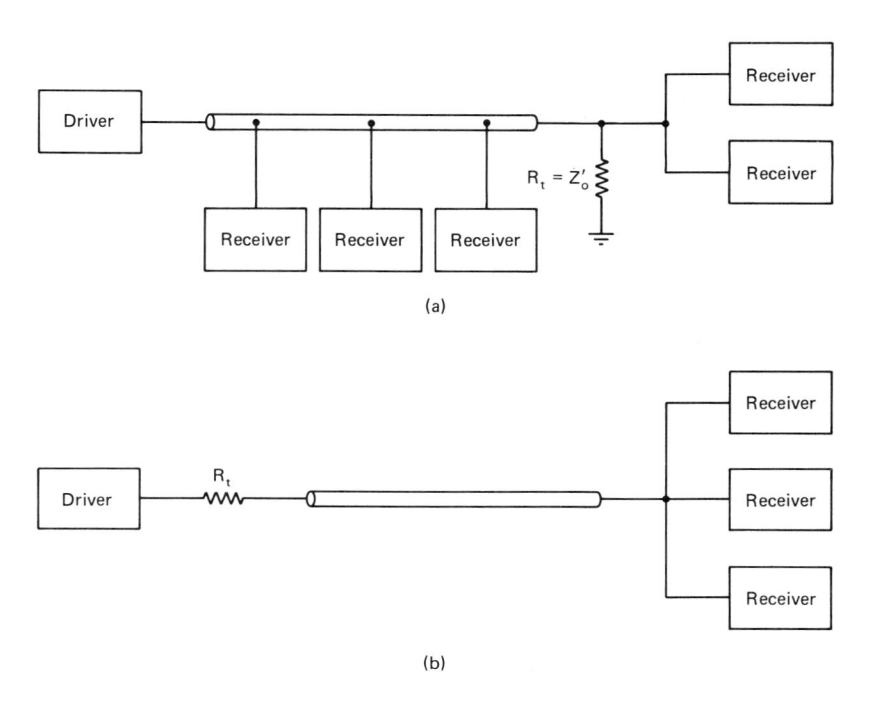

Figure 7.6 Transmission line: (a) parallel output matching and (b) series input matching.

where t_{pd} and C_o are the delay and the distributed capacitance of the unloaded line, respectively, and C_d is the additional distributed capacitance component given by the total lumped capacitance divided by the line length. The characteristic impedance must then change to:

$$Z'_o = Z_o \left(1 + \frac{C_d}{C_o}\right)^{-1/2} \qquad (7.7)$$

which implies that the matching resistance must be Z'_o and not Z_o.

In the series matching configuration, the combination of the matching resistance R_t and the output impedance of the driver must equal the characteristic impedance Z_o of the transmission line.

When a chip uses internal feedback from a particular output, that output should not be used to drive an interconnection line because of the likely effects from reflections. Also, when the interconnection network is formed using microstrip conductors, sharp bends and squared off corners should be avoided because changing the line width results in characteristic impedance (Equation 7.1) variations yielding reflections. Tapered lines with smoothly-changing widths should be used when there is a need to connect points or lines of different impedance. Also the use of vias should be kept to a minimum by careful design of the interconnection network and proper selection of the number of layers in the PCB.

Another interconnection issue that the designer has to resolve is crosstalk which is the coupling of high-frequency components from one interconnection line to another. Crosstalk can reduce the system dynamic noise margin and even cause malfunction. When a signal is coupled from one line to another, current flows in both directions in the recipient line. Assuming both lines propagate in the same direction, backward crosstalk is defined as the current flowing in the direction of the output driver transmitting the signal, and forward crosstalk is the current flowing in the direction of the loads receiving the signal. Forward crosstalk propagates on the passive line, toward the load, at the same speed as the signal in the active line. The backward crosstalk is generated as soon as the signal begins propagating on the active line and is reflected from the input of the passive line. This crosstalk component continues to be generated until the signal reaches the end of the active line. Therefore, the crosstalk lasts, with reflection at the input, for twice the propagation delay in the interconnection and dominates the forward component. Line length, propagation delay, line spacing, line attenuation, and input end reflection coefficient all influence the final value of the crosstalk signal at the passive line load. The crosstalk between two transmission lines can be determined by using the following expressions:[7]

$$\frac{V_B}{V_S} = \frac{K_c + K_L}{4} \qquad (7.8)$$

$$\frac{V_F}{V_S} = \frac{K_c - K_L}{2} \left[\frac{t_{pd}}{t_r} \right] \qquad (7.9)$$

where V_S is the signal amplitude in the active line, V_B is the amplitude of the backward crosstalk, V_F is the amplitude of the forward crosstalk, K_c is the capacitive coupling coefficient, K_L is the inductive coupling coefficient, t_{pd} is the signal propagation delay, and t_r is the signal rise time. Figure 7.7 illustrates typical signal and crosstalk waveforms, as well as variations of the coupling coefficients for different line spacing and board dielectric materials with different dielectric coefficients.[7] The coupling coefficients can be reduced by increasing the line separation or by bringing the ground plane closer to the interconnection lines. The capacitive and inductive coupling coefficients are equal for a uniform dielectric material thus making forward crosstalk equal to zero, according to Equation (7.9). However, this assertion is not valid for composite dielectrics often used in PCBs, when vias are present and the signal planes are located between a set of ground planes. Backward crosstalk between striplines can be kept at low, acceptable levels by using a ratio of line spacing to line width that is greater than one. Forward crosstalk can be minimized by using one interconnect plane between a pair of ground planes, shortening the length of interconnect lines, and using board materials with a lower dielectric constant.

Sec. 7.2 Insertion Issues and Trade-offs 309

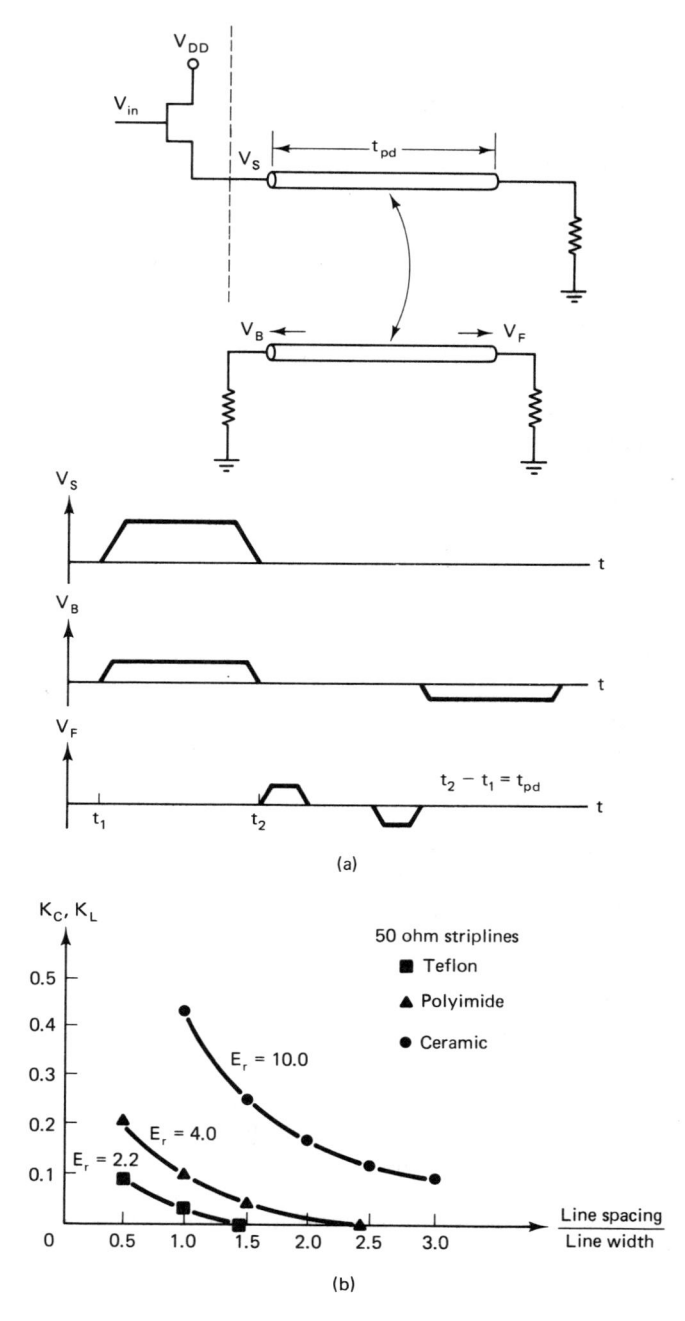

Figure 7.7 Forward and backward crosstalk: (a) signal waveforms and (b) coupling coefficients for different line configurations.

It is clear that the designer must make architectural trade-offs to take advantage of the high-performance of GaAs components at the system level if the system is still to maintain acceptable dynamic noise margins and integration levels. System architectures most effectively utilizing the inherent performance characteristics of GaAs components will have to minimize the amount, as well as the length, of high-speed interconnects and their implementation should consider innovative board materials with low dielectric coefficients. Furthermore, the designer, when using GaAs components, must consider the difficulty of debugging a design with scope-probes and wire-wrap guns as is often the case with Si-based design. The difficulty of modifying a prototype should be used as an incentive for extensive design simulations that include the interconnects and their characteristics. Based on the simulation results, proper tolerances should be added to the design to account for worst-case specifications in manufacturing.

7.2.3 System Timing

System timing is a global design characteristic that affects system performance and reliability. The factors that determine system timing are clocking discipline, clock generation, and clock distribution. The clocking disciplines (used here broadly to define timing control) define the timing relation of various operations performed by the system and can be synchronous or asynchronous. In the case of synchronous design, the sequence of states and the time to establish a state are tightly coupled with a clock signal which is used as a system-wide reference. As a reference sequence the clock defines the successive instants at which system state changes may occur, while as a time reference the clock period defines the time available for a system state to be established. In the case of asynchronous design, there is no global controlling mechanism for sequencing state transitions at a specific time lockstep. In this case, sequence and time are related inside the system components triggered by a control signal at the component input, generated when certain conditions are valid. The establishment of a new state in the component is indicated by an output signal and no time constraints are set for reaching this new state. Proper system operation is guaranteed by assuring only that all system events occur in proper sequence without any requirements for event occurrence at a particular time.

The selection of a clocking discipline can affect both the system performance and the system reliability. In synchronous designs the clock period defines the maximum throughput that can be achieved by the system. Since the establishment of a new state has to be completed within the time defined by the clock period, it becomes apparent that interconnection delays and clock skews for the critical path determine the system clock cycle and are the limiting factors for the system throughput. Clock skews most frequently caused by propagation delay differences on different clock distribution paths can also cause system malfunction. Malfunction is most often the result of failure of memory devices (i.e., flip-flops, registers, etc.) to sample input data, or loss of data due to overlapping clock

Sec. 7.2   Insertion Issues and Trade-offs   311

phases. Both of these events are attributed to clock skews and they were discussed in detail in Chapter 3. On the other hand, synchronous design is, in most cases, straightforward, well supported by CAD tools such as logic and timing simulators, and most importantly it is well understood for test pattern generation for functional verification testing. Furthermore, all the built-in-test techniques discussed in Chapter 6 can be employed only by synchronous designs.

In asynchronous designs, timing constraints are absent from the system level. The system designer, however, has to ensure correct interconnection of asynchronous components which are also assumed to operate correctly. The component interconnection in this case reflects the expected sequence of events during system operation. Since propagation delay within a component depends only on the component's performance and communication of the output does not require a predefined clock, the overall performance of an asynchronous structure is higher than the performance of a synchronous structure. However, the design of asynchronous logic is more complex than synchronous logic, and its support by CAD tools is either poor or nonexistent. In fact, most logic simulators indicate circuit malfunction when simulating an asynchronous circuit, when there is actually no fault in the operation of the circuit. The most complex issue in employing this clocking discipline, however, is functional verification testing which, even for moderate size systems, can be made impossible. This testing difficulty arises because certain design or component errors can cause the system to behave in a stochastic manner, thus making test pattern generation an impossible task. Test application and employment of built-in-test for fault isolation are also very complicated tasks when dealing with asynchronous designs. Therefore, asynchronous timing should be used only when there is no synchronous alternative in achieving some required performance. A synchronous design discipline should be used in most system design cases, especially when fault coverage requirements are very high for the test patterns generated for the system functional verification.

Clock generation is another important issue that should be addressed during system design. The system clock will, in most cases, originate from an electronic oscillator circuit, the period of which is controlled by a crystal. It is essential that the clock signal be kept free of distortion as it is distributed to various circuit boards. Local clock generation on a board should use the system clock as input to ensure both local and global synchronization. It is also recommended to use GaAs components with on-board clock generation capability (from a single phase incoming clock). The main goal here is minimization of clock skews and maintenance of required rise times. The capability of on-board clock generation provides design versatility in the sense that no special provisions are needed for using components requiring different numbers of clock phases in their operation, as long as their internal clocks are produced from the system clock. As discussed in Chapter 3, GaAs circuits can be used with either a single-phase clock or a two phase clock depending on the circuit configuration. Special attention should be placed on providing the capability to control the system clock generation circuitry by external means. This is essential for testing the system by allowing clocking of the system by the test set up, which also provides the test vectors. If this capability is not provided, then the test set up must be synchronized with the system clock, which is far from trivial, and, occasionally, an impossible task.

Clock distribution is probably the most critical design task because it deals with the transmission of the clock where it is needed, and therefore it is prone to all problems concerning high-speed interconnections. The clock distribution network is required to keep the clock noise-free and introduce the minimum clock skew possible. It may be necessary to transmit the clock signal differentially from the clock generation circuitry to its various destinations in the system. The differential signal can be received by a differential line receiver on a circuit board, and then broadcast to short distances as a single-ended signal which can be used directly by various components for generating a single- or multiple-phase on-board clock. An additional precaution is to feed the clock at the center of the circuit board, then use several differential distributions before final conversion to a single-ended waveform that drives only a few components at partitions on the circuit board. The number of components driven by each signal must be constrained to meet clock rise time specifications. Furthermore, the clock distribution network should be designed so that it is free from high-frequency ringing. The length of the clock paths from the clock generation circuitry to the clock destination must be equal to avoid clock skews due to different interconnection delays. Special attention is required when the clock generation circuits reside on a board that also

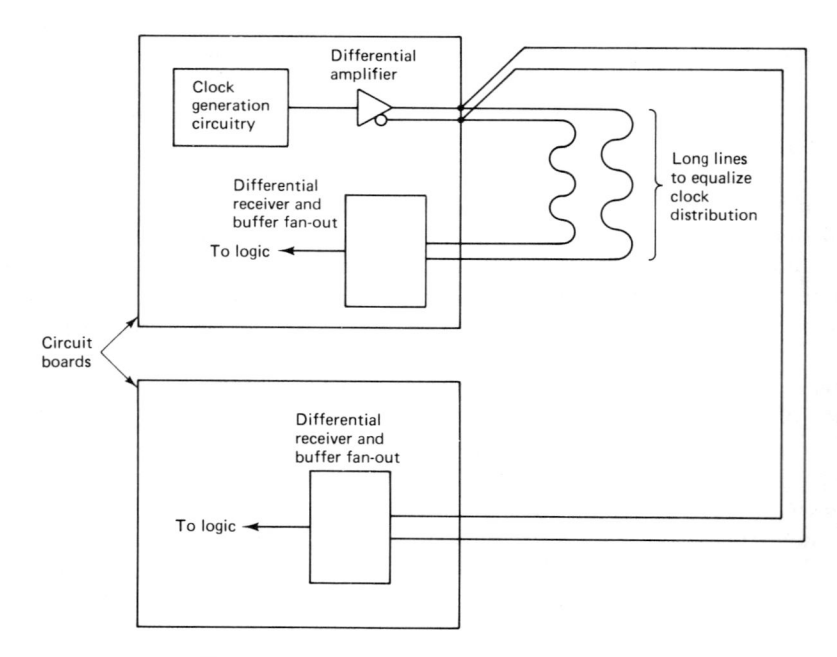

Figure 7.8 Clock distribution to different boards.

Sec. 7.2   Insertion Issues and Trade-offs   313

contains logic driven by the clock. Figure 7.8 illustrates how distribution should be handled in this case to ensure equalized clock paths. In general, however, it is preferable if the clock generation is separated from the logic it drives, because this minimizes crosstalk and clock contamination by noise.

7.2.4 System Testing

In Chapter 6 there was an extensive discussion on testing issues and testing techniques concerning GaAs chips and boards. The same testing issues are present once these components are incorporated into a system. Testing of systems capable of operating at high speeds requires test equipment that may be unavailable (as discussed in Chapter 6 concerning GaAs chips). Therefore, as in the case of chip testing, system testing issues concerning the at-speed application of tests and the capture of the test response has to be both addressed and resolved at the design stage. By doing so at this stage, we provide the capability to functionally verify the system operation and its adherence to design specifications. This capability is extremely important because system producibility and supportability are based on the ability to verify the system function at the time of manufacturing, and to maintain the system in the field. Both of these activities depend on the ability to perform fault detection and isolation tests on the system.

Another important reason for considering testing issues at the board level and above is that testing facilities must propagate throughout the system hierarchy to facilitate system functional verification. For example, if the chips used in the system design include some built-in test scheme, this capability can only contribute to board testing if the board design includes appropriate circuitry that can exercise the testing features of the chips. This implies additional circuitry and interconnects on-board, which is already noted as one of the design complications for GaAs based systems. However, this trade-off is necessary because the benefit of at-speed functional verification may not be achievable using external test equipment. That is not to say that external test equipment will not play a role in testing GaAs based systems. External test equipment must be used to format the tests, supply the test data to the on-board test facilities, and postprocess the test responses. All these functions can be performed at low clock rates. However, the test application and the test response capture should be performed by on-board test circuitry that can interface directly with the GaAs components. This interface must be at the clock rate the system is required to function at during normal operation. A system view of test hierarchy and the system interface to Automatic Test Equipment (ATE) is shown in Figure 7.9. Each board in the system includes a chip, called a Test Node (TN), which interfaces with all chips having built-in test capabilities or board-level Built-In Test (BIT) modules. The function of the TN is to provide test stimulus, control a test, and capture test responses, from a

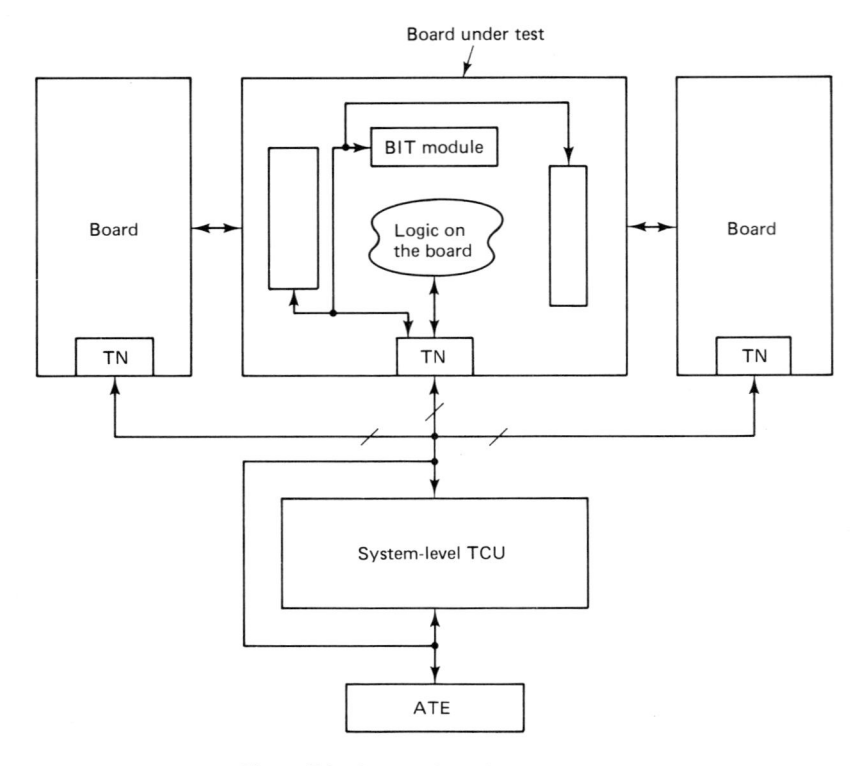

Figure 7.9 System view of test hierarchy.

chip or a collection of chips on a board. The TN communicates, at a higher level of hierarchy, with the Test Control Unit (TCU) which can be a board used solely for the purpose of supplying appropriate test information to each TN in the system, receiving test responses, and communicating with external ATE for test evaluation. The TCU can be viewed as an information bandwidth reduction unit between the high-speed TNs and the slower ATE. When the system is in the test mode, the TCU supplies the required clock and control signals, and the test input data to the TNs. Under test mode, the normal system clock is substituted by the test clock supplied by the TCU so that synchronization of system clock and test patterns is guaranteed. The test response received by each TN is transferred by TCU to the ATE for analysis.[8]

An overall architecture and organization of the test node is shown in Figure 7.10. Several variations of this basic architecture are possible and they have to be evaluated in specific situations where system, board, and chip design details are available. The purpose of the discussion here is to present the feasibility of using the TN, and its general operational characteristics. Although different options can be used for the actual implementation of a TN, the feasibility of implementing an

Sec. 7.2   Insertion Issues and Trade-offs

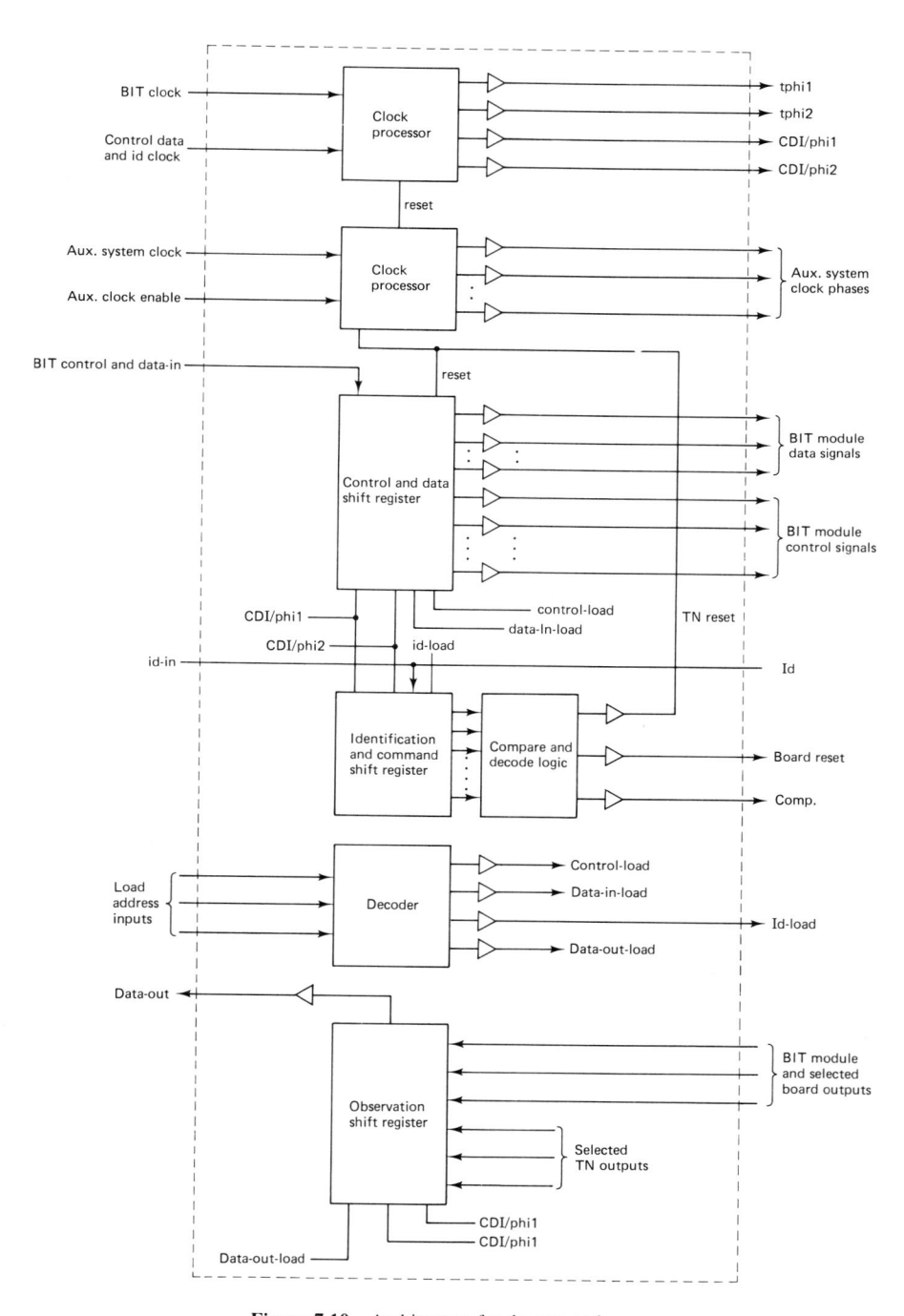

Figure 7.10 Architecture for the test node.

option should be considered in light of the technology constraints and the system design requirements.[8]

Control and data shift register. The main purpose of this shift register is to provide the required control and data signal to BIT modules on a board. This information is loaded serially to the TN, through the BIT control and "data-in" port, by the TCU. The "data-in-load," "control-load," "TN reset," and CDI clock phases are signals generated internally at the TN and control the function of this register. The contents of the register are transferred through the TN outputs to the BIT modules on the board. Note that the architecture of the TN does not assume a specific BIT configuration as long as the BIT module can be controlled by the signals provided by the TN. Notice that the TN can interface with different BIT modules, which implies that the outputs of this register will have different signal assignment for different BIT modules.

Identification and command shift register. The main purpose of this circuit is to determine whether or not the board, with its associated TN, is being addressed by the TCU. The circuit also generates the "TN reset" and "board reset" signals. The circuit logic can be implemented in a shift register, two comparators, and a decoder. The "board reset," "comp," and "TN reset" signals are then generated by comparing the appropriate bits of the shift register with hardwired identification (id) values, and by decoding the command code. All the identification and command codes are supplied bit-serially by the TCU. The identification code indicates which board is to be tested and which BIT modules are connected to the TN. Therefore, the number of bits in this code defines the maximum number of boards in a system and the maximum number of BIT modules on a board that can be used with this approach.

Observation shift register. The purpose of this shift register is to convert the parallel test output data available at the TN-board-logic interface to a serial format for transmission through the "data-out" port to the TCU. The "data-out load" signal, along with the CD clocks, controls the parallel latching of the inputs into the observation register and the output serial shift. In addition to test outputs, the observation register is also used to provide selected TN outputs to the TCU so that correct operation of the TN in generating those outputs can be established.

Load address decoder. The purpose of this circuit is to control the operation of the TN by interpreting the test instructions issued by the TCU. The decoding function of this circuit can be designed to allow independent activation of each load signal, as well as activation of certain combinations of various load signals as needed.

Sec. 7.2   Insertion Issues and Trade-offs   **317**

Clock processors. These circuits are used to disassociate the normal system operation clock from the system clock during test and to provide the necessary clocking to BIT modules during test. These circuits are externally provided with clock signals (by the TCU) to produce the required clock phases which can be transmitted differentially on the board. However, the number of clock phases generated and the clock distribution are general design issues and should be both addressed and resolved for specific system designs based on functional specifications and design performance and test requirements.

A hierarchical-testing example. In Section 6.4.3, a board-level BIT technique and its utilization for fault detection and isolation was discussed along with the design of a module (i.e., testing switch) required for the implementation of this technique. To illustrate the function of the TN and TCU at the system level, as well as the importance of hierarchical testing, we will present an example that uses the testing switch from section 6.4.3, the TN, and the TCU to perform fault detection on a board which contains three ambiguity groups of chips. Figure 7.11 illustrates a hypothetical board configuration. It is assumed that the ambiguity groups (AGs) are synchronous and are clocked by a nonoverlapping, two-phase system clock. All memory elements in the AGs as well as in the testing switches (SW) are assumed to be static which implies that they retain their state in the absence of the clock. Although this example discusses the fault detection procedure only, Figure 7.11 also illustrates the necessary placement of testing switches to achieve fault isolation to an AG. When fault isolation is considered, then the testing procedure is applied by the TCU through the TN by selecting the appropriate testing switches for each ambiguity group instead of for the whole board.

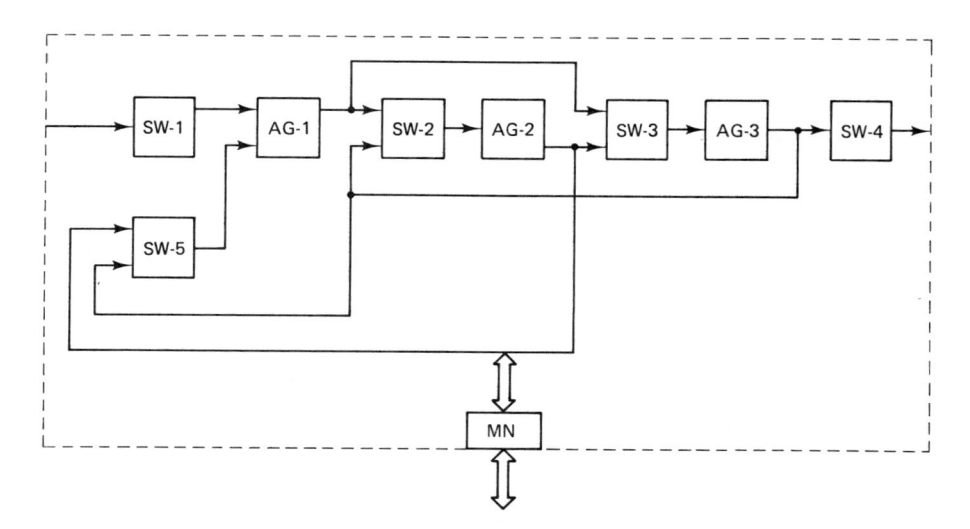

Figure 7.11   Board configuration for the hierarchical testing example.

The following steps will be generally initiated by the system level TCU and carried out through the TN for fault detection on the board:[8]

1. Activate the "auxiliary clock enable" line and hold the auxiliary clock and BIT clock lines inactive.
2. Shift in id and command word containing the appropriate subsystem id, board id, command code indicating the specific TN and board reset, and AG reset code. This word has to be shifted in through the "id-in" input under the control of the CDI clock. Arbitrary values may be used for the AG id and the BIT module id for this step. After an appropriate number of CDI clock cycles, apply the load address bits so that only the "id-load" signal is active for one CDI clock cycle. This step ensures that the necessary TN and board reset is accomplished. Activate the "data-out" signal, which configures the observation register in the TN to a parallel-latch mode of operation for one CDI clock cycle. Then serially shift out the register contents and verify the TN and board reset bits that appear on the "data-out" line at the TN-TCU interface. This step verifies the function of the TN in providing reset signals for itself and the board.
3. Repeat step 2 with another id and command word containing the command code to deactivate the TN and board reset signals. Also, the AG id corresponding to a given AG and the module id corresponding to a given testing switch are used in the id and command word. This step results in the "enable" of a given testing switch module being active, which in turn sets up the testing switch to receive control and data signals from the TN.
4. Shift in control and data word through the "BIT control and data" line containing the appropriate code to reset all the testing switch modules on the board.
5. Repeat step 4 with another control and data word to enable a test path within the chosen testing switch module in step 3. Then apply the BIT module clock for one clock cycle so that the control signals are latched in the chosen testing switch.
6. Shift in control and data words containing the appropriate control and data bits required for initialization of the test pattern generator and the signature analyzer within the chosen testing switch. At the end of the initialization process, the signature analyzer in the chosen testing switch is configured for parallel data compression mode of operation.
7. Repeat step 3 with "all-zero" BIT module id, which results in the "enable" of the chosen testing switch being inactive. The purpose of this step is to prevent further alteration of the latched control signals in the chosen testing switch.
8. Apply the BIT clock for a predetermined number of clock cycles. During this step, the test patterns generated in the testing switch are compressed by

Sec. 7.2   Insertion Issues and Trade-offs   319

the signature analyzer within the testing switch. At the end of this step, hold the BIT clock inactive.

9. Repeat step 4 with the control word that disables shifting-in of all the signature analyzers in testing switches and thus locks the resulting signature in the chosen testing switch.

10. Repeat step 3 with the appropriate id, which results in the "enable" of the chosen testing switch being active.

11. Repeat step 4 with the control word that enables serial shifting in the signature analyzer of the chosen testing switch and also activates the "data-out-load" signal. Then apply the BIT clock synchronously with the CDI clock. Under this mode, the signature output of the chosen testing switch is available through the TN observation register to the TCU. The TCU can compare the obtained signature with the expected one, obtained *a priori* via simulation. This step checks the functionality of the testing switch to a good extent which is essential for fault isolation purposes, before the testing switch is used for testing the corresponding AG(s).

12. Repeat steps 3 to 11 enough times with the AG id and the BIT module id, so as to complete functional verification of all the testing switch modules on the board.

13. Repeat step 4 so that all the testing switch modules are reset again. This step ensures that all the testing switches on the board are in normal "pass-through" mode of operation.

14. Repeat step 3 and step 6 with the AG id corresponding to AG-1 and the BIT module id corresponding to SW-1. At the end of this step, the test pattern generator in SW-1 will be configured to supply pseudorandom test vectors to the AG-1 primary inputs.

15. Repeat step 3 and step 6 with the AG id corresponding to AG-3 and the BIT module id corresponding to SW-4. At the end of this step, the signature analyzer in SW-4 will be configured to accept the AG-3 primary outputs for data compression.

16. Repeat step 7 to deactivate the "enable" of SW-4.

17. Apply the BIT clock and the auxiliary system clock in synchrony for a predetermined number of clock cycles. During this step, pseudorandom vectors generated in SW-1 are applied to the AG-1 primary inputs, and the AG-3 primary outputs are compressed in SW-4. The testing switches SW-2, SW-3, and SW-5 are in "pass-through" mode at this time. At the end of this step, deactivate the BIT clock and the auxiliary system clock.

18. Repeat steps 9 to 11 with the AG id corresponding to AG-3 and the BIT module id corresponding to SW-4, and determine the status of the board, either in the TCU or in external ATE.

An incorrect signature in SW-4 at the end of step 18 indicates a faulty board, while a correct signature implies a nonfaulty board. However, the entire test procedure may have to be repeated several times and the consistency of the results must be verified. Furthermore, there may be a need for performing additional tests involving multiple boards to check completely the interconnections between the board under test and the other boards. However, administering these additional tests differs from the above test procedure only in the id values that need to be used in different steps for the board, AG, and the BIT modules. It involves configuring the testing switch modules on the periphery of the boards to either the test pattern generator or signature analyzer mode of operation, applying the BIT clock and/or auxiliary system clock for a predetermined number of clock cycles, and verifying the relevant signatures, as illustrated in the above test procedure for a single board.

Note the importance of the TN and the TCU in accessing the BIT facilities implemented on the board by the testing switches that provide the capability of performing tests at the system level. Providing the means for the testing features to communicate through the system design hierarchy is essential for system testing and should be carefully considered during the definition of the system architecture.

7.2.5 A Final Note on Insertion Issues

As the preceding discussion in this section indicates, two major issues must be resolved before the inherent high performance of GaAs insertion into system design can contribute to increased system performance. These two issues are intercomponent communication, and system functional verification testing. These two aspects must be addressed early in the system design phase and be resolved through proper trade-offs which can result in a very high performance, and yet producible, system. Although these issues are seemingly unrelated, they actually affect each other in the system implementation phase. In addition, resolution of these issues may affect other system design requirements such as reliability and cost, which have to be considered in developing a new system.

The main point of this discussion is to make the reader aware that the resolution of such issues requires a collective design approach. Such an approach looks at the system design as a global task that considers all these issues and their complex interaction early in the design stage. The design of high-performance systems employing GaAs components must be an integrated multidisciplinary task with a level of interaction among different disciplines uncommon for Si-based systems. The system designer can no longer afford to concentrate on the system architecture, assuming that a host of other tasks will be resolved by specialists in these fields later on during the design in case certain omissions have been made. These other tasks include chip design and test, packaging and intercomponent communications, and system functional verification testing. Classical troubleshooting approaches and fixing of prototypes by breaking board connections and rerouting signals by wire-wrapping simply will not work. Design of high-speed systems that can be verified to perform their intended function at their specified full speed must be performed by an integrated team that covers all required design disciplines at all levels of the system hierarchy. Failure to establish such a team and ensure communication and collective trade-offs, affecting the system as a whole, will virtually ensure poor system performance, if not failure, of the design project. A system design approach which portrays this design-team approach and addresses all pertinent issues during the early stages of the design is shown in Figure 7.1.

7.3 ARCHITECTURAL APPROACHES AT THE SYSTEM LEVEL

From the previous discussion, it is apparent that if GaAs components are forced to "fit" into a Si-based system, there is a good chance that the final system will have less than optimum performance. This result may occur because considerations that are not particularly necessary for implementations in Si are mandatory if any of the GaAs technologies are employed. Therefore, the requirement of a wholistic design approach that considers a target GaAs technology for implementation should be followed from the definition of system architecture until functional verification testing of a prototype.

Systems architectures employing GaAs components must be defined so that technology-related issues are resolved, and full advantage of the high speed of GaAs components can be achieved. The most critical technology related design issue to be impacted by the system architecture is the intercomponent communications presented in Section 7.2.2. Continuous and uniform data flow, on very short paths, among chips on the same board and between logic boards will allow for the fastest overall throughput. Maximization of the ratio of time for on-chip arithmetic and logic operations over time for chip I/O operations will also result in the fastest overall system throughput. All board interconnections among GaAs chips should be 50 Ω terminated transmission lines. These terminations are usually implemented with resistor networks that dissipate considerable power affecting system performance (due to temperature increase) and reliability. This is an additional reason for the architecture design to minimize the number of unique interconnects and operate the available lines at their maximum bandwidth. Minimum capacitive-load interconnections allow transmission of signals with fast edges without placing excessive stress on the output drivers of GaAs chips. Therefore, architectures attractive for GaAs should feature single source, single destination, or low fanout interconnection schemes.

This section presents generic architectural approaches with advantages for GaAs implementation at the system level. Although these approaches are by no means suitable only for GaAs implementations, they lead to structures with inherent characteristics that allow maximum exploitation of the individual component performance.

7.3.1 Pipeline Techniques

Pipelining is a form of embedding parallelism or concurrency in a digital system that is composed of several resources called segments that may be operating simultaneously. In most cases, the segments in the pipeline are specialized in performing a special set of subtasks so that a task has to flow from segment to segment to be completely executed. This flow is normally synchronous. The segment specialization results in the inherent economy of pipelined systems compared to parallel systems, of equivalent performance, that utilize general purpose resources. Another form of pipeline may employ segments that perform the same operations on arrays of data that flow through the segments under synchronous control.

Computer designers have been using pipelining techniques in their efforts to exploit overlapped modes of operation. The fastest "supercomputers" of the 80s, for example the Cray-1 and the CDC-255, use pipelines of 12 or more processors to execute vector operations on arrays of data.[9] A special purpose pipeline of 113 processors is employed by the Cytocomputer, which is specialized to execute image processing operations.[10] Although the processors in this machine are much simpler than Cray-1, the overall structure of 113 processors is one of the longest pipeline structures known to date.

Pipelining can be applied at different levels of design abstraction using different design and control strategies. Pipelining at the system level can be exemplified in the design of the instruction processing unit (IPU). The IPU can be decomposed into various functional segments, such as instruction fetch, instruction decode, address generation, etc., which each take one clock cycle to be performed. Therefore, after a stream of instructions enters this pipeline, the pipeline starts outputting one task (instruction) per cycle. At the subsystem level, pipelining is typically applied to arithmetic units performing functions such as multiplication, division and square-root. At either level, a pipelined module often serves only a single dedicated function. In such a case, the pipeline is defined as a unifunctional pipe with static configuration. If on the other hand a pipelined module can serve a set of functions, each with a distinguishable configuration, the pipeline is defined as a multifunctional pipe.

During system design, trade-offs are made concerning performance requirements and performance characteristics of the intended system architecture and implementation technology. It is critical to recognize the performance characteristics of an architecture based on important system performance measures. By doing so, the efficiency of the implementation technology on this performance, results in an optimum design where this "optimum" is defined by the design specifications. One of the most important performance measures of a system is its throughput rate which is defined as the number of valid outputs produced (or the number of instructions executed) per unit time. This measure directly reflects the processing power of a processor system—the higher the throughput, the more powerful the system. Pipelining is an architectural approach to improving system throughput. Let us consider the example in Figure 7.12, which is a typical pipelined instruction processor with four modules, namely, instruction fetch (IF), instruction decode (ID), operand fetch (OF), and execution (EX). For a nonpipelined processor, the execution time of an instruction will be $T_{np} = t_1 + t_2 + t_3 + t_4$; therefore, the processor throughput is $1/T_{np}$ instructions per second. In the pipeline case, assume $t_s = max(t_1, t_2 t_3, t_4)$ and t_s is the execution time of the slowest module in the pipeline. Then the pipeline throughput is $1/t_s$, because every $T_p = t_s$ time units an instruction can leave the pipeline after its execution, assuming that instructions are independent. A direct comparison shows that $T_p < T_{np}$ and hence the throughput of a pipeline processor can be much larger than that of a nonpipelined processor. For the example in Figure 7.12, if $t_1 = t_2 = t_3 = t_4$, then the comparison shows a fourfold throughput improvement for the pipeline configuration. An important observation at this point is that the throughput of a pipeline is determined by its slowest element, which is often called a "bottleneck."

Another performance measure for a system is its *utilization factor*, which reflects how effective a processing scheme is and is defined as the percentage of "busy" periods with respect to a certain time span. In the case of a pipelined system, however, there is a slight complication with the above definition because a pipelined system consists of several modules, some of which may be "busy" while

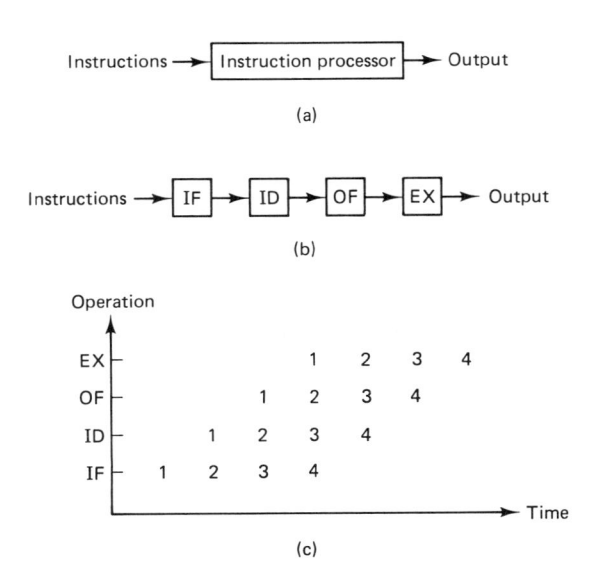

Figure 7.12 (a) Instruction processor; (b) its pipelined configuration; and (c) a space-time diagram indicating the flow of operations in the pipeline.

others are idle. This complication is addressed in Reference 11, where a weighted uniform space-time span index is proposed for evaluating the utilization factor of a pipelined system as an entity. This factor is expressed as

$$\eta = \text{utilization factor of linear pipe}$$

$$= L \frac{(\sum_{i}^{n} a_i t_i)}{\sum_{i}^{n} a_i \left[\sum_{i}^{n} t_i + (L-1)t_j\right]} \quad (7.10)$$

where t_j is the speed of the slowest module; t_i is the speed of the ith module, a_i is a weight factor associated with the space-time span of the ith module as determined by factors such as cost-speed; L is the number of computational tasks or instructions provided to the pipeline in a certain period of time; and n is the total number of modules in the pipeline. In case all modules have the same speed, the above equation simplifies to:

$$\eta = \frac{L}{n + (L-1)} \quad (7.11)$$

It should be noted here that in deriving the above expressions it has been assumed that a continuous supply of tasks (or instructions) is available to the pipe. In practice, however, execution may be discontinued due to precedence constraints, branching, interrupts, etc. As Equation 7.10 indicates, the utilization factor is inversely affected by the speed of the slowest module in the pipe. A significant improvement over a space-time span can be achieved by increasing the speed of the slowest module, so that it is comparable to the speed of the other modules in the pipe.

Data and control flow from one pipe segment to another synchronously under the control of a clock. The propagation delay through each segment and possible signal skews must therefore be carefully considered during design to avoid improper operation in a high-speed system. A maximum clock rate is an important performance measure used to indicate the upper boundary on the throughput achievable in a pipelined system. The maximum clock rate will be limited by the propagation delay of the logic in a pipe-segment and the propagation delays of the interconnects between pipe-segments. In all cases, there are three conditions that must be met to ensure signal balancing and proper operation:

1. The clock must be wide enough to guarantee that the data is properly stable at the output of every segment in the pipe.
2. The clocking discipline should not allow data to pass through more than one pipe segment within a clock phase.
3. The output data from a pipe segment must propagate through the interconnect and be valid at the input of the next segment before the clock that samples the input becomes valid.

Sec. 7.3 Architectural Approaches at the System Level

The discussion on pipelined system configurations points to some inherent features that are very desirable for GaAs implementation. The synchronous operation, the feed-forward close-neighbor interconnection, and the simple control scheme (i.e., clock) are desirable features when GaAs implementation is considered. However, several issues must be resolved before GaAs components are used to utilize any one of the pipeline segments. The first concern is if the overall structure of a pipeline is to be implemented in GaAs or only in selected segments. One should consider design optimization criteria such as the cost-speed product. Some factors to be considered when using this criterion are the scale of integration of available components, component speed, interconnection delays, and component cost. In case the overall pipeline structure is implemented in GaAs, an approach towards design optimization will be to implement the individual segments using bit-serial techniques (as discussed in Chapter 4) so that the system function can be implemented with as few components as possible. This will result in reducing segment throughput when compared to word-parallel techniques, hence easing the data propagation requirements over the interconnect, which can result in better overall system performance. In case selected pipeline segments are implemented in GaAs, then the pipeline bottleneck (i.e., slowest segment) should be considered as the first candidate. Achieving the same throughput at every pipeline segment will result in the highest overall system throughput and segment utilization factor.

One major design issue when using GaAs components in pipeline structures is clock distribution and segment synchronization. Although segment interconnections are very appealing when considering GaAs implementation, global synchronization and minimal clock skews may be difficult to achieve. If not achieved they can result in system malfunction. The difficulty in clock distribution with minimal skew is closely related to the number of components to which the clock has to be distributed and the clock speed.

One application area where pipelining is used extensively is high-throughput signal processing. One of the most popular algorithms in this area that lends itself to pipeline implementation is the Fast Fourier Transform (FFT). The FFT is actually a large class of computational techniques used to compute efficiently the Discrete Fourier Transform (DFT) given by the expression:

$$X_k = \sum_{n=0}^{N-1} x(n) W_N^{nk} \qquad k = 0, 1, \ldots, N-1 \qquad (7.12)$$

where $x(n)$ is the time domain sequence and $W_N = e^{-j2\pi/N}$ is a set of coefficients called twiddle factors. The basic computation in the originally proposed Cooley-Tukey FFT algorithm[12] is called a radix-2 butterfly and it is expressed as:

$$\begin{aligned} X_k &= X_e(k) + W_N^k X_0(k) & k = 0, 1, \ldots, (N/2) - 1 \\ X_{k+N/2} &= X_e(k) W_N^k X_0(k) & k = 0, 1, \ldots, (N/2) - 1 \end{aligned} \qquad (7.13)$$

Assuming that a time domain sequence has N samples, and N is a power of 2, the computation of the FFT of this sequence requires the computation of $N/2$ butterflies per pass over $\log_{2N}$ passes. This computation is illustrated for a sixteen point FFT in Figure 7.13. As illustrated, in addition to the basic computation indicated by Equation 7.13, an FFT processor has to provide for data reordering between passes. Therefore, a pipelined processor can be designed with two basic functional segments for performing the FFT. One segment will perform the basic computation followed by a segment performing the required data reordering. One such pipeline configuration is shown in Figure 7.14. The address generator can be a counter (with an offset in this case) and the memory can store the data in sequence (as though it were a sequential access drum memory). This pipeline

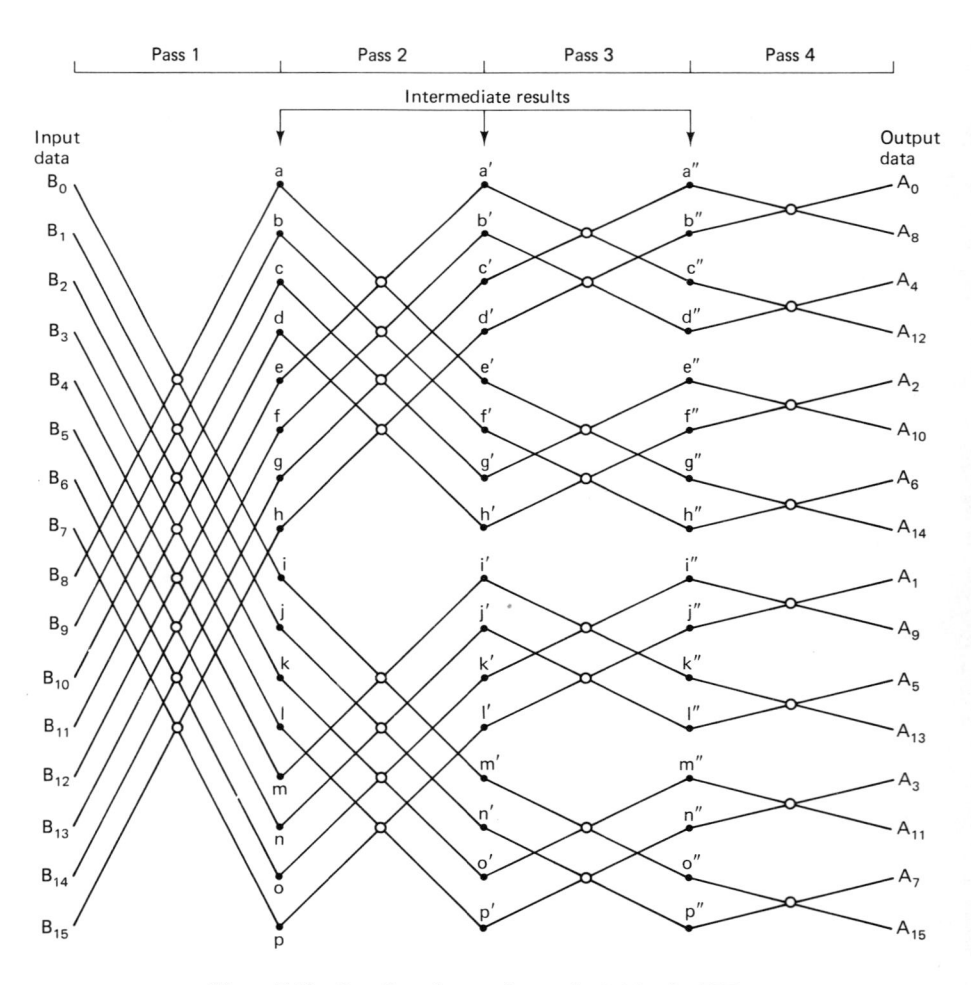

Figure 7.13 Data flow diagram for a radix-2, 16-point FFT.

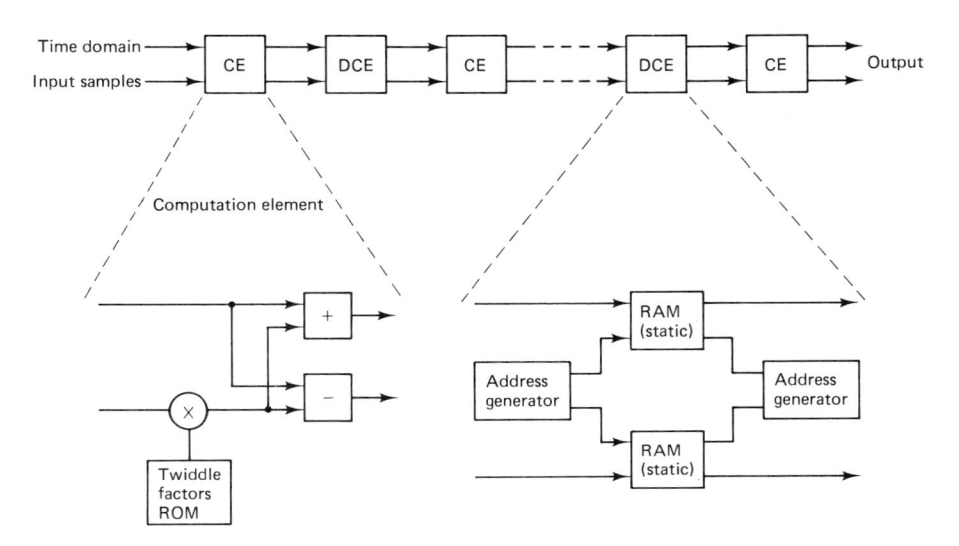

Sec. 7.3 Architectural Approaches at the System Level

Figure 7.14 Pipelined FFT processor configuration.

configuration can be implemented in its entirety with GaAs components or with a mixture of GaAs and Si components. In the former case, the basic computational element can be integrated on a single chip if bit-serial arithmetic is used, and DCFL technology is chosen for its implementation. This approach will ease the performance requirements for the commutator element (assuming single port memory) and will result in a low chip-count implementation. In the latter case, the commutator element can be implemented using a multi-port (4-port) Si-memory and the computational element using a single GaAs chip with bit-serial architecture.

This latter configuration may have a better speed-cost product, depending on the number of bits in the time domain samples and the required accuracy for the output. One point that should always be considered when designing a signal processor, however, is the throughput of the A/D converter. Although not shown in Figure 7.14, the A/D converter is actually part of this pipelined configuration and it can be its slowest segment. Therefore, when design trade-offs are made concerning the pipeline configuration and its implementation, the performance of the A/D converter should also be taken into account.

7.3.2 Systolic Arrays

Systolic arrays are networks of processors which rhythmically compute and pass data through the network. Systolic arrays are more fully described in Reference 13 as effective architectures for performing basic matrix computations. The volume of literature devoted to the subject has expanded rapidly and the concept has been successfully applied to a wide range of problems.

Systolic arrays provide a model of computation which combines the concepts of pipelining, parallelism, and interconnection structures. Several characteristics of systolic array configurations make them attractive for GaAs implementation:

1. All connections between array elements are nearest neighbor and hence can be very short. Also, all data communication paths are single source, single destination, hence decreasing the line loading and thus easing the off-chip driving requirements for the output drivers.
2. Each array element has, in general, a simple structure of moderate circuit complexity that can be accommodated by a technology with LSI capabilities. Furthermore, the simple structure of an array element allows for its exclusive simulation at high clock rates, so that a fault-free design can be measured.
3. The array configuration uses a minimal number of different types of processing elements (usually two or three) thus making interface and synchronization relatively easy to achieve.
4. The operation of the processing elements, as well as the operation of the array, are pipelined, thus allowing synchronous processing at high system clock rates.

Typically a systolic array is configured as an algorithmically specialized system in the sense that its design reflects the data flow and computational requirements of a specific algorithm. In this case, the designer must first perform a mapping from an algorithm to a systolic array structure and must consider the implementation if design requirements are to be met at the lowest cost. The function of the basic computational element in the systolic array is determined by the algorithm under consideration and the network configuration by the data flow. Therefore, the consideration of implementation technology will determine the number of components required to implement the basic computational element based on an architecture that meets the performance requirements for the overall system operation. In some cases, it may be desirable to design systolic arrays that are capable of performing more than one algorithm for a multitude of applications. There are two approaches in designing such an array and actual implementations may be based on a compromise between the two. One approach is to add hardware components that facilitate reconfiguration of the array topology and interconnection network, so the systolic array can emulate different configurations required by different algorithms. The other approach is to accomplish this reconfiguration for different algorithms using software. This case requires the availability of languages capable of expressing parallel computation and, of course, the availability of suitable compilers and operating systems. Considering the level of integration of GaAs components and system integration issues concerning interconnects, GaAs systolic arrays will be mainly focused on implementation of specific algorithms for applications with very high throughput requirements. An example of that is an adaptive multisteering beamforming controller presented in Reference 14. However, GaAs components can be used in conjunction with Si components in implementing "general purpose" systolic arrays. In this case, the GaAs components will be inserted into the "slow" parts of the array.

Although the basic computational element in a systolic array will perform a computation required by the algorithm the array implements, typically in most systolic array configurations the basic computation performed is of the type $Z = Z + X \cdot Y$. The basic computational element in this case is called the inner product step processor and its block diagram is shown in Figure 7.15. It should be noted here that all operands entering the computation are also available as outputs and can be used by other computational elements in a network. Input and output data are latched in the computational element and the sole control during operation is the system clock which provides for global synchronization. A valid output is available during every clock cycle after some latency which is incurred for system initialization. A systolic array is composed of interconnected basic computational elements. The basic network organization of interest here is the mesh-connected scheme in which all connections from a computational element are to neighboring computational elements. Different mesh-connected systolic array configurations are shown in Figure 7.16. Elements on the boundary of the array may have connections to external memory and they may be designated to handle the array I/O operations. These elements have a different architecture than the computational elements in the array, and their function is mainly I/O data formatting and handling. The structures shown in Figure 7.16 have all the characteristics that make systolic arrays attractive for GaAs implementation. However, one major issue that the designer has to resolve, for successfully implementing such structures, is clock distribution. In fact, the stringent requirements for the clock characteristics of GaAs components (i.e., rise and fall times, skew tolerance) may limit the number of computational elements in the array to a level that can be handled reliably by the clock generation circuitry and the clock distribution network. An approach that can resolve the clock distribution issue is the utilization

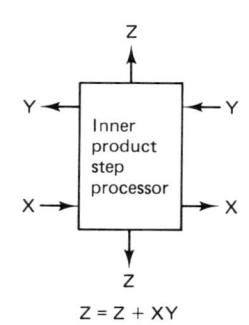

Figure 7.15 Block diagram of a systolic array basic computational element.

$Z = Z + XY$

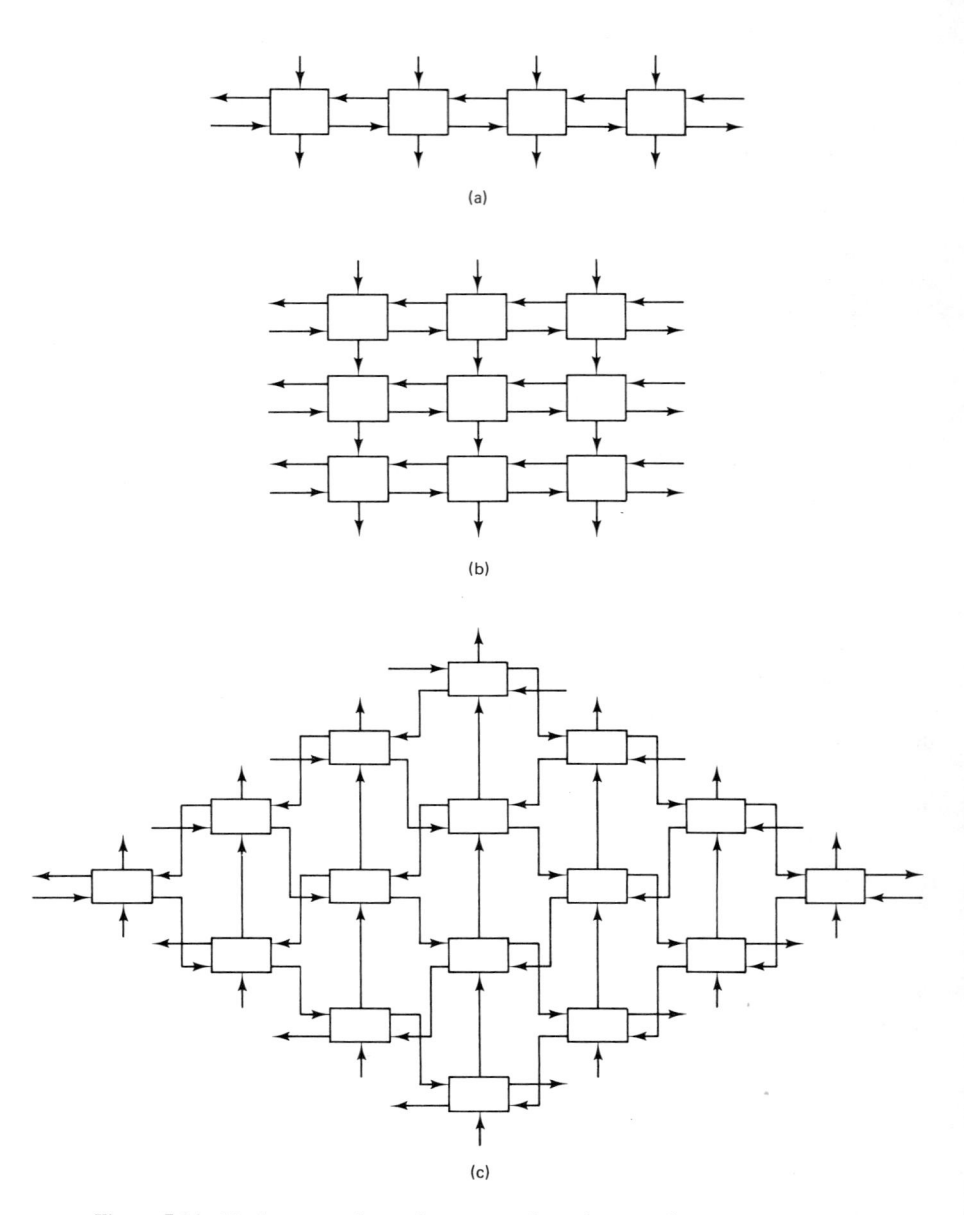

Figure 7.16 Mesh-connected systolic array configurations: (a) linear; (b) orthogonal; and (c) hexagonal.

of asynchronous control in the array. This approach will require a more complex hardware interface between processing elements and, in the case of general purpose systolic arrays, operating system software overhead.

Sec. 7.3　Architectural Approaches at the System Level

The systolic array concept was introduced as a natural approach to solving problems involving matrix computations. This was demonstrated in Reference 13 for matrix multiplication, matrix-vector multiplication, and LU-decomposition (i.e., factorization of a matrix A into lower and upper triangular matrices L and U respectively). Kung and Leiserson have shown in Reference 13 that a linearly connected array can be used to multiply an $N \times N$ band matrix with bandwidth W by a vector of N elements using $W/2$ inner product step processors. Figure 7.17 illustrates the functional block diagram of the array and the matrix computation. As the matrix and vector shift into the array, each processing element performs one computation of each step of the recurrence that is used to compute the entire multiplication. Data moves exactly the same way during the entire operation requiring the control of the clock that provides the needed synchronization. The time to complete the multiplication is $O(2N + W)$ in this case compared to the $O(NW)$ time needed for a uniprocessor using a sequential algorithm to perform the same computation. The symbol $O(n)$ implies "in the order of n."

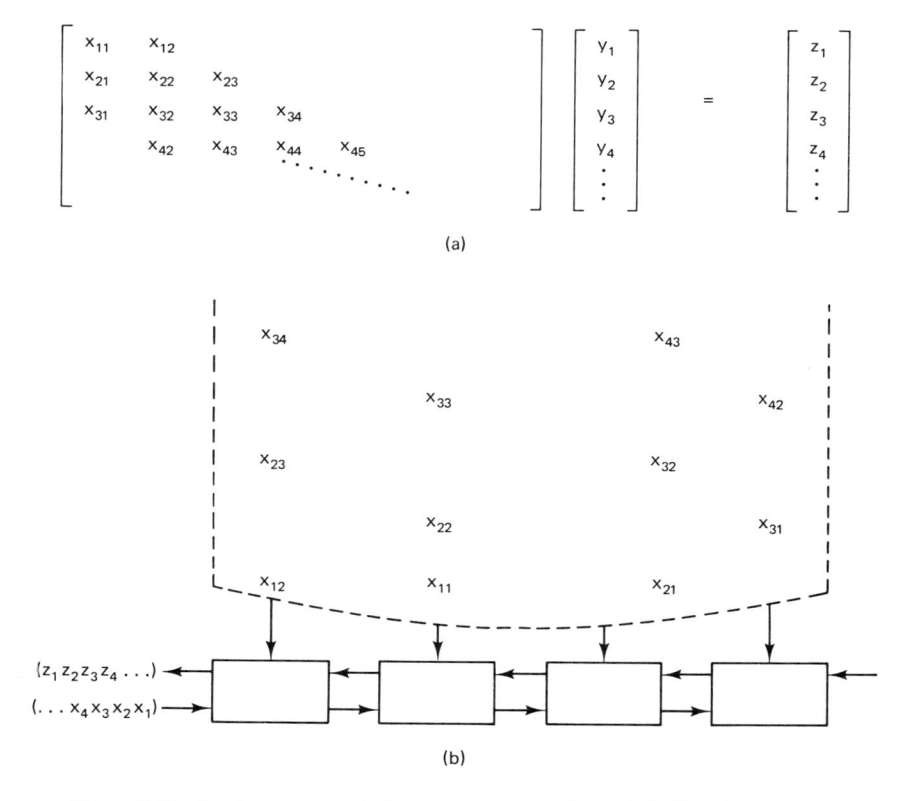

Figure 7.17 Matrix-vector multiplication: (a) computation and (b) linearly connected systolic array performing the computation.

The multiplication of two $N \times N$ matrices of bandwidth W_1 and W_2 was demonstrated in Reference 13 with a hexagonally connected systolic array of $W_1 W_2$ inner product step processors. The elements of the matrices enter and leave the systolic network in three directions and all operations are synchronous under the control of a global clock. Each processing element performs its basic operation at each cycle and computations are overlapped with input/output operations. Figure 7.18 illustrates the data flow in the array structure for matrix multiplication. The multiplication can be performed in $[3N + \min(W_1, W_2)]$ cycles with this structure.

Two important observations can be made from the above discussion. One concerns the application domain of systolic arrays while the other concerns the suitability of their architecture to GaAs implementation. Matrix operations can be used to solve many problems in the system application areas discussed in Section 7.1. Convolution, filtering, correlation, and discrete Fourier transforms are some examples that have been investigated for systolic implementation.[15,16,17] GaAs systolic arrays may provide the computational throughput required for some system applications based on these operations.

The feasibility of implementing systolic arrays in GaAs lies in the simplicity of the global control scheme used and the short interconnect requirements. In addition, the modularity of systolic array structures and the low complexity of the basic computational element are within the integration capabilities offered by GaAs technology. The goal should be to integrate at least a single inner product step processor on a single chip. This can be achieved by using appropriate technology and on-chip architecture. For example, it may be possible to use parallel arithmetic circuits when implementing the processor in DCFL, but it may be required to use bit-serial arithmetic if the technology of implementation is BFL or SDFL. The integration of the basic processor on a single chip should be a design goal because the basic computation will be performed without crossing chip boundaries, which is highly desirable when using GaAs components.

A much simpler structure for the array element can result if the basic processor is designed as a single bit processor and the array operation is pipelined at the bit level. Bit-level systolic arrays were introduced in Reference 18 with additional results presented in References 19 and 20. In this case, the basic processor in the array comprises a full adder, a number of latches, and some simple control logic. The function of each circuit in the processor is determined by the array interconnection topology and the resulting data flow structure. Bit-level systolic arrays performing convolution, correlation, and discrete Fourier transforms are demonstrated in Reference 20. They are architecturally similar to the orthogonally connected systolic array configuration in Figure 7.16. The important features of this type of circuit are the minimum interconnection requirements between close neighbors in the array (single wires), and the enhanced modularity since the basic array element can be used for designing structures for any wordlength. These features are attractive for GaAs implementation especially if it is determined that a single chip implementation of the basic processor is not feasible for some applications.

7.3.3 Special Purpose Architectures

Based on the preceding discussion, it is evident that a different class of system architectures may be best suited for implementation in GaAs than system architectures commonly used in Si implementations. The main feature of these architectures is the retention of all operands in a chip for the maximum possible duration while executing iterative and recursive algorithms. This minimizes the number of components used to perform a function and the number of intercomponent transversals of the data before a computation is completed.

This section presents a set of special purpose system architectures that can exploit the advantages of both Si and GaAs technologies to achieve high performance. These system architectures are designed specifically for signal processing applications where GaAs technology is believed to make a significant impact. The signal processing functions considered for efficient implementation by these architectures are: Fast Fourier Transform (FFT), Pulse Compression (PC), Moving Target Indicator (MTI), Doppler Processing (DP), and Constant False Alarm Rate (CFAR). These functions require high throughput computations and high bandwidth storage and in most cases in present systems cannot be performed in real time at the rate of the incoming signal. The application area where these functions are most commonly used is radar signal processing where real time processing is desirable and GaAs is expected to contribute to this end. The algorithms developed for performing these functions use well defined data structures that allow sequential as well as parallel performance of in-place computations.[21] All data used in the computations (i.e., input data and stored coefficients) are ordered and processed in a predefined order, thus allowing the use of simple memory access control (i.e., first-in, first-out).

In the remainder of this section a brief description of the signal processing functions mentioned above is presented, with the exception of FFT, which was discussed in Section 7.3.1. The presentation of the signal processing functions is followed by the description of the special purpose signal processor architectures that can efficiently perform these functions.

Pulse compression (PC). Figure 7.19 illustrates the pulse-compression function when performed via fast convolution.[21] The input data sequence is first processed via a forward FFT of size N. The result of this transformation is next multiplied by the replica spectrum of the reference signal represented by N complex samples, and the result of this operation is finally acted upon by an inverse FFT of size N. It is assumed throughout this discussion that all computations involve complex data and that the input data sequences are finite but arbitrarily long. For the importance of this function and related mathematical details, the reader is referred to References 22 and 23.

$$B \cdot C = D$$

$$\begin{bmatrix} b_{11} & b_{12} & & & \\ b_{21} & b_{22} & b_{23} & & \\ b_{31} & b_{32} & b_{33} & b_{34} & \\ & b_{42} & & & \ddots \end{bmatrix} \begin{bmatrix} c_{11} & c_{12} & c_{13} & & & \\ & c_{21} & c_{22} & c_{23} & c_{24} & \\ & & c_{32} & c_{33} & c_{34} & c_{35} \\ & & & c_{42} & & \ddots \end{bmatrix} = \begin{bmatrix} d_{11} & d_{12} & d_{13} & d_{14} & \\ d_{21} & d_{22} & d_{23} & d_{24} & \\ d_{31} & d_{32} & d_{33} & d_{34} & \\ d_{41} & d_{42} & & & \ddots \end{bmatrix}$$

(a)

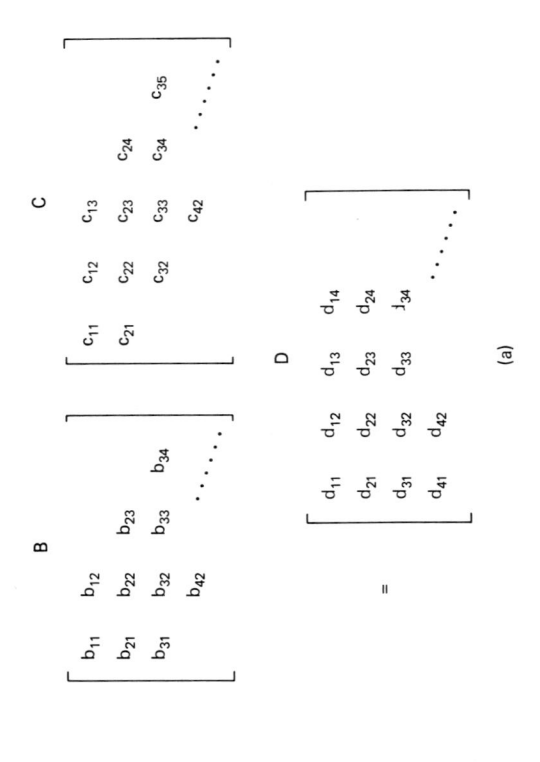

Figure 7.18 Matrix multiplication: (a) computation and (b) hexagonally connected systolic array performing the computation.

334

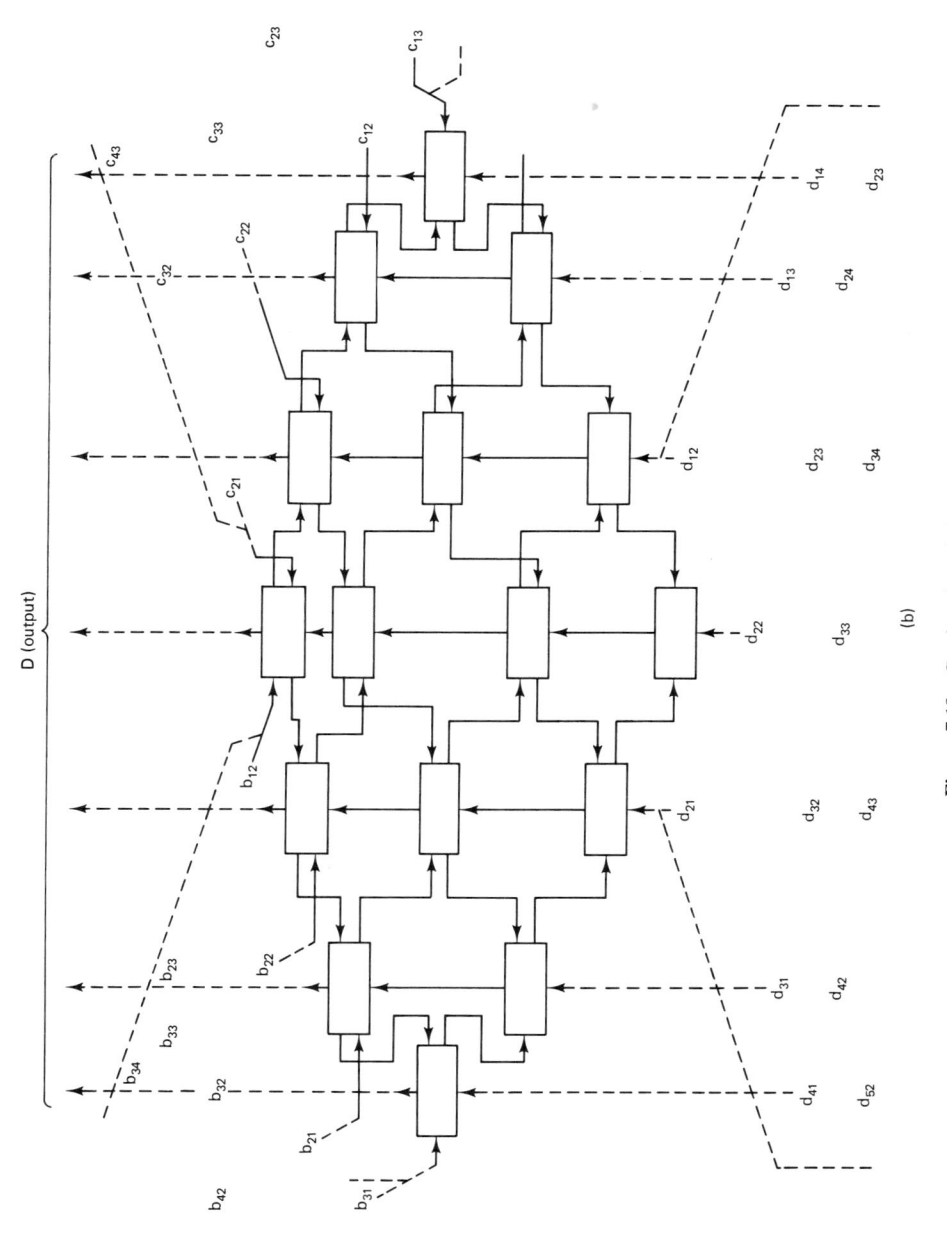

Figure 7.18 Continuation of Figure 7.18

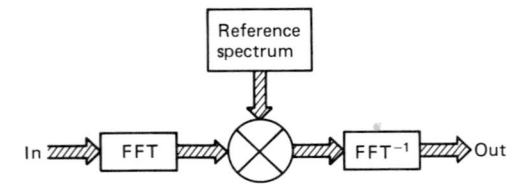

Figure 7.19 Block diagram of the pulse-compression function.

The implementation of the pulse-compression system is highly dependent on the length of the input data stream. Very long input sequences cannot be processed in one step as a single pulse-compressor load because of limitations in the size of FFT one can implement. Therefore, such sequences must be segmented into subsequences of optimal length in a way that minimizes overall processing time.

There are two common approaches for convolving very long sequences: the "overlap-and-save" and "overlap-and-add" techniques.[22] Both approaches involve essentially the same amount of computational effort, but for purposes of sizing up the computational requirements of the pulse compression function the "overlap-and-add" method is adopted. Figure 7.20 illustrates the manner in which the input sequence $\{x(n)\}$ is segmented and processed. Note that the lengths of the input sequence segments, $\{x_i(n)\}$, and reference function sequence, $\{h(n)\}$, are N_2 and N_1, respectively. If the size of FFT mechanized is N points, this method requires that N_2 be chosen so that

$$N \geq (N_1 + N_2 - 1)$$

Ideally, we wish to choose N_2 so that $N = N_1 + N_2 - 1$. However, other overall system requirements may dictate the choice of N_2 (and, for that matter, of N_1), thus forcing us to reconsider even the choice of FFT size, N.

To illustrate how one may perform sequence segmentation, assume an input sequence $\{x(n)\}$ of length $L = 5000$ data points. Furthermore, let the available FFT size be $N = 1024$ points and the reference function require $N_1 = 120$ points for its proper representation. The objective in this case should be to subdivide the input sequence of 5000 data points to as few subsequences as possible. One possible choice is to form five (5) segments of 905 points each and a sixth segment of 475 points. For FFT purposes the segments and the reference sequence must be zero-extended to $N = 1024$ points. In the last convolution, the points to be discarded are the last $N - (N_1 + 475 - 1) = 1024 - (120 + 474) = 430$ points. Figure 7.21 illustrates this operation.

The computational effort generated by a single pulse-compressor load is as follows:

1. $2 \times (N/2 \cdot \log_2 N)$ butterflies (radix-2) and
2. N complex multiplications

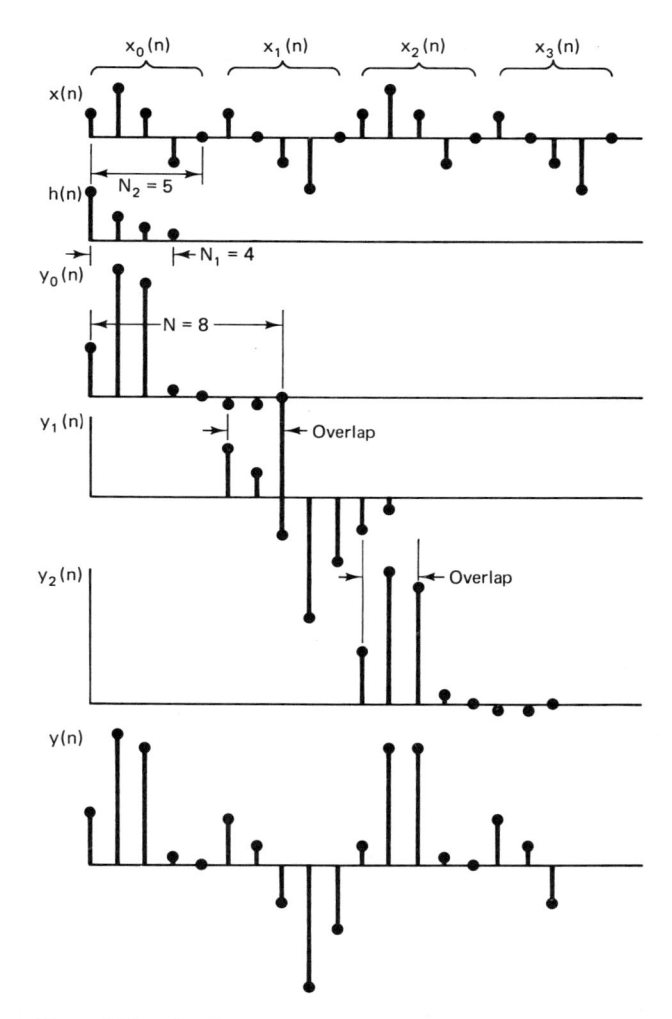

Sec. 7.3 Architectural Approaches at the System Level   337

Figure 7.20 "Overlap-and-add" input segmentation and processing.

Moving target indicator and doppler processing (MTI/DP).

Figure 7.22 illustrates the manner in which digital MTI/Doppler processing is implemented. It involves first pulse cancellation with $K(=4)$ pulses (input data sequences) followed by a forward FFT of size $N(=64)$. Figure 7.22 shows a batch of input data sequences, corresponding to several successive observation intervals, that are ready for processing. Each row represents an input data sequence collected over a complete observation interval with row 1 being the first interval and row r the last one. The MTI/Doppler processing function operates on data found along columns in the two-dimensional data arrays while the pulse-compression

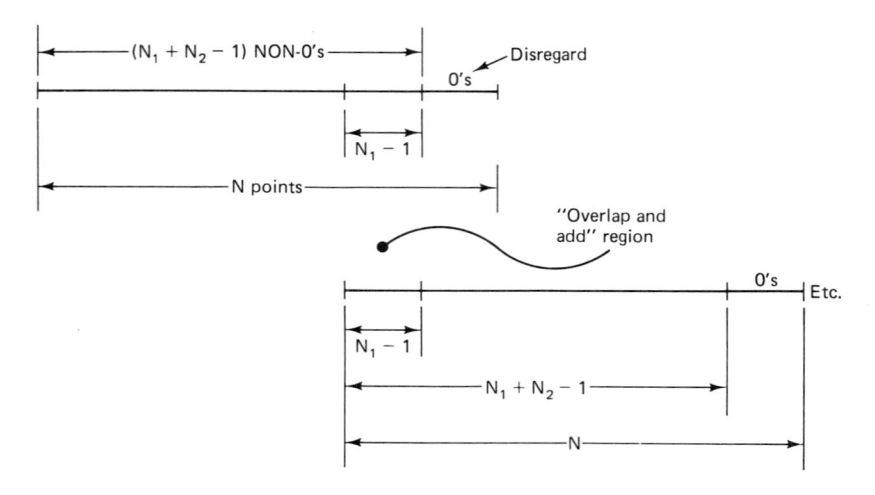

Figure 7.21 "Overlap-and-add" correction.

function requires data arranged along rows. This peculiarity in data accessing is known as "corner turning" and has the potential of introducing serious processing inefficiencies. For a thorough discussion concerning the physical importance and mathematical details of the MTI/Doppler processing functions, the reader might check Reference 23.

The weights associated with an (n + 1)-pulse canceller are given by Expression (7.14):[23]

$$W = (-1)^{i-1} \frac{n!}{(n - i + 1)!\,(i - 1)!} \tag{7.14}$$

where $i = 1, 2, \ldots, n + 1$. Other choices for weights are also possible.

The computational load generated by the MTI/Doppler processing functions includes the following components:

1. **MTI Via a 4-Pulse Canceller**
   It requires r memory accesses per column; four complex multiplications per point produced by the canceller; and $(r-3)$ memory accesses per column to place the results into memory.

2. **Doppler Processing**
   It requires a 64-complex point FFT per column. The size of this FFT is dictated by the number of Doppler channels desired and the choice of 64 is more than adequate for most radar applications.

Many of the computational times are reduced to times required to process a column. Therefore, total processing time will depend very heavily on the number of columns of data involved in the two-dimensional data arrays shown in Figure 7.22.

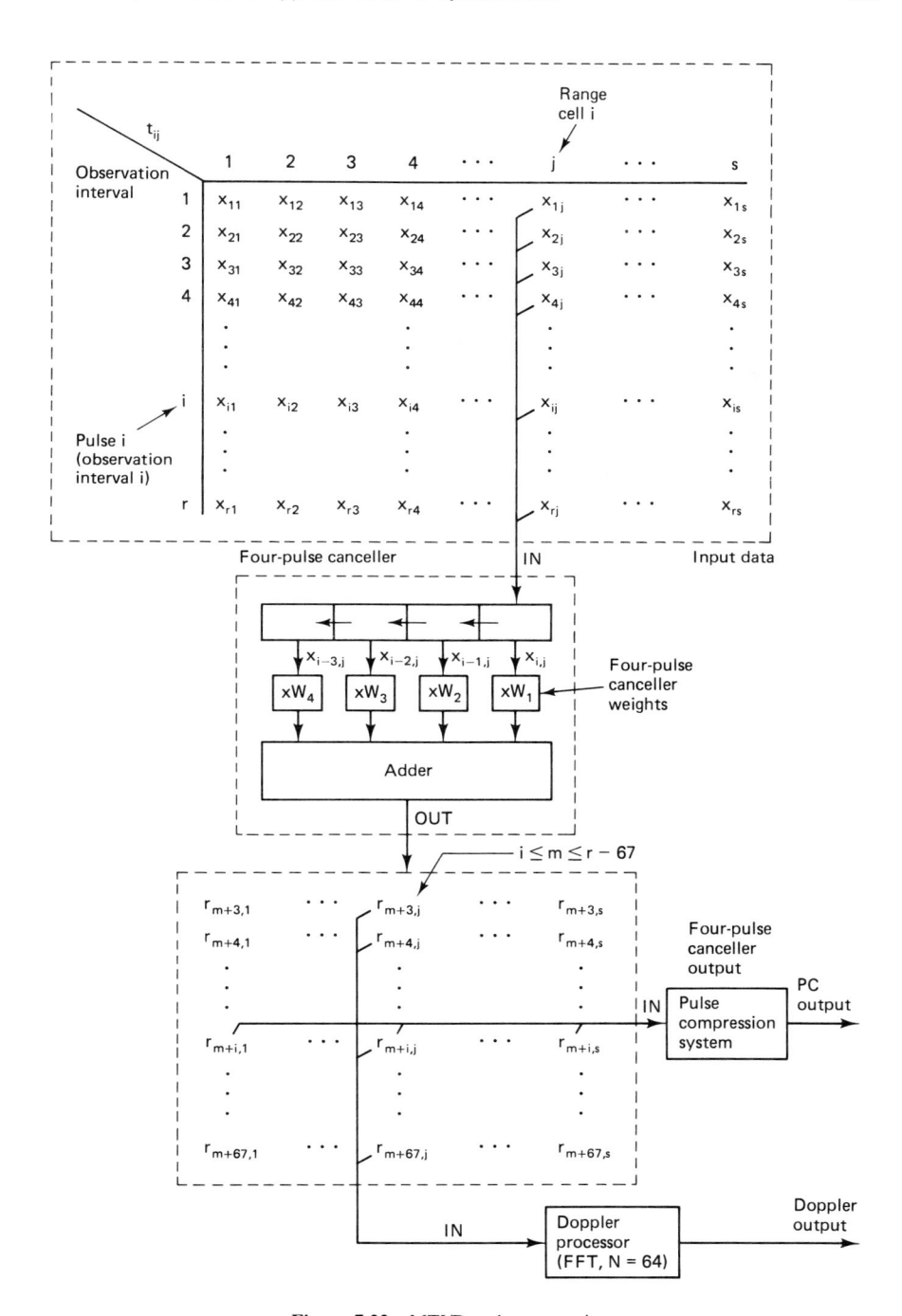

Sec. 7.3 Architectural Approaches at the System Level 339

Figure 7.22 MTI/Doppler processing.

Constant false alarm rate (CFAR). Figure 7.23 illustrates the manner in which the digital CFAR function is implemented. Its computation consists of averaging a predetermined number of data samples, k, before and after the data point of interest, and comparing the two averages (magnitude-wise) and choosing the larger of the two to set a threshold. The data point of interest is then compared against the threshold value set. For more information concerning the physical meaning and mathematical details of this function, Reference 23 should be consulted.

The computational effort involved consists of $\{2 \cdot (k - 1) + 1\}$ complex additions, four real multiplications and one subtraction for each data point of interest. Assuming s data points per observation period (i.e., time between generation of input data sequence otherwise known as interpulse period), the total computational load is practically s times that estimated.

Having briefly discussed the signal processing operations of interest and their computational requirements, we move on to present signal processor architectures capable of performing the postulated set of functions under specific input signal conditions. This allows comparisons of the various architectures under similar conditions and makes processor characteristics such as speed, circuit count, and power requirements important parameters in establishing a figure of merit for each architecture. It should be noted, at this point, that the presented architectures can be implemented with components whose design and performance has been announced in the literature or whose design is feasible with the capabilities offered by GaAs technologies today.

Although the degree of design detail offered throughout this section makes comparisons among various architectures very meaningful, it lacks the level of detail necessary for accurate system sizing (i.e., determination of circuit count, physical size, and power consumption) and throughput projections. This is con-

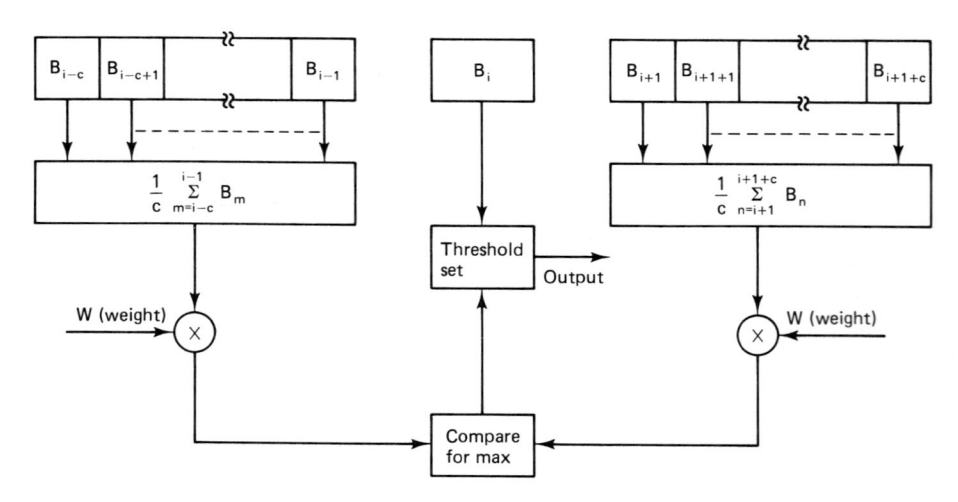

Figure 7.23 CFAR processing block diagram.

Sec. 7.3   Architectural Approaches at the System Level

sistent with the main objectives of this section, which are to establish the relative merits of various architectures rather than their absolute levels of complexity and performance and to demonstrate the effects of GaAs insertion into system design.

With regard to programming the proposed signal processor architectures, it is assumed that programming is carried out at a high level through the use of a minimal instruction set. The instruction set includes I/O instructions for moving data in and out of the processor as well as data-manipulative instructions for performing the signal processing functions. These high-level instructions differ from conventional instruction sets in that each instruction configures the processor to perform the respective signal processing function and leaves it in that configuration until another instruction is invoked.[24]

For illustration purposes, the following assumptions are made for all signal processor architectures to be presented:

1. 16-bit complex operands (i.e., 16-bit real, 16-bit imaginary)
2. 2's complement, fixed point arithmetic
3. up to 1024 complex point FFTs using the radix-2 algorithm
4. convolution of long input sequences is performed in the frequency domain using the overlap-and-add technique.

Although these assumptions are made to allow quantification of the implementation requirements for different architectures, they also hold true for a wide range of signal processor applications.

Architecture-1. This architecture is configured to perform the radix-2, FFT algorithm relying entirely on GaAs components. Figure 7.24 illustrates in block diagram form Architecture-1. The arithmetic processor consists essentially of a hardwired circuit designed to perform efficiently the radix-2 complex butterfly. This processor is further augmented with a "multiplier bank" which, in conjunction with the radix-2 butterfly unit, facilitates the implementation of various signal processing functions such as convolution, MTI/Doppler and CFAR. The multiplier bank can receive FFT output, act upon it, and write the results into data memory. Using this scheme, FFT computations and spectrum multiplications can be pipelined in a way that spectrum multiplication is completely overlapped with the last FFT pass when convolution or other filtering operations are carried out.

The multipliers used in this architecture are 16-bit × 16-bit array multipliers by Fujitsu (or equivalent) with 10.5 ns multiplication time.[25] The adder is the 32-bit adder by NEC (or equivalent) with 4 ns addition time.[26] The memory is assumed to be configured with NTT 4 K SRAM chips (or equivalent) whose access time is less than 4 ns.[27] Although these circuits were announced as proto-

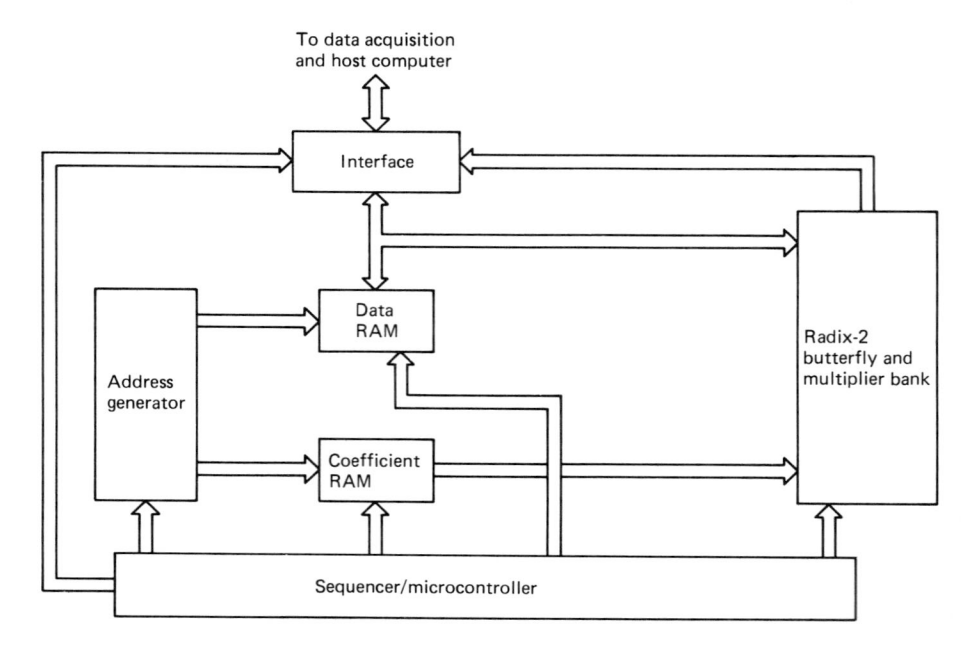

```
                    To data acquisition
                    and host computer
                            ↕
                      ┌──────────┐
                      │ Interface│
                      └──────────┘
```

Figure 7.24 Functional block diagram of Architecture 1.

types, their announcement indicates the evolution of GaAs technology toward producing in quantities circuits of high complexity.

The hardware requirements for implementing Architecture 1 are given in Table 7.1. Four of the multipliers are used to implement the radix-2 butterfly unit, and the remaining eight (8) form the "multiplier bank." The latter facilitates multiplication of two complex samples with appropriate filter coefficients, concurrently. The eleven (11) adders are distributed as follows: six (6) in the radix-2 butterfly, four (4) in the "multiplier bank" and one (1) is provided to facilitate the "overlap-and-add" computation.

TABLE 7.1 Estimate of Hardware Requirements of Architecture-1

1. Twelve (12) multiplier chips
2. Eleven (11) adder chips
3. Twenty-eight (28) single-port, 1 K × 4-bit memory chips
4. One (1) address generator chip
5. One (1) controller chip

   Total number of chips = 53

Sec. 7.3   Architectural Approaches at the System Level   343

The power requirements of Architecture-1 are given in Table 7.2, assuming the published power dissipation for the above chips. The speed characteristics of the various chips used in Architecture 1 are given in Table 7.3.

Since there are two memory operations (i.e., one **read** and one **write**) associated with each of the two operands in every radix-2 FFT butterfly computation, there will be 16 ns consumed in memory accesses. This is in addition to **multiply** and **add** times of 10.5 ns and 4 ns, respectively. However, a timing scheme with an overlapping fetch and execution periods as shown in Figure 7.25 is possible, and the result is a radix-2 butterfly every 20 ns. It should be pointed out that twiddle factors (or filter coefficients) are accessed independently through a bus other than the one used to access data samples. Furthermore, they are only read, and thus they are never a factor in determining system throughput.

Based on a 250-MHz clock and a radix-2 butterfly time of 20 ns, one obtains throughput performance for Architecture 1 as given in Table 7.4. The throughput estimates for this architecture and the architectures that follow are derived based on a collection of expressions presented in detail in Reference 27.

Architecture-2. This architecture is a variation of Architecture-1 that attempts to increase throughput and reduce complexity by exploiting the "ordered" nature of computations involved in the signal processing functions

TABLE 7.2 Estimate of Power Requirements of Architecture-1

1. Multiplier (GaAs)	12 chips × 0.9532 watt/chip = 11.44 watts
2. Adders (GaAs)	11 chips × 1.2 watts/chip = 13.20 watts
3. Memory (SRAM, GaAs)	28 chips × 0.9 watt/chip = 25.20 watts
4. Controller (GaAs)	1 chip × 1 watt/chip = 1.00 watt
5. Address generator (GaAs)	1 chip × 1 watt/chip = 1.00 watt
	Totals: 53 chips 51.84 watts

TABLE 7.3 Speed Characteristics of Chips Used in Architecture-1

1. Array multiplier, 16 × 16 bit (direct coupled FET logic)	Multiplication time = 10.5 ns
2. Adder, 32-bit (buffered FET logic)	Add time = 4 ns
3. Memory (SRAM), 1K × 4 bit (DCFL)	Access time = 2.8 ns
4. System clock (address generator produces a new address every 4 ns)	250 MHz

344   GaAs Insertion into System Design   Chap. 7

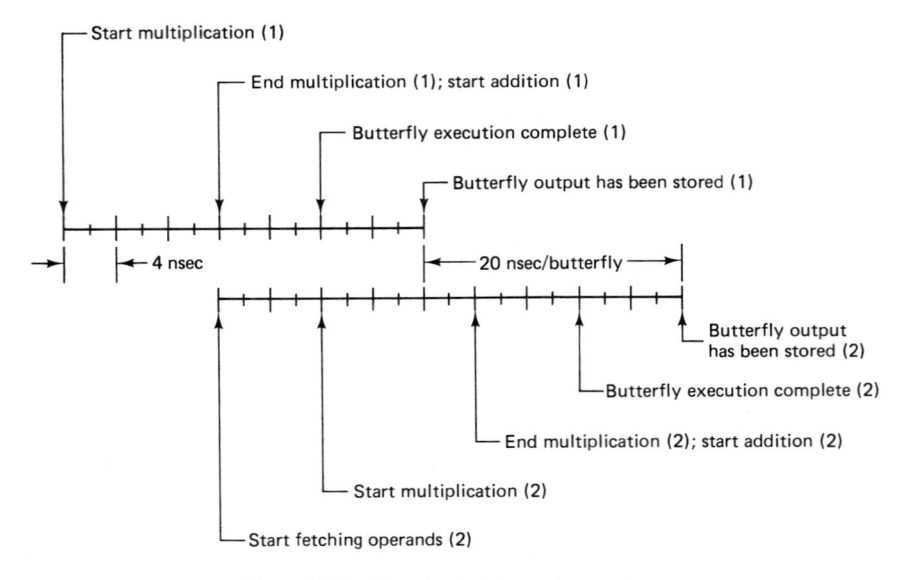

Figure 7.25 Butterfly fetch/execution overlap.

TABLE 7.4 Estimate of Throughput Characteristics of Architecture-1

	Clock	
Function	250 MHz	100 MHz
FFT-1024 complex points	102.40 μs	256.0 μs
Pulse compression	204.8 μs	512.0 μs
MTI/Doppler (per range-cell)	6.53 μs	16.32 μs
CFAR (per range-cell)	0.201 μs	0.502 μs

under consideration. Figure 7.26 shows a block diagram of Architecture-2. As shown, the RAM of Architecture-1 has been substituted with FIFO memory for both the data and coefficients. The FIFO is a two-port memory and does not require addressing for its operation. Therefore, this architecture does not include an address generator. In terms of number of components, this architecture requires one chip less (i.e., address generator) than Architecture-1 assuming that a 256 × 16 FIFO can be integrated on a single chip (i.e., storage equivalent to 4 K bits). The throughput, however, of this architecture will be higher than the throughput of Architecture-1 since read and write operations can occur concurrently. This feature becomes particularly critical when the computation time and the memory access time are almost equal.

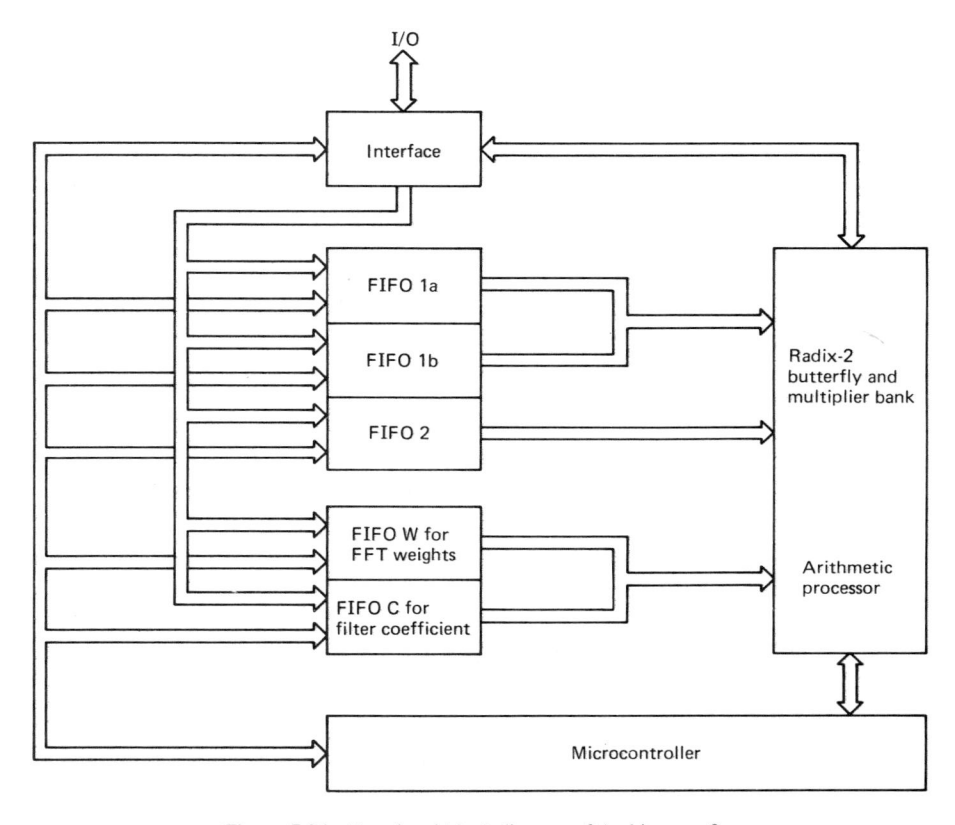

Sec. 7.3　Architectural Approaches at the System Level　　　　345

Figure 7.26　Functional block diagram of Architecture 2.

One attractive characteristic of this architecture is the simplicity of data shuffling between FFT passes using a memory "corner turning" technique discussed in Reference 28. This technique is implemented using FIFOs 1a and 1b (Figure 7.26) in an interleaved fashion and instead of address generation, requires simple control of the "load" and "unload" functions of the FIFOs (see Section 3.7.2) which is provided by the processor microcontroller. To illustrate the application of this technique, let us consider the computation of a 16-point FFT with a data flow diagram presented in Figure 7.13. For this computation, the data and intermediate results exchanged between FIFOs 1a and 1b are shown in Figure 7.27 for the different steps in the algorithm.

A further reduction in the number of components used, and also in power dissipation, can be achieved if the radix-2 butterfly circuit is implemented in a single chip using bit-serial arithmetic. The complexity of such a circuit is comparable to a 16×16 array multiplier. The single-chip butterfly circuit may have lower throughput than the circuit in Architecture-1 depending on the on-chip clock frequency and number of I/Os. With the current capability of GaAs

		First in $\downarrow$							Last in $\downarrow$
Before first pass	FIFO 1b	Empty							
	FIFO 1a	B_0	B_1	B_2	B_3	B_4	B_5	B_6	B_7
	FIFO 2	B_8	B_9	B_{10}	B_{11}	B_{12}	B_{13}	B_{14}	B_{15}
After first pass	FIFO 1b	a	i	b	j	c	k	d	l
	FIFO 1a	Empty							
	FIFO 2	e	m	f	n	g	o	h	p
After second pass	FIFO 1b	Empty							
	FIFO 1a	a'	e'	i'	m'	b'	f'	j'	h'
	FIFO 2	c'	g'	k'	o'	d'	i'	l'	p'
After third pass	FIFO 1b	a"	c"	e"	g"	i"	k"	m"	o"
	FIFO 1a	Empty							
	FIFO 2	b"	d"	f"	h"	j"	l"	h"	p"
After fourth pass	FIFO 1b	Empty							
	FIFO 1a	A_0	A_8	A_4	A_{12}	A_2	A_{10}	A_6	A_{14}
	FIFO 2	A_1	A_9	A_5	A_{13}	A_3	A_{11}	A_7	A_{15}

Figure 7.27 Data and intermediate results stored in the FIFOs during the FFT computation.

DCFL technology it is estimated that a single-chip, bit-serial radix-2 butterfly with 100 MHz throughput is possible.[28]

Architecture-3. This architecture uses both Si and GaAs components exploiting the capabilities of both technologies to achieve high throughput with low power requirements and modest demands on the capabilities of these technologies in terms of integration levels. Figure 7.28 illustrates a functional block diagram of Architecture-3. The arithmetic processor in this case can be implemented in CMOS and using the integration potential of this technology can include a hardwired radix-2 butterfly and a multiplier bank on a single chip. This leads to an arithmetic processor with 4 multipliers in the radix-2 butterfly circuit and 4 multipliers in the multiplier bank section for a total of 8 16-bit multipliers. Using bit-serial arithmetic, these multipliers along with the required adders and control circuitry can be integrated on a single chip.[29] Assuming a 25 MHz clock for the arithmetic processor, then a radix-2 butterfly will be performed every 640 nsec (i.e., 16 clock cycles times 40 nsec/cycle). Performance at this level is possible with appropriate CMOS technology as has been demonstrated with prototype circuits from the Very High Speed Integrated Circuits (VHSIC) program sponsored by the Department of Defense.

High system throughput, with a relatively low-throughput arithmetic processor unit, can be achieved by using a network of arithmetic processors operating in parallel and supplied with data by high bandwidth memory as depicted in Figure 7.28. In this case, a processor load for the basic radix-2 FFT computation is 2-

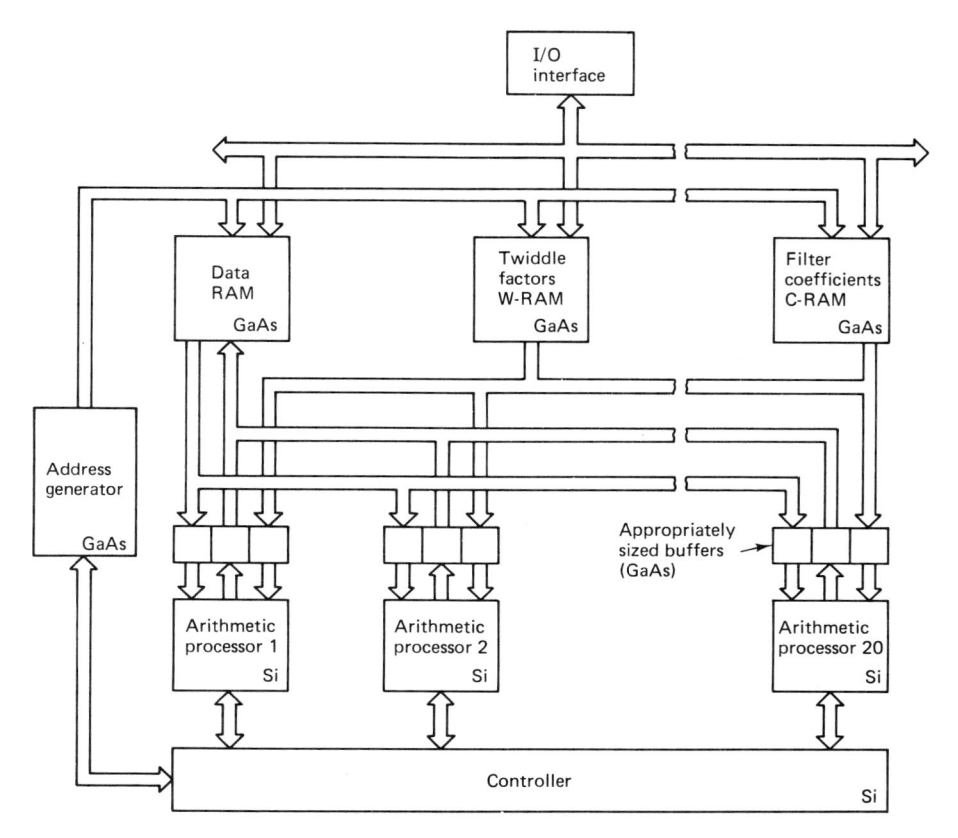

Sec. 7.3   Architectural Approaches at the System Level   **347**

Figure 7.28   Functional block diagram of Architecture 3.

complex operands (data) or 4 16-bit words. Assuming a network of 20 arithmetic processors, the memory has to supply one 16-bit word per 320/80 = 4 nsec. This calculation clearly indicates that the memory used is assumed to be single port (i.e., two memory references, read-write, per computational cycle). This configuration can be supported by the GaAs memory chips assumed for Architecture-1. When GaAs memory is used, however, the address generator also needs to be implemented in GaAs and data buffering becomes necessary to match the speed of the memory and address generator complex to the lower speed of the Si-based section of the system. The buffers have to be also GaAs circuits.

The hardware requirements of Architecture-3 are given in Table 7.5, while its power requirements are given in Table 7.6. Since there are 20 arithmetic processors in this configuration executing a radix-2 butterfly every 642 nsec, the effective throughput for the basic FFT computation is one butterfly/32 nsec. Based on this performance, Table 7.7 provides the throughput characteristics of Architecture-3.

TABLE 7.5 Estimate of Hardware Requirements of Architecture-3

1. Twenty (20) arithmetic processor chips (Si)
2. Twenty-eight (28) single-port, 1 K × 4-bit memory chips (GaAs)
3. Sixty (60) 4 × 32-bit buffer chips (GaAs)
4. One (1) address generator chip (GaAs)
5. One (1) controller chip (Si)

   Total number of chips = 110

TABLE 7.6 Estimate of Power Requirements of Architecture-3

1. Arithmetic processor (CMOS):	20 chips × 0.4 watts/chip = 8.0 watts
2. Memory, 1 K × 4-bit (GaAs):	28 chips × 0.9 watt/chip = 25.2 watts
3. Buffer, 4 × 32-bit (GaAs):	60 chips × 0.1 watt/chip = 6.0 watts
4. Address generator (GaAs)	1 chip × 1 watt/chip = 1.0 watt
5. Controller (CMOS):	1 chip × 0.3 watt/chip = 0.3 watts
	Totals: 110 chips 40.5 watts

TABLE 7.7 Estimate of Throughput Characteristics of Architecture-3

	Clock
Function	25 MHz
FFT-1024 complex points	122.7 μsec
Pulse compression	247.2 μsec
MTI/Doppler (per range cell)	10.2 μsec
CFAR (per range cell)	0.48 μsec

The discussion on the special purpose architectures does not intend to define specific figures of merit for these architectures nor does it intend to present a detailed description of signal processing algorithms and their implementation. The intention is to present some ideas of how GaAs components can be used in signal processor architectures and what the relative performance of such architectures can be expected to be. When detailed design of a processor is considered, however, factors that will affect the final performance are delays due to interconnect and/or bus protocols, specific memory access requirements, and delays in data acquisition in case of real time systems. The number of components will, in general, be slightly higher due to the required "glue" logic for satisfying certain design and/or test requirements. The size of components, the power dissipation, and the packaging technology determine the number of components that can be accommodated on a single board. An effort should be made during design to minimize the number of boards needed to implement a processor architecture because of the negative effect of off-board interconnection delays to system throughput.

PROJECTS

7.1. Figure 7.A illustrates an image processing system that receives image data from a sensor and selects targets based on the processed input data. The raw image is enhanced and then segmented to locate objects or regions of interest. Once an object or region has been located, it is examined for identifying features that can lead to final classification of targets. Any identified target can then be tracked through subsequent images.

Before an image can be segmented, the objects in that image must be detected and roughly classified according to shape and boundary characteristics. Some techniques used to detect objects use a gradient operator to locate potential object boundaries or edges. The gradient operator applied to a continuous function produces a vector at each point whose magnitude gives the magnitude of this maximum change. A digital gradient is often computed by convolving two windows with an image, one window giving the x component g_x of the gradient, and the other giving the y component g_y. This operation can be described by the expressions

$$g_x(i,j) = \text{mask}_x^* I(i,j) \qquad g_y = \text{mask}_y^* I(i,j)$$

where $I(i,j)$ is indicating some neighborhood of pixel (i,j) and * denotes convolution. Figure 7.B illustrates an example of this operation.

The task for this project is a filter design that performs the above operation using the Sobel operators. The incoming image data, in this case, is convolved with four 3×3 masks (i.e., Sobel operators) weighted to measure the differences in intensity along the horizontal, vertical, and left and right diagonals. These four measurements E_H, E_V, E_{D_x}, and E_{D_y} are then combined to estimate edge magnitude and direction. The inclusion of the diagonal measurements allows a more accurate, yet computationally simple, magnitude approximation than could be attained with just the usual horizontal and vertical measurements, and it simplifies the directional computation. The gradient magnitude:

$$MAG = [E_H^2 + E_V^2]^{1/2}$$

can be estimated as[30]

$$MAG = MAX[|E_H|, |E_V|, |E_{D_x}|, |E_{D_y}|] + K[E_\perp]$$

where

$$E_H = \begin{bmatrix} -1 & -2 & -1 \\ 0 & 0 & 0 \\ 1 & 2 & 1 \end{bmatrix} * I \qquad E_V = \begin{bmatrix} -1 & 0 & 1 \\ -2 & 0 & 2 \\ -1 & 0 & 1 \end{bmatrix} * I$$

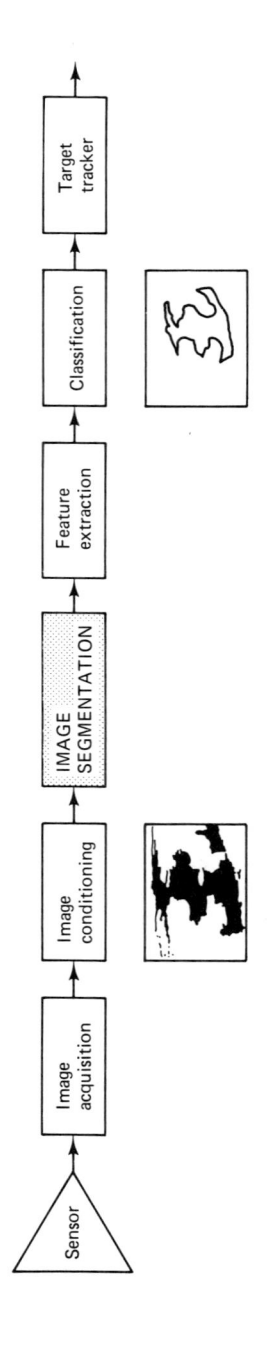

Figure 7.A Image processing system application.

Chap. 7   Projects

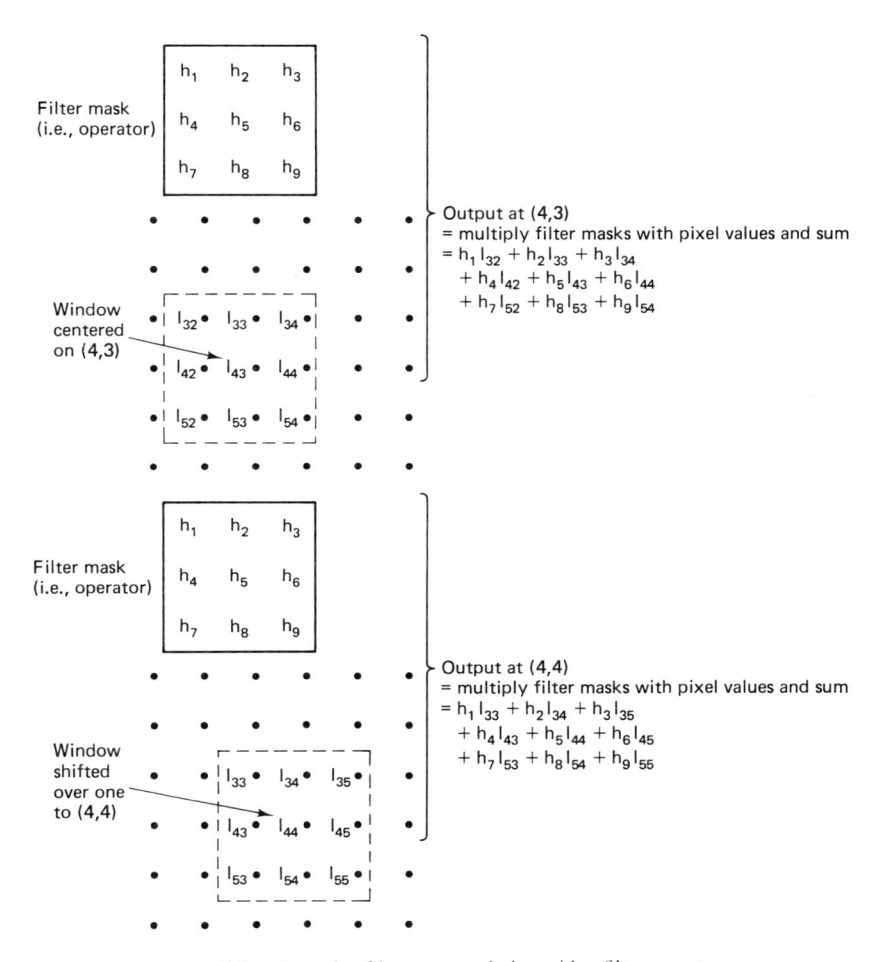

Figure 7.B   Example of image convolution with a filter operator.

$$E_{D_x} = \begin{bmatrix} -2 & -1 & 0 \\ -1 & 0 & 1 \\ 0 & 1 & 2 \end{bmatrix} * I \qquad E_{D_y} = \begin{bmatrix} 0 & 1 & 2 \\ -1 & 0 & 1 \\ -2 & -1 & 0 \end{bmatrix} * I$$

and $E_\perp$ is the measure in the direction perpendicular to the selected maximum measure. The gradient direction

$$\Theta = \arctan\left[\frac{E_V}{E_H}\right]$$

is assigned from a template of eight angle directions based on the selected maximum measure and its corresponding sign.

One pertinent question that affects the design and also the performance of the filter is how one approximates the gradient magnitude computed by Equation 3, or in other words, what is the best value for K. In most cases, the determination of the value of K will be application-dependent. For the filter design of this project, the value of K should be determined as the value that minimizes the magnitude error, which can be computed using error analysis for Euclidean norm approximations.

Compute the optimum value of K for this case. Perform the functional design for the Sobel filter, including the design of its architecture. An application of interest requires a filter throughput of 50 MHz and the filter should be implemented on a single board. Design a Sobel filter that meets these requirements for 256×256 pixel images where each input pixel is 8 bits, and 12-bit accuracy is required at the output. Which parts of the filter are suitable for GaAs implementation and why? The filter is also required to store one frame of the input image and one frame of the output. Design the appropriate memory configuration, and its interface to the Sobel filter, to meet this requirement. In your design use off-the-shelf components where applicable, and indicate their characteristics.

At the end of this project you're required to write a report that justifies your selection of components, design partitioning, and functional and logic design. Provide simulation results that demonstrate your filter is performing its intended function according to its functional specifications. One example of test image data is shown in Figure 7.C.

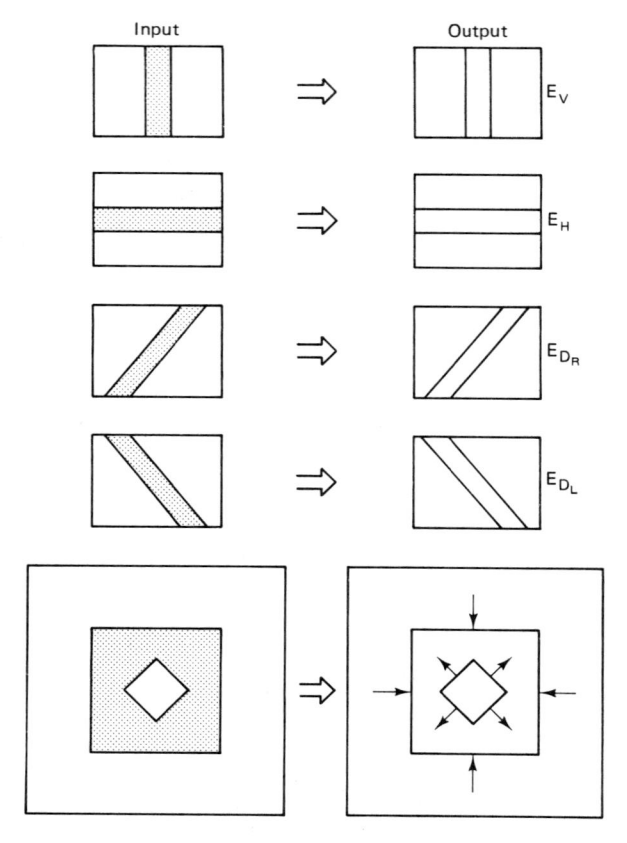

Figure 7.C Example of test image data for the Sobel filter.

What testability features (if any) did you incorporate into your design and how are you going to use these features for functional verification testing? What is a reasonable test plan? Does it make sense (in terms of performance, number of chips, power requirements) to implement the entire design in GaAs? Independently of your design approach the overall objective should be to meet the performance requirements for the filter function using the smallest number of chips while dissipating the least possible power.

REFERENCES

1. D. K. Ferry, *Gallium Arsenide Technology* (ch. 1), Indianapolis: Howard W. Sams & Co., 1985.
2. T. L. Rasset et al., "A 32-bit RISC Implemented in Enhancement-Mode JFET GaAs," *IEEE Computer* 19, no. 10, p. 60, October 1986.
3. E. R. Fox et al., "Reduced Instruction Set Architecture for a GaAs Microprocessor System," *IEEE Computer* 19, no. 10, p. 71, October 1986.
4. V. Milutinovic et al., "Architecture/Compiler Synergism in GaAs Computer Systems," *IEEE Computer* 20, no. 5, p. 72, May 1987.
5. A. Barna, *High Speed Pulse and Digital Techniques,* New York: John Wiley & Sons, 1980.
6. H. Howe, *Stripline Circuit Design,* Dedham, Mass.: Artech House, 1974.
7. T. Gheewala, D. MacMillan, "High Speed GaAs Logic Systems Require Special Packaging," *EDN,* p. 135, 17 May 1984.
8. N. Kanopoulos et al., "Advanced CAD for Systems Testability," RTI technical report 3650-O1F, prepared for U. S. Army Laboratory Command, Fort Monmouth, N.J., June 1987.
9. E. W. Kozdrowicki, D. J. Thies, "Second Generation of Vector Supercomputers," *Computer* 13, p. 71, November 1980.
10. S. R. Sternberg, "Cytocomputer Real-Time Pattern Recognition," *Proc. 8th Pattern Recognition Symposium,* Washington, D.C.: National Bureau of Standards, 1978.
11. C. V. Ramamoorthy, H. F. Li, "Efficiency in Generalized Pipeline Networks," *Proc. AFIPS 1974 National Computer Conf.,* p. 625, 1974.
12. J. W. Cooley, J. W. Tukey, "An Algorithm for the Machine Calculation of Complex Fourier Series," *Math. Comp.* 19, p. 297, 1965.
13. H. T. Kung, C. I. Leiserson, "Systolic Arrays for VLSI," in *Sparse Matrix Proceedings,* edited by I. S. Duff, G. W. Stewart, p. 256, SIAM, 1978.
14. C. E. Hein, R. M. Zieger, J. A. Urbano, "The Design of a GaAs Systolic Array for An Adaptive Null Steering Beam-forming Controller," *Computer* 20, no. 7, p. 92, July 1987.
15. H. T. Kung, "Let's Design Algorithms for VLSI Systems," *Proc. Conf. on VLSI,* California Institute of Technology, p. 65, January 1979.
16. H. T. Kung, L. M. Ruane, D. W. Yen, "A Two-Level Pipelined Systolic Array for Convolutions," in *VLSI Systems and Computations,* p. 255, Rockville, Md.: Computer Science Press, October 1981.

17. P. R. Cappello, K. Steiglitz, "Digital Signal Processing of Systolic Algorithms" in *VLSI Systems and Computations*, p. 245, Rockville, Md.: Computer Science Press, October 1981.
18. J. V. McCanny et al., "Bit Level Systolic Arrays," *Proc. Asilomar Conf. on Circuits Systems and Computers*, p. 124, November 1981.
19. J. V. McCanny, J. G. McWhirter, "Implementation of Signal Processing Functions Using 1-Bit Systolic Arrays," *Electronic Letters* 18, no. 6, p. 241, January 1982.
20. J. V. McCanny, J. G. McWhirter, "Some Systolic Array Developments in the United Kingdom," *Computer* 20, no. 7, p. 51, July 1987.
21. L. R. Rabiner, B. Gold, *Theory and Application of Digital Signal Processing*, Englewood Cliffs, N.J.: Prentice-Hall, 1975.
22. B. A. Bowen, W. R. Brown, *VLSI System Design for Digital Signal Processing*, Englewood Cliffs, N.J.: Prentice-Hall, 1982.
23. M. I. Skolnik, *Introduction to Radar Systems*, New York: McGraw-Hill Book Co., second edition, 1980.
24. P. N. Marinos, N. Kanopoulos, "GaAs Technology Insertion in High Performance Digital Signal Processor Design," *Proc. Asilomar Conf. on Circuits Systems and Computers*, p. 171, November 1986.
25. Y. Nakayama et al., "A GaAs 16x16 Bit Parallel Multipler," *IEEE J. Solid-State Circuits* SC-18, no. 5, p. 599, October 1983.
26. R. Yamamoto, "Design and Fabrication of Depletion GaAs LSI High-Speed 32-Bit Adder," *IEEE J. Solid-State Circuits* SC-18, p. 511, October 1983.
27. M. Hirayama et al., "A GaAs 4Kb SRAM with Direct Coupled FET Logic," *Proc. Int. Symposium Solid State Circuits*, p. 46, 1984.
28. P. Marinos, N. Kanopoulos, "Advances in Signal Processor Architectures," in *Advanced Semiconductor Technology and Computer Systems*, New York: Van Nostrand Reinhold Company, Inc., 1988.
29. N. Kanopoulos, P. Marinos, "A High Performance VLSI Signal Processor Architecture," *Proc. Int. Conference on Computer Design: VLSI in Computers*, p. 526, October 1984.
30. N. Kanopoulos, N. Vasanthavada, R. L. Baker, "Design of an Image Edge Detection Filter Using the Sobel Operator," *IEEE J. Solid-State Circuits*, vol. SC-23, p. 358, April 1988.

INDEX

A

Active elements in interconnection lines, 67
Ad-hoc testing, 254–257
ADC (analog-to-digital converters), 159–161, 163–170
  with pipelining, 327
Adders, 123–131
ALU (arithmetic logic units), 299
Ambiguity groups, 281–282, 287–291, 317–320
Analog-to-digital converters, 159–161, 163–170
  with pipelining, 327
AND gates in flip-flops, 83
AND/NOR functions, 76
AND-OR gates for multiplexers, 141
AND-OR-INVERT function, 73
Anneal in fabrication, 2

Architectural approaches, 321–349
  and chip design, 227–229
Architecture-1, 341–343
Architecture-2, 343–346
Architecture-3, 346–349
Arithmetic circuits
  adders, 123–131
  multipliers, 131–141
Arithmetic logic units, 299
Array personalization, 188
Arrays
  gate, 187–189
  multipliers of, 133–134
  systolic, 327–332
Asynchronous logic and designs, 311
  avoidance of, 260
  counters, 102–105
ATE (Automatic Test Equipment), 313–314

355

B

Backward crosstalk, 308–309
Band-limiting and interconnection lines, 66–68
Base-K counters, 102
BCD (Binary-Coded Decimal) counters, 113–114
BDFL (Buffered Diode FET Logic), 42, 45
BFL. *See* Buffered FET logic
BILBO (built-in logic block observer), 278–280
Binary arithmetic circuits
  adders, 123–131
  multipliers, 131–141
Binary-Coded Decimal counters, 113–114
Biomedical research, 297–298
BIST (built-in self-test), 272–291
  sequence generators for, 118
BIT (built-in-test), 272–273, 313–314
Bit-level systolic arrays, 332
Bit-serial arithmetic circuits
  adders, 129, 131
  multipliers, 136–141
Board level BIST, 280–291
Booth algorithm, 135–136
Breakdown region, FET, 11, 16
Buffer memory, FIFO, 153
Buffered Diode FET Logic, 42, 45
Buffered FET logic, 42
  for AND-OR-INVERT function, 73
  for carry lookahead circuits, 129
  characteristics of, 76–77
  for comparators, 166
  for flip-flops, 81, 87
  and LSSD, 263
  with multiplexers, 141
  for NOR gates, 68–69, 204–206
  for PLAs, 156–158
  propagation delay and power dissipation by, 55–56
Buffers, I/O, 171–172

Built-in logic block observer, 278–280
Built-in self-test, 272–291
  sequence generators for, 118
Built-in-test, 272–273, 313–314
Bulk resistivity, 1

C

CAD. *See* Computer-aided design
Calibration of ADC comparators, 166–167
Capacitance
  and chip packaging, 234
  and interconnection lines, 65, 67
  parasitic. *See* Parasitic capacitance
  and RAM speed, 151
Capacitor-Diode FET Logic, 42–43
  NAND configuration, 72–73
  for RAM, 152
Capacitors
  formation of, 3
  for power supply decoupling, 237
Carry circuits
  lookahead, 128–131
  with multipliers, 133
  ripple, 124–129
Cascading of decimal counters, 113
CDC (cyclic redundancy check) codes, 275
CDFL (Capacitor-Diode FET Logic), 42–43
  NAND configuration, 72–73
  for RAM, 152
Cell-configurable arrays, 188
Cells
  design approach using, 194, 199–200
  placing and routing of, 222–224
  standard, 184–187
Ceramic packages, 238–242
CFAR (constant false alarm rate), 340–341

Index

Channel current, FET, 14
Channel half-width, FET, 15
Charge storage mechanisms, 26, 28–29
Chemical Vapor Deposition, 2
Chips, design of
　computer-aided, 189–203
　example of, 218–227
　methodologies for, 183–189
　packaging, 235–244
Circuit extraction, 203
Circuits
　selection of, for design, 203–211
　simulation of, 192–193
　See also Logic circuits
CLEAR capabilities
　for counters, 105, 110–113
　for shift registers, 90–91
Clock processors, 315, 317
Clocks and clock circuitry
　and chip design, 212–213
　with FIFO memory, 153
　with flip-flops, 80–87
　with pipelining, 324
　with shift registers, 89–90, 94
　and system design, 310–313
　with systolic arrays, 329
　and testing, 261–262
　See also Skew, clock
CMOS circuits, interfacing of, 171, 174
　and chip design, 217
Collective design approach, 320–321
Column decoders, 146
Common mode coupling and chip packaging, 236
Communications applications, 298
Comparators for ADC circuits, 163–166
Complementary clock dividers, 98–101
Computer-aided design, 30, 189
　for circuit extraction, 203
　for circuit simulation, 192–193
　for layout, 193–201
　for logic simulation, 190–192
　for net list extraction, 202

for rule checking, 201–202
and schematic capture, 189–190
for switch level simulation, 202–203
Computer-aided tomography, 297
Computer simulation
　models for, 32–34
　for noise margins, 62–63
　See also Computer-aided design
Computers, scientific, 299
Concurrency, pipelining as, 322
Configurations for logic gates, 68–75
Connection problems, 302–310
Constant false alarm rate, 340–341
Contact photolithography, 5
Control circuits
　counters for, 98
　for RAM, 145, 152
　with shift registers, 94
Control and data shift registers, 315–316
Controllability for testing, 260–262
Convergence performance, 30
Conversion of data, 159–160
　analog-to-digital, 163–170
　digital-to-analog, 161–163
Corner turning technique, 345
Counters, 98–102
　combinations of, 114
　lockout with, 118
　modulus, 115–118
　nonbinary, 111–114
　ripple, 102–105
　synchronous, 105–111
Coupling coefficients, 308
Cross overs and diodes, 44
Cross-point switches, 176–177
Crosstalk, 308
　and chip packaging, 236
　and interconnection lines, 66
Cryptography, sequence generators for, 118
CT (computer tomography) scanners, 297
Current density, channel, 14

Current-voltage characteristics of
    JFETs, 11–13
Custom chip design, 184, 199
CVD (Chemical Vapor Deposition), 2
Cyclic redundancy check codes, 275
Cytocomputer, 322

D

DAC (digital-to-analog converters),
    159–163
D-algorithm, 249
Data, encryption and decryption of, 123
Data acquisition systems, 159
Data conversion, 159–160
    analog-to-digital, 163–170
    digital-to-analog, 161–163
Data latency, 156
DC testing, 253–254
DCFL. *See* Direct-Coupled FET Logic
Decade counters, 111–112
Decision time with ADCs, 165
Decoder circuits
    with counters, 105, 107, 111
    for RAM, 146, 152
Decoupling, power supply, and chip
    packaging, 237
Decryption of data, 123
Delay converter time, 165
Demultiplexers, 142–144
Depletion-mode MESFETs
    in logic gate design, 40–46
    propagation delay and power dissipation by, 50–56
Depletion Mode Process, 1–8
Depletion region, FET, 9
Design for testability, 257
Design partitioning, 300–302
Design principles, 182
    and architectures, 227–229
    chip methodologies, 183–189
    for chip packaging, 238–244

and circuit extraction, 203
and circuit simulation, 192–193
for interfacing, 214–218
for layout, 193–201
logic simulation, 190–192
net list extraction, 202
rule checking, 201–202
schematic capture, 189–190
selection of circuits, 203–211
special considerations for, 211–213
switch level simulation, 202–203
for testing, 246, 257–272
DFT (design for testability), 257
DFT (Discrete Fourier Transform),
    325–326
Differential amplifiers for ECL compatibility, 171
Differential linearity, 160–161
Digital circuits, interconnection lines
    in, 63–68
Digital Signal Processing, 295–297
Digital-to-analog converters, 159–163
Diode models, 24–27
    with SPICE, 32–33
Diodes in D-MESFET circuits, 44
Direct-Coupled FET Logic, 46–48
    for carry lookahead circuits, 129
    characteristics of, 76–77
    configuration using, 70–71, 74
    for flip-flops, 81, 87
    with multipliers, 133–135, 141
    power dissipation by, 58
    for RAM, 148
    for ripple carry adders, 124, 126
Disable-shift capability, 290
Discrete Fourier Transform, 325–326
Discriminators, 175–176
Divide-by-two circuits, 98–100
D-MESFET (depletion-mode MESFET)
    in logic gate design, 40–46
    propagation delay and power dissipation by, 50–56
Down counters, 105, 115–117

Index

Drain, FET, 9, 115–116
Drift velocity, 16–17
Drivers, I/O, 171–174
DSP (Digital Signal Processing), 295–297
D-type flip-flops, 81–82
Dual-phased clock system for chip design, 212
Dynamic complementary frequency divider, 101
Dynamic flip-flops, 83–85
Dynamic RAM, 146
Dynamic shift registers, 94–95

E

ECG signals, applications for, 297–298
ECL circuits, interfacing of, 171–173
  and chip design, 217–218
EDAC (error detection and correction), 221, 298
Edge detectors, 174
EEG signals, applications for, 297–298
Electric field configuration, 36
Electrostatic discharge, protection from, 171–174
E-MESFET. *See* Enhancement-mode MESFETs
Encryption of data, 123
Enhancement-mode MESFETs, 6
  for comparators, 167
  in logic gate design, 41, 46–49
  propagation delay and power dissipation by, 56–59
Epoxy in packaging, 242
Error detection
  with chip design, 218–219
  real-time, 123
Error detection and coding, 221, 298
Errors
  counter decoding, 105, 107
  quantization, 160
  and signature analysis, 277
ESD (electrostatic discharge), protection from, 171–174
Evaluation of tests, 252–253

F

Fabrication, 1–8
FAN algorithm, 249
Fan-in
  with counters, 110
  with DCFL circuits, 71
  and power dissipation, 55
Fan-out
  and chip design, 208, 223
  with counters, 101
  with D-MESFET circuits, 45
  with E-MESFET circuits, 46–47
  with flip-flops, 85
  and logic families, 76
  and power dissipation, 55
  with shift registers, 92
Fast Fourier Transform, 228–229, 325–326
Faults
  chip design for detection of, 220
  isolation testing for, 290–291
  simulation of, 252–253
  undetectable, 248
Feedback
  avoidance of, 260
  and chip testing, 281
  reflections from, 307
Feedback shift register, 120
Feedforward ADCs, 169–170
Feed-forward interconnection, 325
Feed-Forward Static Logic, 43
FFSL (Feed-Forward Static Logic), 43
FFT (Fast Fourier Transform), 228–229, 325–326
Fiber optic communications systems, 298

FIFO (First-In, First-Out) memory, 153–156
Flash ADC, 167–168
Flicker noise, 30–31
Flip-flops, 80–87
  See also Counters; Shift registers
Forward crosstalk, 308–309
Frequency dividers
  counters as, 98–101
  demultiplexers for, 143–144
Full adders, 123–124
  for multiplication, 133
Function generators, multiplexers as, 141
Functional verification testing, 311

G

Gain bandwidth product with ADCs, 166
Gate arrays, 187–189
Gates, FET, 9, 11
  fabrication of, 2–4, 6
  See also Logic gate design
Glitches, 160
Global feedback paths, avoidance of, 260

H

Half-adders, 133
HBT (heterojunction bipolar transistors), 8, 23–24
HEMTs (high electron mobility transistors), 7–8, 22–23
  for RAM, 152
Heterojunction bipolar transistors, 8, 23–24
Hierarchical-testing example, 317–320

High electron mobility transistors, 7–8, 22–23
  for RAM, 152
High pin-count packages, 239–240
High-resolution ADCs, 163
High-speed gates, logic families for, 76
High-speed memory, 151
High-speed testers, 252–254
High-throughput processing, 295–297
Hold ADC mode, 166
Hybrid packages, 242–244
Hysteresis with ADCs, 165

I

Identification and command shift register, 315–316
Image processing, 295–297
Implanted resistors, formation of, 2
Impurities in fabrication, 1
Incomplete scan, 271–272
Initialization of stored-state circuits, 260
Inner product step processor, 329, 332
Instruction processing units, 322
Instruction sets, 301–302
Integral linearity of ADC, 160
Intercomponent communications, 321
Intercomponent connection problems, 302–310
Interconnection lines and patterns
  and cell structures, 185–186
  and chip design, 212
  in logic gate design, 63–68
Interfacing
  and chip design, 214–218
  of GaAs circuits, 171–174
I/O drivers, 171–174
Ion implementation, 2
IPU (instruction processing units), 322

Index **361**

J

Jam loading of shift registers, 91
JFET (junction FETs), 9–16
  models for, 27–30
Johnson counters, 103
Junction FETs, 9–16
  models for, 27–30

K

Karnaugh maps, 156

L

Latency, data, 156
Layers
  with chip design, 194–198
  and interconnection lines, 67
Layout and chip design, 193–201
Least Significant Digit, 113
Left-shift registers, 92
Level-sensitive scan design, 263–267
Level shifting in logic gate design, 41–42
LFSR (linear feedback shift registers), 120, 274–277
Libraries, standard cell, 185–187
Linear converters, 159–160
Linear feedback shift registers, 120, 274–277
Linear region, FET, 11
Load address decoder, 315–316
Loading
  and packaging, 235–236
  of shift registers, 91
Local clock generation, 311
Local oscillators, counters for, 98
Lockout with counters, 118
Logic circuits
  arithmetic, 123–141

counters. *See* Counters
cross-point switch, 176–177
data conversion, 159–170
discriminators, 175–176
edge detectors, 174–175
flip-flops, 80–87
I/O drivers, 171–174
memory, 144–156
multiplexers/demultiplexers, 141–144
programmable logic arrays, 156–159
sequence generators, 118–124
shift registers, 87–98
Logic families
  interfacing of, and chip design, 214–218
  in logic gate design, 76–77
Logic gate design, 40–41
  configurations for, 68–75
  D-MESFET approaches to, 42–46
  E-MESFET approaches to, 46–49
  interconnection lines in, 63–68
  logic family selection in, 76–77
  and noise margins, 59–63
  propagation delays and power dissipation in, 49–59
Logic simulation, 190–192
Logical redundancy, avoidance of, 258–259
Lookahead carry circuits, 128–131
Low pin-count packages, 240
LSD (Least Significant Digit), 113
LSSD (level-sensitive scan design), 263–267

M

Mainframe computers, 299
Master-slave flip-flops, 81–82, 86
  with SET and CLEAR, 90
Matrix computations, 331–332

Maximum length sequences, 122–123
MBE (molecular beam epitaxy), 7–8, 23
Memory circuits
 and custom design, 184
 FIFO, 153–156
 RAM, 144–153
MESFET (metal semiconductor FET)
 devices, 2, 9, 16–19, 182
 compared to HEMTs, 22
 SPICE modeling of, 32–34
Mesh-connected systolic arrays, 329–330
Metalization steps, 3–5
Microstrip structures, 241, 303
MOCVD (metalorganic chemical vapor deposition), 7–8
 with HBTs, 23
Modeling
 for computer simulation, 32–35
 diode, 24–27
 JFET, 27–30
 noise, 30–32
Modulus counters, 115–117
 mod-5, 113–114
 mod-6, 114
 mod-9, 114
 mod-10, 111–112
 mod-16, 104, 106
 mod-K, 102, 111
Molecular beam epitaxy, 7–8
 with HBTs, 23
Monotonic ADC operation, 160–161
MSD (Most Significant Digit), 113
MTI/DP (moving target indicator and doppler processing), 337–339
Multilayer ceramic packages, 238–242
Multiplexers, 141–144
Multiplication, matrix, 331–332
Multipliers, 131–141
 with Architecture-1, 341–343

N

NAND function, 71
 in BFL and SDFL configurations, 72–73
N-bit adders, 124
Net list extractions, 202
Noise and noise margin
 and chip design, 211–212
 and connections, 305
 in D-MESFET circuits, 45
 and flip-flops, 85
 in logic gate design, 59–63
 models for, 30–32
 and production yield, 77
 quantization, 160
Nonbinary counters, 111–114
Nonlinear converters, 159
Nonvolatile memory, 156–159
NOR circuits, 76
 in BFL configuration, 68–69, 72–73
 in DCFL configuration, 70–71
 design of, 204–206
 in flip-flops, 83
 for ripple carry adders, 124–129
 in SDFL configuration, 68–70, 72–73
 simulation of, 206, 211
Normally-off FETs, 19–21

O

Observability for testing, 260–261
Observation shift register, 315–316
Offset with ADCs, 165
OR function, 74–75
 wired, 73, 262
OR-NAND gates with flip-flops, 87
Organization of memory circuits, 144–145
Overlap, clock phase, 212

Overlap-and-save approach, 336
Overshoot, 17–19

P

Packaging of chips, 235–244
Parallel ADC circuits, 163–164
Parallel data and shift registers, 87, 91–92, 94–97
Parallel signature analysis, 276
Parallelism, pipelining as, 322
Parasitic capacitance, 226
    layout to prevent, 211–213
    modeling of, 204–205, 208–209
    and RAM speed, 151
Partitioning, design, 300–302
Pass devices with shift registers, 94–97
Path-Oriented Decision Making, 249
Pattern generation for testing, 249–250, 260, 283, 285
PC (pulse compression), 333, 336
PCM (process control monitor), 225–226
Periods, sequence, 119
Phase/frequency discriminators, 175–176
Photolithography, 5
Pinch-off voltage, 11–14
Pipeline structures and techniques, 322–327
    ADC, 169–170
    bit-serial multiplier, 136–141
    for speed, 229
PLA (Programmable Logic Arrays), 156–159
Placement of cells, 223–224
PODEM (Path-Oriented Decision Making), 249
Polarity hold addressable latch, 270
Power consumption and dissipation
    by D-MESFET circuits, 42–43
    and interconnection lines, 65
    by logic families, 76
    and logic gate design, 49–59
    by RAM, 152
    by shift registers, 94
Power distribution for chip design, 211
Power supplies
    decoupling of, 237
    for E-MESFET circuits, 46
    for FETs, 41–42
Presettable counters, 110–111, 115
Process control monitor, 225–226
Production yields, 76–77
Programmable Logic Arrays, 156–159
Propagation delay
    and chip design, 228
    and chip packaging, 235
    and circuit simulations, 192–193
    and clock skew, 310–311
    with counters, 105–106, 110
    by interconnection lines, 63–68
    in logic gate design, 49–59
    with ripple adders, 128
    with shift registers, 89
    and synchronous circuits, 80–81
Pseudo complementary buffers, 47
Pseudorandom sequences, 122–123
Pulse compression, 333, 336
Pulse stretchers, 174–175

Q

Quantization error, 160
Quick Chip, 188

R

R-2R ladder DAC, 161–162
Race conditions, 212–213
RAM (random-access memory), 144–153
Random-access scanning, 269–271

Random sequences, 122–123
Random test patterns, 250
Read operations
 with Architecture-2, 344
 FIFO, 155
 RAM, 145, 229
Real-time error detection, 123
Real-time processing, 296–297
Recessed gates, 6
Recoders with multiplier circuits, 137–138
Reconfigurable cell array, 188
Redundancy, 248
 avoidance of, 258–259
Reference voltages for ADC circuits, 163–164
Reflections from connections, 305–306
Refresh cycles
 for dynamic flip-flops, 85
 for dynamic RAM, 146
Relative accuracy of ADCs, 160
Reliability and chip layout, 201
RESET for testing, 260
Resistors
 with E-MESFET circuits, 46
 formation of, 2
 for interconnection line termination, 64
 and transmission line matching, 236
Resolution
 of ADCs, 165
 of DACs, 163
Right-shift registers, 92
Ring counters, 102–103
Ringing
 from connections, 305–306
 and packaging, 236
Ripple carry adders, 124–129
Ripple counters, 102–105
Rise times of logic families, 171
Routing of cells, 223–224
Row decoders, 146
Rule checking, design, 201–202

S

S-A (stuck-at) type fault model, 247–249
SAG (self-aligned-gate) fabrication, 6–7
SAINT (Self-Aligned Implantation for N-Layer Technology) FETs, 151
SAR (synthetic aperture radar), 297
Satellites, EDAC for, 298
Saturation current, FET, 11–13, 15–16, 20
Saturation region, FET, 11
Scaling, 35–37
 and chip design, 208
Scan path testing technique, 267–268
Scan/set logic for testing, 268–269
Scanners, computer tomography, 297
Scanning with LSSD, 263–267
SCFL (Source-Coupled FET Logic), 47–49
Schematic capture, 189–190
Schottky barriers in MESFETs, 16
Schottky Diode FET Logic, 42, 44
 characteristics of, 76–77
 configuration using, 68–70, 72–73
 for flip-flops, 87
 for ripple carry adders, 128
Schottky diodes, formation of, 2
Schottky gates in SAG fabrication, 6
Scientific computers, 299
SDFL. *See* Schottky Diode FET Logic
SDHT (selectively doped heterojunction transistors), 22–23
Segment synchronization, 325
Self-Aligned Implantation for N-Layer Technology FETs, 151
Self-aligned-gate fabrication, 6–7
Self-testing, 272
Semidynamic shift registers, 94, 96
Sense amplifiers, 148, 150
Sequence generators, 118–123
Serial data and shift registers, 87–90

Index

Series-parallel feedback ADCs, 168–169
Series-parallel feedforward ADC, 169–170
SET capabilities for shift registers, 90–91
Settling time with DACs, 160–161
Setup converter time, 165
Shielding and chip packaging, 236
Shift-and-add multiplication, 133
Shift Register Latch, 263
Shift registers, 87–98, 315–316
   with sequence generators, 118–123
Shift-right, shift-left register, 92–94
Shot noise, 30–31
Signal coupling and chip packaging, 236
Signal processing, 295–297
Signature analysis, 274–278, 288
Silicon-based packaging, 242
Simulation
   of circuits, 192–193
   of faults, 252–253
   of logic, 190–192
   for logic gate design, 55
   switch level, 202–203
Single-phase clocks
   for dividers, 98
   for flip-flops, 81–82
Size
   of chip packaging, 235
   of memory circuits, 144
Skew, clock
   with flip-flops, 81–83
   with pipelining, 324
   from propagation delays, 310–311
Source, FET, 9
Source-Coupled FET Logic, 47–49
Special purpose architectures, 333–349
Speed of circuits, 8, 203–204, 227, 294
   pipelining for, 229
SPICE circuit simulator, 32–34, 193
   for logic gate design, 55
   for noise models, 31

Spikes, 160
SRL (Shift Register Latch), 263
ST (self-testing), 272
Standard cells, 184–187
Standing waves, voltage, 236
Static flip-flops, 83
Static RAM, 146–149
Static shift registers, 94
Stored-state circuits, avoidance of, 260
Stray capacitance and chip packaging, 234
Stripline structures, 241, 303–304
Structured design for testability, 262–272
Stubs and chip packaging, 236
Stuck-at type fault model, 247–249
Successive-approximation ADC, 168
Supercomputers, 299, 322
Switch level simulation, 202–203
Switches
   cross-point, 176–177
   testing, 283–284, 287, 317–320
Synchronization with testing, 252, 254
Synchronous operations
   with counters, 105–111
   with flip-flops, 80–87
   with pipelining, 325
Synthetic aperture radar, 297
System applications, 295–299
System architecture approaches, 321
System testing, 313–320
System timing, 310–313
Systolic arrays, 327–332

T

T-bar gate structure, 7
TCU (Test Control Unit), 314, 317–320
TEGFET (two-dimensional electron gas FETs), 7–8, 22–23

Temperature
  and chip design, 211, 215
  and chip packaging, 237
  and JFET saturation current, 29
Termination
  of interconnection lines, 64
  of transmission lines, 305–307
Test Control Unit, 314, 317–320
Test Node chip, 313–314, 317
Testing, 245
  approaches to, 253–257
  BIST for, 272–291
  of chips, 225–227
  and control of clock, 311
  designing for, 257–272
  issues in, 246–252
  system, 313–320
Testing switches, 283–284, 287, 317–320
Thermal conversion, 182
Thermal noise, 30–31
Thermometer code, 164
Thin film resistors, formation of, 2
Threshold control and production yield, 76–77
Threshold sensitivity and chip design, 215
Threshold voltage
  with FETs, 19–20
  with RAM, 147–148
Throughput
  with Architecture-2, 344
  with pipelining techniques, 322–323, 325
Time-division multiplexing, 143–144
Timing
  counters for, 110
  system, 310–313
TMR (Triple Modular Redundancy) design technique, 218–227
TN (Test Node) chip, 313–314, 317
Tomography, computer-aided, 297

Track ADC mode, 166
Transconductance
  in HEMTs, 22
  of JFETs, 16
Transfer functions of interconnection lines, 64–65
Transient response for SDFL inverter, 52
Transient spikes, 160
Transition detectors, 174
Transmission lines
  in chip packaging, 236
  termination of, 305–307
Triple Modular Redundancy design technique, 218–227
Truncation with multiplier stages, 137–138, 141
TTL circuits, interfacing of, 171, 174
  and chip design, 217
Tungsten silicide gate metalization, 6
Twisted-ring counters, 103
Two-dimensional effects, 17
Two-dimensional electron gas FETs, 7–8, 22–23
Two-phase clocks
  for dividers, 98–101
  for flip-flops, 81–82
Two's complement numbers, multiplication of, 133–135

U

UBFL for flip-flops, 85–87
UFL (unbuffered FET logic), 42–43
  propagation delay and power dissipation by, 55–56
Unity gain point, 59
Up counters, 103–106
Utilization factor of linear pipes, 323–324

Index

V

Variable modulus counters, 115–117
Velocity overshoot, 17–19
Velocity saturation, 20–21
Vias, 2–3
 and diodes, 44
Voltage standing waves, 236
Voter circuits, 219–220, 226–227
VSW (voltage standing waves), 236

W

Wafer etching, 2
WIRED-AND function, 74–75
WIRED-OR function, 73
 avoidance of, 262
WOMBAT program, 202
Write operation
 with Architecture-2, 344
 FIFO, 229
 RAM, 145, 229

Y

Yields, production, 76–77

SOUTHEASTERN MASSACHUSETTS UNIVERSITY
TK7874.K345 1989
Gallium arsenide digital integrated circ

3 2922 00044 017 9

DATE DUE

320561